Ionic Liquids and Deep Eutectic Solvents: Greener Approaches for Sustainable Chemistry

Ionic Liquids and Deep Eutectic Solvents: Greener Approaches for Sustainable Chemistry

Editors

Slavica Ražić
Aleksandra Cvetanović Kljakić
Enrico Bodo

Basel • Beijing • Wuhan • Barcelona • Belgrade • Novi Sad • Cluj • Manchester

Editors

Slavica Ražić
Department of Analytical
Chemistry - Faculty of
Pharmacy
University of Belgrade
Belgrade
Serbia

Aleksandra Cvetanović
Kljakić
Faculty of Technology
University of Novi Sad
Novi Sad
Serbia

Enrico Bodo
Chemistry Department
University of Rome "La
Sapienza"
Rome
Italy

Editorial Office
MDPI
St. Alban-Anlage 66
4052 Basel, Switzerland

This is a reprint of articles from the Special Issue published online in the open access journal *Molecules* (ISSN 1420-3049) (available at: https://www.mdpi.com/journal/molecules/special_issues/Liquids_Solvents).

For citation purposes, cite each article independently as indicated on the article page online and as indicated below:

Lastname, A.A.; Lastname, B.B. Article Title. *Journal Name* **Year**, *Volume Number*, Page Range.

ISBN 978-3-7258-0866-3 (Hbk)
ISBN 978-3-7258-0865-6 (PDF)
doi.org/10.3390/books978-3-7258-0865-6

© 2024 by the authors. Articles in this book are Open Access and distributed under the Creative Commons Attribution (CC BY) license. The book as a whole is distributed by MDPI under the terms and conditions of the Creative Commons Attribution-NonCommercial-NoDerivs (CC BY-NC-ND) license.

Contents

About the Editors vii

Preface ix

Orfeas-Evangelos Plastiras and Victoria Samanidou
Applications of Deep Eutectic Solvents in Sample Preparation and Extraction of Organic Molecules
Reprinted from: *Molecules* **2022**, *27*, 7699, doi:10.3390/molecules27227699 1

Slavica Ražić, Tamara Bakić, Aleksandra Topić, Jelena Lukić and Antonije Onjia
Deep Eutectic Solvent Based Reversed-Phase Dispersive Liquid–Liquid Microextraction and High-Performance Liquid Chromatography for the Determination of Free Tryptophan in Cold-Pressed Oils
Reprinted from: *Molecules* **2023**, *28*, 2395, doi:10.3390/molecules28052395 18

Alena Koigerova, Alevtina Gosteva, Artemiy Samarov and Nikita Tsvetov
Deep Eutectic Solvents Based on Carboxylic Acids and Glycerol or Propylene Glycol as Green Media for Extraction of Bioactive Substances from *Chamaenerion angustifolium* (L.) Scop.
Reprinted from: *Molecules* **2023**, *28*, 6978, doi:10.3390/molecules28196978 33

Alexandra Cristina Blaga, Elena Niculina Dragoi, Alexandra Tucaliuc, Lenuta Kloetzer and Dan Cascaval
Folic Acid Ionic-Liquids-Based Separation: Extraction and Modelling
Reprinted from: *Molecules* **2023**, *28*, 3339, doi:10.3390/molecules28083339 50

Maria Myrto Dardavila, Sofia Pappou, Maria G. Savvidou, Vasiliki Louli, Petros Katapodis, Haralambos Stamatis, et al.
Extraction of Bioactive Compounds from *C. vulgaris* Biomass Using Deep Eutectic Solvents
Reprinted from: *Molecules* **2023**, *28*, 415, doi:10.3390/molecules28010415 63

Julius Choi, Alberto Rodriguez, Blake A. Simmons and John M. Gladden
Valorization of Hemp-Based Packaging Waste with One-Pot Ionic Liquid Technology
Reprinted from: *Molecules* **2023**, *28*, 1427, doi:10.3390/molecules28031427 82

Xinyu Cui, Yani Wang, Yanfeng Wang, Pingping Zhang and Wenjuan Lu
Extraction of Gold Based on Ionic Liquid Immobilized in UiO-66: An Efficient and Reusable Way to Avoid IL Loss Caused by Ion Exchange in Solvent Extraction
Reprinted from: *Molecules* **2023**, *28*, 2165, doi:10.3390/molecules28052165 94

Vanessa Piacentini, Andrea Le Donne, Stefano Russo and Enrico Bodo
A Computational Analysis of the Reaction of SO_2 with Amino Acid Anions: Implications for Its Chemisorption in Biobased Ionic Liquids
Reprinted from: *Molecules* **2022**, *27*, 3604, doi:10.3390/molecules27113604 114

Imran Ali, Gunel T. Imanova, Hassan M. Albishri, Wael Hamad Alshitari, Marcello Locatelli, Mohammad Nahid Siddiqui and Ahmed M. Hameed
An Ionic-Liquid-Imprinted Nanocomposite Adsorbent: Simulation, Kinetics and Thermodynamic Studies of Triclosan Endocrine Disturbing Water Contaminant Removal
Reprinted from: *Molecules* **2022**, *27*, 5358, doi:10.3390/molecules27175358 124

Dhari K. Luhaibi, Hiba H. Mohammed Ali, Israa Al-Ani, Naeem Shalan, Faisal Al-Akayleh, Mayyas Al-Remawi, et al.
The Formulation and Evaluation of Deep Eutectic Vehicles for the Topical Delivery of Azelaic Acid for Acne Treatment
Reprinted from: *Molecules* **2023**, *28*, 6927, doi:10.3390/molecules28196927 . . . . **140**

Dorota Mańka and Agnieszka Siewniak
Deep Eutectic Solvents as Catalysts for Cyclic Carbonates Synthesis from CO_2 and Epoxides
Reprinted from: *Molecules* **2022**, *27*, 9006, doi:10.3390/molecules27249006 . . . . **154**

Dennis Woitassek, Till Strothmann, Harry Biller, Swantje Lerch, Henning Schmitz, Yefan Song, et al.
Tunable Aryl Alkyl Ionic Liquid Supported Synthesis of Platinum Nanoparticles and Their Catalytic Activity in the Hydrogen Evolution Reaction and in Hydrosilylation
Reprinted from: *Molecules* **2023**, *28*, 405, doi:10.3390/molecules28010405 . . . . **171**

Xiaokang Wang, Yuanyuan Cui, Yingying Song, Yifan Liu, Junping Zhang, Songsong Chen, et al.
Studies on the Prediction and Extraction of Methanol and Dimethyl Carbonate by Hydroxyl Ammonium Ionic Liquids
Reprinted from: *Molecules* **2023**, *28*, 2312, doi:10.3390/molecules28052312 . . . . **191**

Olga Terenteva, Azamat Bikmukhametov, Alexander Gerasimov, Pavel Padnya and Ivan Stoikov
Macrocyclic Ionic Liquids with Amino Acid Residues: Synthesis and Influence of Thiacalix[4]arene Conformation on Thermal Stability
Reprinted from: *Molecules* **2022**, *27*, 8006, doi:10.3390/molecules27228006 . . . . **206**

About the Editors

Slavica Ražić

Slavica Ražić is a full professor and the Chair of the Department of Analytical Chemistry at the Faculty of Pharmacy, University of Belgrade. She is a member of the Executive Boards of the European Chemical Society (EuChemS) and the Division of Analytical Chemistry (DAC-EuChemS). Slavica is also a titular member of the Analytical Chemistry Division of the International Union of Pure and Applied Chemistry (ACD-IUPAC) and its representative in the Interdivisional Committee on Green Chemistry for Sustainable Development. Her current research interests related to this Special Issue are in the development of analytical methods for environmental and natural food samples, assisted by chemometric methods of analysis and in line with the principles of green and sustainable chemistry, unconventional and environmentally friendly solvents and extractions in sample preparation, as well as the physico-chemical characterization of the extracts obtained using suitable separation and spectroscopic analysis techniques.

Aleksandra Cvetanović Kljakić

Dr. Aleksandra Cvetanović Kljakić is a principal research fellow at the Faculty of Technology in Novi Sad. Her research focuses on the extraction and isolation of bioactive molecules of natural origin and their biological and chemical characterization. She has received numerous national and international awards for her work and exceptional results. Her current research interests related to this Special Issue are in the extraction and isolation of bioactive molecules, green extraction solvents, green chemistry, NADES solvents, subcritical water, supercritical carbon dioxide, and biological characterization of bioactive molecules.

Enrico Bodo

Dr. Enrico Bodo is a full professor at the Department of Chemistry of the University of Rome, La Sapienza. He has been an elected member of the Board of the Faculty of Science of the University of Rome, La Sapienza, and is currently an elected member of the Board of the Division of Theoretical and Computational Chemistry of the Italian Chemical Society. His scientific area is in physical chemistry (i.e., material modeling, ab initio molecular dynamics, classical molecular dynamics, and ionic liquids). His current research interests related to this Special Issue are in the application of theoretical and computational methods, such as molecular dynamics, to the description of the properties of ionic liquids, and solvents.

Preface

Conventional solvents are still used in laboratory practice but are increasingly being replaced by environmentally friendly solvents. Remarkable progress has been made in the field of unconventional, i.e., green, solvents. This Special Issue aims to provide a comprehensive overview of the "state of the art" in recent advances in the field of the two classes of green solvents, ionic liquids (Ils) and deep eutectic solvents (DES), through various sustainable chemical approaches. Both ILs and DES have the potential to be environmentally friendly alternatives to conventional solvents as designer solvents due to their highly tunable physical, chemical, and physicochemical properties. In line with the principles of green chemistry, the emergence of deep eutectic solvents, in particular natural deep eutectic solvents (NADES) and ionic liquids (ILs), has opened new avenues with applications in various fields. This Reprint is divided into three parts, although overlaps are inevitable due to the multidisciplinary approach to this undoubtedly attractive and challenging topic. In the first part, the authors focus on the sample preparation of various complex matrices (natural and food products, biomass, pharmaceuticals, etc.) in general and on the extraction of targeted bioactive compounds, essential molecules, and pollutants. The next sections summarize research results related to biomass and nanocomposite-based adsorption technologies for reuse, water treatment, pollutant removal, IL- and DES-based green synthesis, catalysis, and improvement of chemical and technological processes. The latest findings on the recycling/recovery of more environmentally friendly solvents, the use of advanced analytical techniques and methods, and adherence to green chemistry principles are fully recognized.

Slavica Ražić, Aleksandra Cvetanović Kljakić, and Enrico Bodo
Editors

Review

Applications of Deep Eutectic Solvents in Sample Preparation and Extraction of Organic Molecules

Orfeas-Evangelos Plastiras [1] and Victoria Samanidou [2],*

[1] UMR CNRS 8181, Unité de Catalyse et Chimie du Solide, Université de Lille, 59925 Lille, France
[2] Laboratory of Analytical Chemistry, Department of Chemistry, Aristotle University of Thessaloniki, 54124 Thessaloniki, Greece
* Correspondence: samanidu@chem.auth.gr; Tel.: +30-231-099-7698

Abstract: The use of deep eutectic solvents (DES) is on the rise worldwide because of the astounding properties they offer, such as simplicity of synthesis and utilization, low-cost, and environmental friendliness, which can, without a doubt, replace conventional solvents used in heaps. In this review, the focus will be on the usage of DES in extracting a substantial variety of organic compounds from different sample matrices, which not only exhibit great results but surpass the analytical performance of conventional solvents. Moreover, the properties of the most commonly used DES will be summarized.

Keywords: deep eutectic solvents; extraction; organic compounds; green sample preparation; analytical chemistry

1. Introduction

Over the last decades, conventional solvents and extraction techniques—for example, methanol as a solvent or liquid–liquid extraction as a technique—tend to be replaced by new, simpler, inexpensive, and environmentally friendly approaches. The movement of "Green Analytical Chemistry" (GAC) and its 12 principles, presented by P. Anastas in 1998, has especially inspired scientists to switch to greener and alternative solutions for extracting compounds from different matrices [1]. Thus, developing new and sustainable solvents to match the above criteria is of high interest to many researchers globally [2–8]. A new group of organic salts that possess melting points below 100 °C, called ionic liquids (ILs), have emerged and demonstrate properties that could replace volatile organic solvents. However, ionic liquids pose some problems, such as toxicity, stability, biodegradability, and expense regarding their synthesis.

That is where deep eutectic solvents (DES) emerged in 2001 as the perfect candidate for replacing ILs, demonstrating cheaper and easier synthesis, while their environmentally friendliness is much more evident [9–12]. Deep eutectic solvents have low melting points due to low lattice energy formed by their large and asymmetrical ions. Habitually, they are formed by the combination of metal salt, called the hydrogen bond donor (HBD), with a quaternary ammonium salt, the hydrogen bond acceptor (HBA). Thus, hydrogen bonds are formed by charge delocalization, and the mixture's melting point is lower than the separate components of the DES [13,14]. Hydrophilic and hydrophobic DES and natural deep eutectic solvents (NADES) can, without a doubt, be applied to analytical chemistry in different fields to extract molecules from a variety of samples [4,8,15–29].

Older reviews on deep eutectic solvents focus, mainly, on the extraction or microextraction techniques [2,4], highlighting their low cost, simplicity, and environmental safeness, on their synthesis and properties [17,20,27,28], or on the variety of samples that they can be applied to [8], such as gas, biodiesel, or metals. The aim of this review is the introduction of deep eutectic solvents and natural deep eutectic solvents and understanding how can they be applied in the extraction of a plethora of compounds, focusing mainly on the different

group of organic compounds that can be extracted. Additionally, the synthesis and the properties of these solvents will be discussed briefly. In Figure 1, the main scheme of how DES are used in extraction is presented.

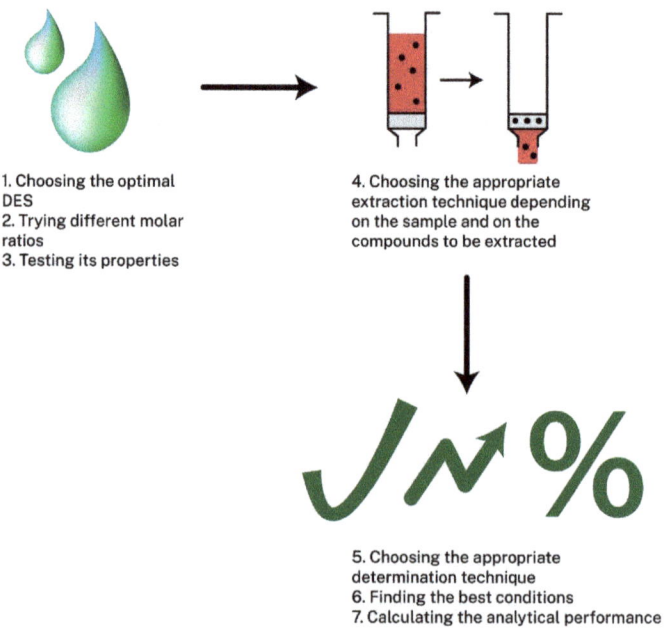

Figure 1. Representative scheme of the typical development of an extraction method by using deep eutectic solvents.

2. Synthesis and Properties of Deep Eutectic Solvents

2.1. Synthesis of Deep Eutectic Solvents

The most common way to synthesize DES is by mixing two or three cheap and safe salts and heating them up in order to obtain a homogenous solution. Molten salts at the ambient temperature can also be prepared through the mixture of metal salts with quaternary ammonium salts [30–32]. There are four main types of deep eutectic solvents: type I, where a metal chloride is combined with a quaternary salt, type II, where a hydrated metal chloride is merged with a quaternary salt, type III, where an HBD is put together with an HBA quaternary salt, and type IV, where an organic HBD compound reacts with a metal chloride [2,33]. Table 1 summarizes the four types of the DES, while in Figure 2, the most frequently used HBA and HBD are presented.

Table 1. The four types of Deep Eutectic Solvents and their formulas.

Type	Formula	Terms
I	$Cat^+X^- + zMCl_x$	M = Zn, Sn, Ga, In, Al, Fe
II	$Cat^+X^- + zMCl_x \cdot yH_2O$	M = Cr, Cu, Fe, Ni, Co
III	$Cat^+X^- + zRZ$	Z = COOH, $CONH_2$, OH
IV	$MCl_x + RZ$	M = Al, Zn and Z = $CONH_2$, OH

Cat^+ = any phosphonium, ammonium or sulfonium cation, X^- = a Lewis base, often a halide anion, MCl_x = metal chloride, RZ = organic compound.

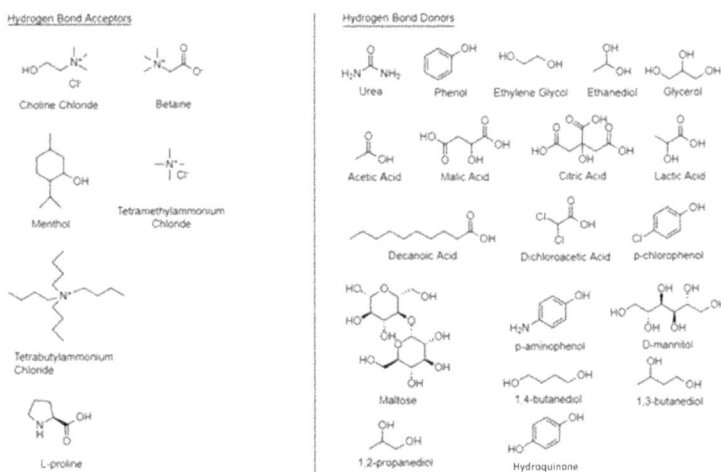

Figure 2. Structures of some common HBAs and hydrogen bond donors HBDs used in the synthesis of deep eutectic solvents.

Choline chloride (ChCl) is the most widely used HBA due to its low cost, low toxicity and biodegradability than has the advantage of reacting with safe and inexpensive HBDs, such as carboxylic acids, urea, or glycerol [32,34]. Additionally, type III acidic deep eutectic solvents are synthesized from quaternary organic salts with Brønsted acids (for example, citric acid) and type IV acidic DES from Lewis acids (for instance, zinc chloride or bromide) [35].

2.2. Properties of Deep Eutectic Solvents

Physicochemical properties of deep eutectic solvents derive from the various interactions between an HBA and HBD, which depend on the molar ratios, the organic compounds themselves, and the nature of the interactions (π-π and/or hydrogen bonding, anion exchange, weak non-covalent interactions) [36]. The properties that need to be considered when using DES and/or NADES are density, acidity, conductivity, viscosity, volatility, the melting and freezing point, and surface tension. Additionally, the toxicity, cost, thermal stability, and biodegradability should not be passed over [2,37]. The value of pH that results from the combination of HBA and HBD can drastically affect the extraction efficiency of the target compounds, so this parameter ought to not be overlooked as well [38]. As described by El Achkar et al. [39] in a thorough review about the properties of DES, viscosity is an important parameter to be studied, among others, since it severely impacts the extraction efficiency of a developed method. Water content is an ambiguous parameter since, for some, it might seem as an impurity, while others rely on it and add water on purpose to expose the reliability of the DES. Molar ratio and temperature of extraction can impact the isolation of organic compounds as well, and they are studied in almost all developed methods of extraction. Consequently, by knowing the parameters that can affect the extraction with the usage of DES, one can exploit their high tunability [40].

3. Extraction of Organic Compounds

3.1. Pesticides, Fungicides and Herbicides

Choline chloride (ChCl) with phenol, ethylene glycol, or 4-chlorophenol were used as deep eutectic solvents, in different molar ratios, from Abolghasemi et al. [41] so as to extract triazole fungicides from vegetable samples and fruit juice. The team used headspace single-drop microextraction (HS-SDME) as an extraction technique and gas chromatography coupled with a flame ionization detector (GC-FID) to determine the compounds. By examining a lot of parameters and choosing the optimal ones, they were able to achieve a limit of detections (LODs) of 0.82–1.0 µg/mL and relative standard deviations (RSD) of 3.9–6.2% when ChCl and 4-chlorophenol (1:2, respectively) were used as the DES media.

Farajzadeh et al. [42] synthesized a new DES of high density—consisting of menthol and dichloroacetic acid (1:2 molar ratio)—that was utilized as the extraction solvent of different pesticides originating from honey samples. Dispersive liquid–liquid microextraction (DLLME) was chosen as the technique, while the determination was done by GC-FID. By using these techniques and the aforementioned DES under the best conditions, they were able to detect and quantify the pesticides in the range of 0.32–1.2 ng/g and 1.1–4.0 ng/g, respectively, with enrichment factors (EF) 279–428 and absolute recoveries obtained from the extraction being between 56 and 86%. In another paper, Farajzadeh et al. [43] successfully extracted pesticides from vegetable and fruit juice samples by liquid phase microextraction (LPME) with the help of ChCl and p-chlorophenol as the DES. Interestingly, they used different temperatures to disperse the DES into the aqueous phase, thus improving the extraction. EFs were once again very high, in the range of 280–465, with absolute recoveries of 56–93%, and very low LODs and LOQs were achieved (0.13–0.31 and 0.45–1.1 ng/mL respectively).

Pesticides can also be found in wastewater, which poses health risks to the public. According to Florindo et al. [44], a variety of hydrophobic DES were picked to evaluate their removal efficiency of four neonicotinoid pesticides from aqueous environments. The highest extraction efficiency (up to 80%) was accomplished with DL-menthol and organic acids as DES. Another possible source of exposure to pesticides is milk, so the team of Jouyban et al. [45] developed a method to determine some of them by forming the DES in matrix and using them in the liquid phase extraction coupled with the DLLME of solidified organic droplets. ChCl and decanoic acid (molar ratio 1:2) were the best candidates for this extraction, providing LODs of 0.9–3.9 ng/mL, LOQs of 4.8–6.9 ng/mL, EFs of 320–445, and recoveries of 64–89%.

Musarurwa and Tavengwa [46] highlighted the importance of using DES in the liquid–liquid micro-extraction (LLME) so as to successfully extract pesticides from different food samples. One work of Nemati et al. [47] focused on the usage of solidified droplets in DLLME, coupled with stir bar sorptive extraction (SBSE) with the help of water-miscible DES, to extract acidic pesticides from tomato samples. Very high enrichment factors were introduced by this method (2530–2999) and pretty low LODs and LOQs were achieved (7–14 ng/L and 23–47 ng/L, respectively). In another work, Nemati et al. [48] effectively extracted pesticides from milk samples by using tetrabutylammonium chloride (TBAC) and dichloroacetic acid (1:1) as the elution solvents in their DLLME method that they developed based on organic polymers. Once more, low LODs (0.09–0.27 ng/mL) and LOQs (0.31–0.93 ng/mL) were obtained, with EFs of 162–188 and extraction efficiencies of 81–94%.

L-menthol with decanoic acid were used as hydrophobic DES by Lin et al. [49] so as to extract five fungicides in tea and fruit juices by freezing and, therefore, solidifying the DES to use them in ultrasound-assisted DLLME (UA-DLLME). After a thorough examination of the parameters involved in the efficiency of the extraction and, then, using the optimum ones, the relative recoveries were 72–109%, RSDs were 13.5–14.8%, and LODs were 0.75–8.45 µg/mL. Tea samples caught the eye of Torbati et al. [50], who decided to use water-miscible DES to extract and preconcentrate herbicides by LLE that were coupled with in-syringe DLLME in a narrow tube. This method allowed them to accomplish LODs of 2.6–8.4 ng/kg, with EF in the range of 350–445 and absolute recoveries of 70–89%.

3.2. Flavonoids, Phenolic and Other Bioactive Compounds

Skarpalezos and Detsi [51] emphasized the significance of DES when selected as the extraction means for flavonoids from different natural sources, including isoflavones, flavones, anthocyanins, flavonois, flavonones, and flavan-3-ols. Choi et al. [52] showed how important it is to use greener approaches to the usage of solvents for the extraction of bioactive compounds from a plethora of natural products, suggesting ILs and DES as the perfect candidates for this task. Plants house a range of bioactive compounds that can be isolated with innovative extraction techniques that involve both DES and natural deep eutectic solvents (NADES), according to Ivanović et al. [53]. DES and NADES can also be used to extract bioactive compounds from agricultural by-products as well, as Socas- Rodríguez et al. [54] highlighted in a recent review. Moreover, a comparison of conventional organic solvents with DES was thoroughly documented by Nakhle et al. [55] regarding, once again, the microextraction of bioactive compounds.

Bi et al. [56] were able to extract two flavonoids, amentoflavone and myricetin, by using ChCl with 1,4-butanediol (1:5 molar ratio) in water and heating at 70 °C for 40 min to get the optimal results. The LODs that were achieved were 0.07 and 0.09 µg/mL, respectively, for the two compounds, and RSDs were 2.72 and 3.06. Nam et al. [57] chose *Flos sophorae*, a common and traditional Chinese medicine, in order to extract flavonoids that are present in this flower, such as kaempferol, quercetin, and isorhamnertin glycosides. By synthesizing a DES of glycerol and L-proline (2:5), by freeze-drying and by applying ultrasound-assisted extraction (UAE), they were able to extract these organic compounds. Other experiments with common solvents, such as methanol, and other extraction techniques—for example, SPE with a C18 cartridge—were evaluated and compared to the usage of DES. It was found that the ultrasound-assisted extraction with DES gave lower absolute recoveries than the SPE method. Nevertheless, the advantages of safety, low-cost, and environmental friendliness that DES provide surpassed the lower extraction efficiency.

Genistein, daidzin, daidzein and genistin, four isoflavones, were successfully extracted with the application of NADES, according to Bajkacz et al. [58]. It was the first time that a solution consisting of 30% ChCl and citric acid with 1:1 molar ratio was used in the ultrasound-assisted extraction of isoflavones from soy products. The determination of the efficiency of the extraction was done by ultra-high performance liquid chromatography coupled with an ultraviolet detector (UHPLC-UV). Enrichment factors reached up to 598, with recoveries between 64.7 and 99.2%. LODs ranged from 0.06 to 0.14 µg/g and RSDs ranged from 1.0 to 5.9%. Meng et al. [59] used the same extraction technique—UAE, with NADES consisting of ChCl:1,2-propanediol (1:4 molar ratio)—so as to extract flavonoids from Pollen Typhae. Compared to methanol and 75% aqueous ethanol, NADES exhibited higher recoveries of 86.87–98.89% for kaempferol, quercetin, isohamnetin, and naringenin, rendering NADES a useful tool for this type of extractions.

Ternary DES, denoting deep-eutectic solvents consisting of the combination of three solvents, were used by Tang et al. [60] to extract two flavonoids from *Ginkgo biloba*. ChCl with oxalic acid and ethylene glycol, at 1:1:3 molar ratio, gave the best extraction efficiencies for myricetin and quercetin under the optimal conditions, which consisted of heating to 60 °C for 30 min. The extracted amount from this plant was 1.11 and 1.40 mg/g of sample for each compound respectively. Flowers of *Abelmoschus Manihot* (Linn.) *Medicus* were picked from Wan et al. [61] as the source of bioactive flavonoids that could be extracted with the aid of ChCl and acetic acid (1:2 molar ratio). The target compounds of interest were myricetin, isoquercitrin, and hyperoside, and they were successfully extracted in mild conditions (30 °C, 30 min, 35:1 mg/mL solid-to-solvent ratio), providing 1.11, 5.64, and 11.57 mg of each compound, respectively, per gram of the flower. Quercetin, isohamnetin, and kaempferol were extracted from vegetables with the benefit of betaine-D-mannitol as the ideal DES by Dai and Row [62].

Five Chinese herbal medicinal plants, *Notoginseng Radix* et *Rhizoma*, *Salviae Miltiorrhizae Radix* et *Rhizoma*, *Epimedii Folium*, *Rhei Rhizoma* et *Radix*, and *Berberidis Radix* were selected from the team of Duan et al. [63] to extract flavonoids, anthraquinones, alkaloids, saponins, and phenolic acids as bioactive compounds by applying UAE with DES. A variety of combinations between ChCl, L-proline, and betaine as HBA were examined and tailored for the needs of each extraction. NADES based on ChCl and malic acid caught the eye of Radošević et al. [64], who extracted a variety of bioactive organic compounds from plants, such as anthocyanins and phenolic compounds. Their cytotoxicity was also evaluated in vitro toward two human tumor cell lines, MCF-7 and HeLa. Bioactive compounds, mainly scoparone, quercetin, and rutin, were successfully extracted from Herba *Artemisiae Scopariae* with ILs and DES from Ma et al. [65]. Amounts of 554.32, 899.73, and 10,275.92 µg/g of each compound were able to be extracted under the ideal conditions.

Anthocyanins contained in blueberries were extracted with ternary NADES by da Silva et al. [66]. ChCl:glycerol:citric acid, in molar ratios of 0.5:2:0.5, was used as the NADES that provided similar or higher extraction efficiencies from other conventional organic solvents. Wine lees are another known source of anthocyanins, which Bosiljkov et al. [67] were able to extract by UAE in almost 30 min with the help of the environmentally friendly ChCl-malic acid deep-eutectic solvent. Lactic acid-glucose and ChCl-1,2-propanediol are perfect candidates for the green extraction of anthocyanins from *Catharanthus roseus*, as indicated by Dai et al. [68].

Phenolic compounds are of high importance and interest from scientists all around the world, as they are great antioxidants. Ruesgas-Ramón et al. [69] pinpointed the utilization of DES in the extraction of that group of organic compounds. Redha [70] motivates researchers and suggests to use DES when extracting phenolic compounds from natural sources, as these solvents take into account the principles of Green Analytical Chemistry (GAC). In another study, Dai et al. [71] were able to isolate phenolic metabolites of different polarities from *Carthamus tinctorius* by applying natural deep-eutectic solvents, highly suggesting their usage in the extraction of bioactive compounds from natural sources. Bubalo et al. [72] applied both microwave-assisted and ultrasound-assisted extraction (MAE, UAE) with ChCl-oxalic acid as the DES that was able to extract phenolics from samples of grape skin. Petunidin-3-O-glucoside, delphinidin-3-O-glucoside, cyanidin-3-O-glucoside, malvidin-3-O-glucoside, peonidin-3-O-glucoside, quercetin-3-O-glucoside, and (+)-catechin were the target compounds that were able to be extracted, with LODs ranging from 0.05–0.37 mg/L and impressively low RSD of 0.30–0.96%. García et al. [73] were drawn by the DES as well, especially ChCl-xylitol and ChCl-1,2-propanediol, which they used to extract apigenin, oleacin, oleocanthal, tyrosol, hydroxytyrosol, luteolin, and 1-acetoxypinoresinol from virgin olive oil, increasing the yield to 20–33% and 67.3–68.3%, respectively, for each DES mentioned when compared to conventional organic solvents. Chanioti et al. [74] used NADES for the isolation of phenolic compounds from olive pomace with the usage of non-ordinary extraction techniques, such as microwave-assisted extraction (MAE), UAE, homogenate extraction (HAE), or high hydrostatic pressure-assisted extraction (HHPAE). The NADES that were used for this work were combinations of ChCl with maltose, glycerol, lactic acid, and citric acid. ChCl with lactic acid and citric acid exhibited the best results. Shishov et al. [75] developed a method that involved rotating disk sorptive extraction (RDSE) to extract phenolic compounds from vegetable oil by coating the surface of the disk with ChCl. The DES were formed by the reaction of the coating and the target organic compounds, and then, the phenolic compounds, vanillic acid, tyrosol, gallic acid, thymol, p-coumarinic acid, syringaldehyde, and protocatechuic acid, were eluted in an aqueous phase by the decomposition of the DES. The proposed method exhibited LODs of 10–60 µg/L and absolute recoveries between 66 and 87%.

Alam et al. [76] presented the studies of many scientists around the usage of ChCl-based DES as a possible, greener, and cheaper means of extraction of phenolic compounds from biomass. Orange peel waste was a source of polyphenols that caught the eye of Ozturk et al. [77], who tried and succeeded in extracting them with ChCl-ethylene glycol (1:4 molar ratio). Surprising enough, the peels are high in content of ferulic acid, gallic acid, and p-coumaric acid, which were extracted via solid–liquid extraction (SLE). By-products from agro-foods also contain phenolic compounds that can be isolated by UAE and lactic acid-glucose-water as the DES, according to Fernández et al. [78]. The extraction efficiency was validated by HPLC-DAD, where 14 phenols were evaluated from pear, tomato, and olive by-products. LODs varied from 0.0006 to 0.0891 µg/g, with the time of each analysis taking up to 18 min. El Kantar et al. [79] extracted polyphenols, with lactic acid-glucose as DES, from grapefruit peels by a solid–liquid extraction (SLE) with a 10:1 ratio of liquid to solid. Fu et al. [80] applied pulse-ultrasonication-assisted extraction (P-UAE) with NADES

(ChCl-malic acid) as a greener and more efficient approach in order to extract significant phenolic compounds from peels of *Carya cathayensis* Sarg. Wheat waste biomass was chosen by Cherif et al. [81] as the main source of phenolic compounds to be extracted by UAE and, further, thermal treatment of glycerol-citric acid-glycine as the deep-eutectic solvent. With this combination, they managed to isolate 94.62 mg of ferulic acid equivalents per gram of dry mass. Rutin, a flavonoid, could be solubilized more in NADES, based on ChCl or glycerol, than in water—almost 660–1577 times more—and thus, it could be extracted by tartary buckwheat hull with UAE, as instructed by Huang et al. [82]. Additionally, 9.5 mg of rutin per g of sample were able to be isolated, with the extraction efficiency reaching 95%.

In a review, Bubalo et al. [83] introduced the usage of greener solvents for the extraction of biologically active compounds from plants, of which NADES played an important role. Native Greek medicinal plants, mainly sage, mint, dittany, marjoram, and fennel, were selected as the resource of polyphenols to be extracted by Bakirtzi et al. [84] by novel NADES, synthesized from lactic acid with ChCl, ammonium acetate, sodium acetate, or glycine-water in 3:1, 3:1, 3:1 and 3:1:3 molar ratios, respectively. *Lonicerae japonicae flos*, a common Chinese medicinal plant, was the main sample from which five phenolic acids were extracted (caffeic acid, chlorogenic acid, 4,5-dicaffeoylquinic acid, 3,5-dicaffeoylquinic acid, and 3,4-dicaffeoylquinic acid) by Peng et al. [85]. MAE was the main technique used for this feat, while, as DES, ChCl with 1,3-butanediol was used in 1:6 molar ratio. Recoveries for this work ranged from 79.25 to 86.01% for the five organic compounds of interest. Caffeine, rutin, ferulic acid, naringin, rosmarinic acid, and 7-methylrosmanol were extracted from *Rosmarinus officinalis* L. by ChCl-1,2-propanediol in water, providing a total amount of phenolics of 20,588 µg/g. MAE with ChCl-glycerol (1:2) and 20% water constituted the main extraction means for Gao et al. [86] when they developed the method of isolation of phenolic compounds from the leaves of mulberry (*Morus Alba* L.). A total amount of 8.352 mg of phenolics per gram of leaves was able to be extracted, under 18 min at 66 °C, by using 20 mL of DES per gram of sample. Lastly, they utilized a macroporous resin, which was put in a chromatography column to separate the organic compounds of interest from the DES. Park et al. [87] were able to extract two phenolic compounds, caffeic acid and chlorogenic acid, from Herba *Artemisiae Scopariae* by UAE of tetramethyl ammonium chloride-urea (1:4 molar ratio) as DES, which was further mixed with methanol-water (60:40 *v/v*). As a liquid-to-solid ratio 10:1 was chosen as the optimal one, which exhibited recoveries of 97.3–100.4%, with RSDs lower than 5%. Furthermore, they were able to extract 0.31 mg and 9.35 mg of each compound, respectively, per gram of sample. Last but not least, Liu et al. [88] used betaine-ethanediol (1:4 molar ration) in 30% water, at 60 °C for 30 min, to extract hesperidin, neohesperidin, naringin, and nariturin from *Aurantii Fructus*. Compared to the methanol extracts, the DES extracts contained higher amounts of isolated organic compounds, namely 3.03, 35.94, 83.98, and 8.39 mg/g. respectively. Some of the organic compounds found in the *Ixora javanica* flower have skin-lightening cosmetic and antioxidant properties, and Oktaviyanti et al. [89] developed a method to extract them. ChCl-propylene glycol, with a 1:1 molar ratio, was the ideal candidate for the application in the UAE of these compounds. The extraction time was only 5 min at 57 °C, and 33 mg of quercetin equivalent per gram of dried sample were able to be obtained.

Pontes et al. [90] used ChCl-acetic acid (1:2 molar ratio) as their ideal DES in the extraction of phenolics from olive leaves, and it was able to extract more phenolic compounds than ethanol under the same conditions. Mogaddam et al. [91] were able to develop a method of extracting three phenolic compounds with organic-free solvents based on DES (tetrabutylammonium chloride-hydroquinone) and by exploiting the LLE/DLLME with heating. LODs reported were 0.13–0.42 ng/mL, EFs were 370–445, and extraction recoveries were from 74 to 89%, with RSDs being equal to or lower than 7.4%. Li et al. [92] tried to tackle the problem of phenols being contained in wastewater, which later result in aqueous environments, by developing a method of isolating them with L-proline-decanoic acid during a 60 min period at 50 °C. The reusage of the NADES that was used was studied, and it was found that it could be reused up to six times, with extraction efficiencies being 62% for the first cycle and 57% up to the sixth cycle.

In Table 2, the methods developed for the extraction of the organic compounds from the first two subgroups are summarized and presented.

Table 2. Summary and analytical performance of the methods discussed in Sections 3.1 and 3.2.

Group of Compounds	DES	Molar Ratio	Extraction Technique	Recovery (%)	LOD [1]	Reference
Triazole fungicides	ChCl: 4-chlorophenol	1:2	HS-SDME	94–97	0.82–1.0 mg/L	[41]
Pesticides	Menthol: dichloroacetic acid	1:2	DLLME	56–86	0.32–1.2 ng/g	[42]
Pesticides	ChCl: p-chlorophenol	1:2	LPME	56–93	0.13–0.31 ng/mL	[43]
Neonicotinoid pesticides	Menthol: dodecanoic acid	2:1	LLE	80	N/R [2]	[44]
Pesticides	ChCl: decanoic acid	1:2	DLLME	64–89	0.9–3.9 ng/mL	[45]
Acidic pesticides	ChCl: ethylene glycol	1:2	DLLME-SBSE	76–90	7–14 ng/L	[47]
Pesticides	TBAC: dichloroacetic acid	1:1	DLLME	81–94	0.09–0.27 ng/mL	[48]
Fungicides	Menthol: decanoic acid	1:1	UA-DLLME	72–109	0.75–8.45 µg/mL	[49]
Herbicides	ChCl: butyric acid	1:2	LLE-DLLME	70–89	2.6–8.4 ng/kg	[50]
Flavonoids	ChCl: 1,4-butanediol	1:5	UAE	N/R	0.07, 0.09 µg/mL	[56]
Flavonoids	Glycerol: L-proline	2:5	UAE, SPE	81–87	N/R	[57]
Isoflavones	ChCl: citric acid	1:1	UAE	64.7–99.2	0.06–0.14 µg/mL	[58]
Flavonoids	ChCl: 1,2-propanediol	1:4	UAE	86.7–98.9	0.05–0.14 µg/mL	[59]
Flavonoids	ChCl: oxalic acid: ethylene glycol	1:1:3	LLE	N/R	1.11–1.40 mg/g	[60]
Flavonoids	ChCl: acetic acid	1:2	UAE	N/R	1.11–11.57 mg/g	[61]
Flavonoids	Betaine: D-mannitol	N/R	UAE	91.7–95.8	0.14–0.17 µg/mL	[62]
Flavonoids	Different types	Diff. [3]	UAE	93.8–107.7	0.14–0.22 µg/mL	[63]
Bioactive compounds	ChCl: malic acid	1:1	UAE	N/R	N/R	[64]
Bioactive compounds	ILs, ChCl based DES	Diff.	Reflux	N/R	N/R	[65]
Anthocyanins	ChCl: glycerol:citric acid	0.5:2:0.5	UAE	N/R	0.02 mg/L	[66]
Anthocyanins	ChCl: malic acid	Diff.	UAE	N/R	0.15–0.28 mg/L	[67]
Anthocyanins	Lactic acid: glucose	1:2	Stirring	N/R	N/R	[68]
Phenolic metabolites	Different types	Diff.	N/R	75–97	N/R	[71]
Phenolics	ChCl: oxalic acid	1:1	MAE, UAE	N/R	0.05–0.37 mg/L	[72]
Phenolics	ChCl: 1,2-propanediol	1:1	Vortex	N/R	N/R	[73]
Phenolics	ChCl: lactic acid	1:2	MAE, HUE, UAE	N/R	N/R	[74]
Phenolics	ChCl + phenolics	N/R	RDSE	66–87	10–60 µg/L	[75]
Polyphenols	ChCl: ethylene glycol	1:4	SLE	N/R	N/R	[77]
Phenolics	Lactic acid: glucose	5:1	UAE	86.0–109.6	0.6–89.1 ng/g	[78]
Polyphenols	Lactic acid: glucose	5:1	SLE	N/R	N/R	[79]
Phenolics	ChCl: malic acid	1.5:1	P-UAE	N/R	N/R	[80]
Phenolics	Glycerol: citric acid: glycine	Diff.	UAE	N/R	N/R	[81]
Flavonoids	ChCl: glycerol	1:1	UAE	95	N/R	[82]
Polyphenols	Lactic acid: glycine:water	3:1:3	UAE	N/R	N/R	[84]
Phenolic acids	ChCl: 1,3-butanediol	1:6	MAE	79.2–86.0	N/R	[85]
Phenolics	ChCl: glycerol	1:2	MAE	77.8–83.8	0.15–0.78 µg/mL	[86]
Phenolics	Tetramethyl ammonuium chloride:urea	1:4	UAE	97.3–100.4	N/R	[87]
Flavanones	Betaine: ethanediol	1:4	Heated LLE	97.0–101.6	N/R	[88]
Bioactive compounds	ChCl: propylene glycol	1:1	UAE	N/R	N/R	[89]
Phenolics	ChCl: acetic acid	1:2	Thermo-shaking	N/R	N/R	[90]
Phenolics	TBAC: hydroquinone	1:2	LLE/DLLME	74–89	0.13–0.42 ng/mL	[91]
Phenols	L-proline: decanoic acid	1:4.2	Stirring	57–62	N/R	[92]

[1] Limit of Detection, [2] Not reported, [3] Different molar ratios tested.

3.3. Pharmaceutical Compounds and Preservatives

An antibacterial sulfonamide named sulfamerazine was the compound of interest to be extracted by Liu et al. [93], who applied pipette-tip solid-phase extraction (PT-SPE) with graphene modified by DES (ChCl-ethylene glycol) as the adsorbent. The limit of detections that was calculated from this method were 0.01 µg/mL, with relative recoveries of 91.01–96.82% and RSDs below 3.84%. Paracetamol, a common analgetic and antipyretic drug, was able to be isolated from synthetic urea with an inexpensive, new, and simple method by Dogan et al. [94]. A shaker-assisted deep eutectic solvent microextraction (SA-DES-ME) was used as the extraction technique, with betaine-oxalic acid (1:2 molar ratio) playing the role of the DES. The analytical performance of the proposed method was great, with recoveries varying from 94.2–107.1%, LOD of 14.9 µg/L, and very low RSDs (below 3.3%). Li et al. [95] wanted to isolate thiamphenicol and chloromycetin, two antibiotics, from milk in order to purify it. This was achieved by using molecular imprinted polymers (MIPs) in the solid phase extraction (SPE) of the two organic compounds, along with deep eutectic solvents, with 87.02% and 91.23% being the recoveries of the two antibiotics, respectively.

Antiviral organic compounds that are used as drugs to combat viral-related diseases can also be extracted with the use of DES. Jouyban et al. [96] managed to isolate sofosbuvir and daclatasvir from urine with homogenous LLE by applying the DES, p-aminophenol-tetrabutyl ammonium chloride, as a more environmentally friendly alternative that is as effective as conventional solvents. RSDs were lower than 9.3%, low LODs were achieved with sofosbuvir at 1.3 and daclatasvir at 1.0 µg/L, as well as EFs of 90 and 96, respectively, and lastly, extraction recoveries of 90 and 96%. Quinine was successfully able to be extracted via its hydrogen bonding with hydrophilic alcohol-based NADES by Fan et al. [97]. The same NADES could be reused by performing a back-extraction with 1% v/v aqueous acetic acid. Acetic acid with menthol served as the deep eutectic solvent to isolate diclofenac from aqueous media. The method developed by Kurtulbaş et al. [98] could reach up to 80% removal of the pharmaceutical compound. Antibiotic residues from honey were extracted with a method developed by Shahi et al. [99] that involved the synthesis of nanocomposites of multiwall carbon nanotubes with urea-formaldehyde and their application in SPE. DES with DLLME were also combined with the above method, aiding in the extraction and the obtaining of improved results, such as LODs of 0.32–0.86 ng/g and LOQs of 1.1–2.9 ng/g, respectively, with the RSD calculated as equal to or lower than 9.1%.

Hadi et al. [100] developed a method to extract tocotrienols and tocopherols from palm oil with the usage of ChCl, with carboxylic acids as the means of extraction. A total of 18,525 mg of tocopherols per kg of sample were able to be isolated with this method. Choline chloride with sucrose, ethylene glycol, and ethanolamine were used in the LLE of α-tocopherols by the team of Mohammadi et al. [101]. α-, γ-, and δ-tocopherols were able to be extracted from soybean oil deodorizer distillate by Liu et al. [102] with phenolic DES, ChCl with p-cresol. The total extraction efficiency of all tocopherols reached 77.6%. Ge et al. [103] managed to synthesize, in situ, the hydrophobic deep eutectic solvent comprised of decanoic acid and DL-menthol, and they used it in the liquid–liquid microextraction (LLME) of four parabens in water samples. With this method, LODs of 0.6–0.8 ng/mL and RSDs less than 7.2% were able to be achieved.

3.4. Polycyclic Aromatic Hydrocarbons, Volatile Organic Compounds and Pollutants

Hashemie et al. [104] wrote an extensive review on the usage of ionic liquids and DES in sorptive-based extraction techniques in order to extract environmental pollutants, urging scientists to switch to more greener alternatives for this type of sample preparation. Among these pollutants, PAHs play an important role.

In one interesting work, Shakirova et al. [105] used fatty acids that came from the in situ alkaline hydrolysis of triglycerides in milk so as to synthesize NADES between them, with menthol or thymol being the second component. With this method, they were able to extract thirteen polycyclic aromatic compounds (PAHs) from milk: pyrene, anthracene, phenanthrene, chrysene, fluoranthene, fluorene, naphthalene, benz[*a*]anthracene, benzo[*a*]pyrene, benzo[*b*]fluoranthene, benzo[*k*]fluoranthene, benzo[*ghi*]perylene, and dibenz[*a,h*]anthracene. LODs varied from 0.002–0.09 µg/kg, with 70–91% being the extraction efficiencies and RSDs being lower than 5.2%. Nie et al. [106] chose volatile organic compounds (VOCs) as their compounds of interest to be extracted from tobacco, utilizing MAE with a DES that was coupled with headspace solid phase microextraction (HS-SPME), followed by gas chromatography with a mass spectrometer (GC-MS) for their determination.

3.5. Polysaccharides, Pigments and Terpenes

A famous Chinese plant that has many invigorating health benefits when consumed, *Dioscorea opposita* Thunb, caught the eye of Zhang and Wang [107] who developed a method of extraction of polysaccharides by using UAE with ChCl and 1,4-butanediol, constituting the deep eutectic solvent. Under the best conditions, with the temperature reaching 94 °C, the water content of the extraction with the DES being 32.89% v/v, and the extraction time being less than 45 min, they were able to get the best results. Shang et al. [108] used MAE with the same DES, ChCl-1,4-butanediol with 1:5 molar ratio, in order to be able to extract polysaccharides from *Fucus vesiculosus*, attaining 116.33 mg of the target compounds per gram of sample. Optimal conditions for this isolation were 35 min of extraction at 168 °C, with 32% content of water. Furthermore, the in vitro evaluation assay of their biological activity was done in HeLa cells, highlighting their antioxidant activity. Edible brown seaweed, called *Sargassum horneri*, is another source of polysaccharides that attracted Nie et al. [109]. With their method that consisted of the usage of ChCl-1,2-propanediol with 1:2 molar ratio in 30% of water (v/v), of the heating to 70 °C, and of the solid-to-liquid ratio reaching 1:30 g/mL, they managed to extract them with high efficiencies. Once again, the in vitro biological activity was assessed. Das et al. [110] developed a method of isolating κ-carrageenan from *Kappaphycus alvarezii* that involved hydrated DES, with 10% hydrated ChCl-glycerol (1:2 molar ratio) being the best candidate for the extraction, giving a yield of 60.25%.

Citric acid with glucose, with 1:1 molar ratio, served as the natural deep eutectic solvent for Liu et al. [111] who wanted to extract natural pigments, such as curcumin, demethoxycurcumin, and bis-demethoxycurcumin, from *Curcuma longa* L. Subsequently, they performed SPE to recover the NADES and reuse it. This method exhibited impressively low RSDs, 0.46–0.91%, and LODs ranging from 0.25 to 0.37 mg/L. Patil et al. [112] proposed a green approach of extracting curcuminoids from *Curcuma longa* by UAE with DES, namely ChCl-lactic acid, with a 1:1 molar ratio. The maximum amount of curcuminoids that were able to be extracted was 77.13 mg/g. Curcumin was extracted and preconcentrated from herbal tea and food samples by a LLME emulsification of DES, and its performance was determined by UV-vis, as reported by Aydin et al. [113]. The pH that was selected for this method was the value 4, while other parameters were also studied. Under the best conditions, LODs were 2.86 µg/L, RSDs were lower than 6%, and recoveries varied from 96 to 102%. Zhang et al. [114] extracted astaxanthin from shrimp by-products by applying a UAE with DES and compared this method with the extraction using ethanol. It was found that the proposed method was able to obtain more of the organic compound than with the conventional solvent (146 µg/g with DES and 102 µg/g with ethanol). Rhodamine B, a colorant that is no longer used, by law, due to its carcinogenic properties, was isolated from chili oil by Wang et al. [115] by doing an extraction with ChCl-ethylene glycol and determining the efficiency by ultra-high performance liquid chromatography coupled with a fluorescence detector (UHPLC-FD).

Su et al. [116] applied UAE with two different DES, ChCl-urea and betaine-ethylene glycol, to extract terpene trilactones from the leaves of *Ginkgo biloba*. The latter DES worked best and gave the highest extraction efficiencies that reached 99.37% and 1.94 mg/g.

3.6. Other Organic Compounds

Benzophenone-type UV filters can be extracted and preconcentrated with the benefit of hydrophobic DES by an air-assisted dispersive liquid–liquid mixroextraction (AA-DLLME) from aqueous samples, as reported by Ge et al. [117]. DL-menthol with decanoic acid was used as the deep eutectic solvent and the proposed method was applied successfully, providing 88.8–105.9% relative recoveries. There were two common organic solvents, acetonitrile and butanol, extracted from aqueous solutions, with LLE and DES by Rabhi et al. [118], by forming ternary systems.

An environmentally friendly extraction method that involved ChCl-1,3-butanediol in the ball mill-assisted extraction of tanshinones (tanshinone I, tanshinone II A, and cryptotanshinone) from plants was developed by Wang et al. [119]. Extracted amounts reached 0.181, 0.421, and 0.176 mg/g respectively for each compound, with RSDs lower than 1.9%, recoveries of 96.1–103.9%, and LODs of 5–8 ng/mL being reported. Catechin compounds, such as (+)-epicatechin gallate, (−)-epigallocatechin gallate, and catechin were extracted from Chinese green tea by the usage of deep eutectic solvents. Zhang et al. [120], with this method, demonstrated an eco-friendly pathway that could achieve 35.25, 114.2, and 3.629 mg of an extracted compound, respectively, per gram of sample, with extraction efficiencies ranging from 82.7 to 97.0%.

Zhao et al. [121] extracted the essential oil out of cumin seeds (*Cuminum cyminum* L.) by a MAE that involved NADES and was done in three steps. ChCl-L-lactic acid (1:3 molar ratio) was the ideal candidate for this work, which helped this team to extract 58 certified essential oil components, with the major components consisting of cuminol, moslene, terpineol, and cuminal. The highest yield reported was 2.22% of essential oil from the seeds. MAE, with NADES consisting of ChCl and oxalic acid (1:1 molar ratio), were involved in the extraction of essential oil from turmeric (*Curcuma longa* L.). The yield that was obtained was 0.85%, while 49 organic compounds were determined by GC-MS [122].

Milani et al. [123] chose tetraethylammonium chloride with ethylene glycol, at 1:2 molar ratio, as their deep eutectic solvent in the UAE of rebaudioside A and stevioside, two steviol glycosides found in *Stevia rebaudiana*. Tests with conventional extraction methods showed that the proposed method was three times more efficient.

In Table 3, all the discussed methods from Sections 3.3–3.6 are summarized, and their analytical performances are demonstrated.

Table 3. Summary and analytical performance of the methods discussed from Sections 3.3–3.6.

Group of Compounds	DES	Molar Ratio	Extraction Technique	Recovery (%)	LOD [1]	Reference
Sulfonamide	ChCl: ethylene glycol	1:2	PT-SPE	91.0–96.8	0.01 µg/mL	[93]
Analgetic	Betaine: oxalic acid	1:2	SA-DES-ME	94.2–107.1	14.9 µg/L	[94]
Antibiotics	ChCl: glycerol	1:2	MIPs-SPE	87.0–91.2	N/R [2]	[95]
Antivirals	TBAC: p-aminophenol	1:2	LLE	90–96	1.0–1.3 µg/L	[96]
Antimalarial	Menthol: fenchyl alcohol	1:1	Stirring	101	N/R	[97]
NSAID [3]	Menthol: acetic acid	Diff. [4]	LLE	80	N/R	[98]
Antibiotics	TBAC: butanol	1:1	SPE, DLLME	84–99	0.32–0.86 ng/g	[99]
Tocotrienols, tocopherols	ChCl: malonic acid	1:1	LLE	93.0–99.8	N/R	[100]
α-tocopherols	ChCl: sucrose	1:2	LLE	N/R	N/R	[101]
α-, γ-, δ-tocopherols	ChCl: p-cresol	1:2	Vortex	77.6	N/R	[102]
Parabens	Menthol: decanoic acid	2:1	LLME	69.1–78.5	0.6–0.8 mg/mL	[103]
PAHs	Menthol or thymol with fatty acids	Diff.	Stirring	70–91	2–90 ng/kg	[105]
VOCs	ChCl: urea	1:3	MAE, HS-SPME	N/R	N/R	[106]
Polysaccharides	ChCl: 1,4-butanediol	N/R	UAE	N/R	N/R	[107]
Polysaccharides	ChCl: 1,4-butanediol	1:5	MAE	91.2	N/R	[108]
Polysaccharides	ChCl: 1,2-propanediol	1:2	UAE	N/R	N/R	[109]
Polysaccharides	ChCl: glycerol	1:2	Thermal treatment	60.3	N/R	[110]
Natural pigments	Citric acid: glucose	1:1	SPE	88.5–94.4	0.25–0.37 mg/L	[111]
Curcuminoids	ChCl: lactic acid	1:1	UAE	N/R	N/R	[112]
Curcuminoids	ChCl: phenol	1:4	VA-LLME	96–102	2.86 µg/L	[113]
Pigment	Different types	Diff.	UAE	N/R	N/R	[114]
Pigment	ChCl: ethylene glycol	Diff.	N/R	N/R	N/R	[115]
Terpene trilactones	Betaine: ethylene glycol	Diff.	UAE	99.4	N/R	[116]
Benzophenone-type UV filters	Menthol: decanoic acid	1:1	AA-DLLME	88.8–105.9	0.05–0.2 ng/mL	[117]
Organic solvents	Menthol: capric acid	2:1	LLE	N/R	N/R	[118]
Tanshinones	ChCl: -1,3-butanediol	N/R	BMAE	96.1–103.9	5–8 ng/mL	[119]
Catechins	Different types	Diff.	LLE	82.7–97.0	N/R	[120]
Essential oils	ChCl: L-lactic acid	1:3	MAE	N/R	N/R	[121]
Essential oils	ChCl: oxalic acid	1:1	MAE	N/R	N/R	[122]
Steviol glycosides	Tetraethylammonium chloride: ethylene glycol	1:2	UAE	N/R	N/R	[123]

[1] Limit of Detection, [2] Not reported, [3] Non-steroidal anti-inflammatory drugs, [4] Different molar ratios tested.

4. Conclusions

As discussed herein, deep eutectic solvents and natural deep eutectic solvents can have a vast variety of applications in the extraction and preconcentration of organic compounds. Some of the groups of organic compounds that were isolated by their usage involve pesticides, fungicides, herbicides, bioactive compounds, flavonoids, phenolic and pharmaceutical compounds, pigments, terpenes, polycyclic aromatic compounds, volatile organic compounds and other pollutants, preservatives, or other organic compounds.

This quite big range of possible extracted compounds renders DES an important tool to be exploited. Additionally, they offer both high absolute and relative recoveries when coupled with extraction or microextraction techniques, very low limits of detection and quantification, as well as quite low relative standard deviations. Additionally, their low cost, environmental friendliness, and simplicity attract researchers to use them and to find new combinations that can combat conventional organic solvents that pose risks to the health of the user or to organisms in general. Their properties are still being studied, and hopefully, their commercialization will soon be done.

Author Contributions: Conceptualization, V.S.; methodology, V.S and O.-E.P.; writing—original draft preparation, V.S. and O.-E.P.; writing—review and editing, V.S. and O.-E.P.; supervision, V.S.; project administration, V.S. All authors have read and agreed to the published version of the manuscript.

Funding: This research received no external funding.

Institutional Review Board Statement: Not applicable.

Informed Consent Statement: Not applicable.

Data Availability Statement: Not applicable.

Conflicts of Interest: The authors declare no conflict of interest.

References

1. Armenta, S.; Garrigues, S.; Esteve-Turrillas, F.A.; de la Guardia, M. Green extraction techniques in green analytical chemistry. *Trends Anal. Chem.* **2019**, *116*, 248–253. [CrossRef]
2. Plastiras, O.E.; Andreasidou, E.; Samanidou, V. Microextraction techniques with deep eutectic solvents. *Molecules* **2020**, *25*, 6026. [CrossRef] [PubMed]
3. Paiva, A.; Craveiro, R.; Aroso, I.; Martins, M.; Reis, R.L.; Duarte, A.R.C. Natural deep eutectic solvents—Solvents for the 21st century. *ACS Sustain. Chem. Eng.* **2014**, *2*, 1063–1071. [CrossRef]
4. Cunha, S.C.; Fernandes, J.O. Extraction techniques with deep eutectic solvents. *Trends Anal. Chem.* **2018**, *105*, 225–239. [CrossRef]
5. Espino, M.; Fernandez, M.D.; Gomez, F.J.V.; Silva, M.F. Natural designer solvents for greening analytical chemistry. *Trends Anal. Chem.* **2016**, *76*, 126–136. [CrossRef]
6. Vanda, H.; Dai, Y.T.; Wilson, E.G.; Verpoorte, R.; Choi, Y.H. Green solvents from ionic liquids and deep eutectic solvents to natural deep eutectic solvents. *Comptes Rendus Chim.* **2018**, *21*, 628–638. [CrossRef]
7. Florindo, C.; Branco, L.C.; Marrucho, I.M. Quest for green-solvent design: From hydrophilic to hydrophobic (Deep) eutectic solvents. *ChemSusChem* **2019**, *12*, 1549–1559. [CrossRef]
8. Tang, B.; Zhang, H.; Row, K.H. Application of deep eutectic solvents in the extraction and separation of target compounds from various samples. *J. Sep. Sci.* **2015**, *38*, 1053–1064. [CrossRef]
9. Kissoudi, M.; Samanidou, V. Recent advances in applications of ionic liquids in miniaturized microextraction techniques. *Molecules* **2018**, *23*, 1437. [CrossRef]
10. Pena-Pereira, F.; Namiesnik, J. Ionic liquids and deep eutectic mixtures: Sustainable solvents for extraction processes. *ChemSusChem* **2014**, *7*, 1784–1800. [CrossRef]
11. Romero, A.; Santos, A.; Tojo, J.; Rodriguez, A. Toxicity and biodegradability of imidazolium ionic liquids. *J. Hazard. Mater.* **2008**, *151*, 268–273. [CrossRef] [PubMed]
12. Berthod, A.; Ruiz-Angel, M.J.; Carda-Broch, S. Recent advances on ionic liquid uses in separation techniques. *J. Chromatogr. A* **2018**, *1559*, 2–16. [CrossRef] [PubMed]
13. Smith, E.L.; Abbott, A.P.; Ryder, K.S. Deep Eutectic Solvents (DESs) and their applications. *Chem. Rev.* **2014**, *114*, 11060–11082. [CrossRef]
14. Abbott, A.P.; Capper, G.; Davies, D.L.; Munro, H.L.; Rasheed, R.K.; Tambyrajah, V. Preparation of novel, moisture-stable, Lewis-acidic ionic liquids containing quaternary ammonium salts with functional side chains. *Chem. Commun.* **2001**, *2001*, 2010–2011. [CrossRef] [PubMed]

15. Shishov, A.; Bulatov, A.; Locatelli, M.; Carradori, S.; Andruch, V. Application of deep eutectic solvents in analytical chemistry. A review. *Microchem. J.* **2017**, *135*, 33–38. [CrossRef]
16. Dai, Y.T.; van Spronsen, J.; Witkamp, G.J.; Verpoorte, R.; Choi, Y.H. Natural deep eutectic solvents as new potential media for green technology. *Anal. Chim. Acta* **2013**, *766*, 61–68. [CrossRef]
17. Hansen, B.B.; Spittle, S.; Chen, B.; Poe, D.; Zhang, Y.; Klein, J.M.; Horton, A.; Adhikari, L.; Zelovich, T.; Doherty, B.W.; et al. Deep eutectic solvents: A review of fundamentals and applications. *Chem. Rev.* **2021**, *121*, 1232–1285. [CrossRef]
18. Dai, Y.T.; van Spronsen, J.; Witkamp, G.J.; Verpoorte, R.; Choi, Y.H. Ionic liquids and deep eutectic solvents in natural products research: Mixtures of solids as extraction solvents. *J. Nat. Prod.* **2013**, *76*, 2162–2173. [CrossRef]
19. Tang, W.; An, Y.; Row, K.H. Emerging applications of (micro) extraction phase from hydrophilic to hydrophobic deep eutectic solvents: Opportunities and trends. *Trends Anal. Chem.* **2021**, *136*, 116187. [CrossRef]
20. van Osch, D.; Dietz Chjt van Spronsen, J.; Kroon, M.C.; Gallucci, F.; Annaland, M.V.; Tuinier, R. A search for natural hydrophobic deep eutectic solvents based on natural components. *ACS Sustain. Chem. Eng.* **2019**, *7*, 2933–2942. [CrossRef]
21. Zdanowicz, M.; Wilpiszewska, K.; Spychaj, T. Deep eutectic solvents for polysaccharides processing. A review. *Carbohydr. Polym.* **2018**, *200*, 361–380. [CrossRef] [PubMed]
22. Makos, P.; Slupek, E.; Gebicki, J. Hydrophobic deep eutectic solvents in microextraction techniques-A review. *Microchem. J.* **2020**, *152*, 104384. [CrossRef]
23. Dwamena, A.K. Recent advances in hydrophobic deep eutectic solvents for extraction. *Separations* **2019**, *6*, 9. [CrossRef]
24. Santana-Mayor, A.; Rodriguez-Ramos, R.; Herrera-Herrera, A.V.; Socas-Rodriguez, B.; Rodriguez-Delgado, M.A. Deep eutectic solvents. The new generation of green solvents in analytical chemistry. *Trends Anal. Chem.* **2021**, *134*, 116108. [CrossRef]
25. Yang, Z. Natural Deep Eutectic Solvents and Their Applications in Biotechnology. In *Application of Ionic Liquids in Biotechnology*; Itoh, T., Koo, Y.M., Eds.; Springer: Cham, Switzerland, 2019; pp. 31–59.
26. Chen, J.N.; Li, Y.; Wang, X.P.; Liu, W. Application of deep eutectic solvents in food analysis: A review. *Molecules* **2019**, *24*, 4594. [CrossRef]
27. Zhao, R.T.; Pei, D.; Yu, P.L.; Wei, J.T.; Wang, N.L.; Di, D.L.; Liu, Y.W. Aqueous two-phase systems based on deep eutectic solvents and their application in green separation processes. *J. Sep. Sci.* **2020**, *43*, 348–359. [CrossRef]
28. Li, L.N.; Liu, Y.M.; Wang, Z.T.; Yang, L.; Liu, H.W. Development and applications of deep eutectic solvent derived functional materials in chromatographic separation. *J. Sep. Sci.* **2021**, *44*, 1098–1121. [CrossRef]
29. Tang, B.K.; Row, K.H. Recent developments in deep eutectic solvents in chemical sciences. *Mon. Chem.* **2013**, *144*, 1427–1454. [CrossRef]
30. Abbott, A.P.; Boothby, D.; Capper, G.; Davies, D.L.; Rasheed, R.K. Deep eutectic solvents formed between choline chloride and carboxylic acids: Versatile alternatives to ionic liquids. *J. Am. Chem. Soc.* **2004**, *126*, 9142–9147. [CrossRef]
31. Sereshti, H.; Jamshidi, F.; Nouri, N.; Nodeh, H.R. Hyphenated dispersive solid- and liquid-phase microextraction technique based on a hydrophobic deep eutectic solvent: Application for trace analysis of pesticides in fruit juices. *J. Sci. Food Agric.* **2020**, *100*, 2534–2543. [CrossRef]
32. Zhang, Q.H.; Vigier, K.D.; Royer, S.; Jerome, F. Deep eutectic solvents: Syntheses, properties and applications. *Chem. Soc. Rev.* **2012**, *41*, 7108–7146. [CrossRef] [PubMed]
33. Abbott, A.P.; Capper, G.; Davies, D.L.; Rasheed, R.K.; Tambyrajah, V. Novel solvent properties of choline chloride/urea mixtures. *Chem. Commun.* **2003**, *39*, 70–71. [CrossRef] [PubMed]
34. Florindo, C.; Oliveira, F.S.; Rebelo, L.P.N.; Fernandes, A.M.; Marrucho, I.M. Insights into the synthesis and properties of deep eutectic solvents based on cholinium chloride and carboxylic acids. *ACS Sustain. Chem. Eng.* **2014**, *2*, 2416–2425. [CrossRef]
35. Qin, H.; Hu, X.T.; Wang, J.W.; Cheng, H.Y.; Chen, L.F.; Qi, Z.W. Overview of acidic deep eutectic solvents on synthesis, properties and applications. *Green Energy Environ.* **2020**, *5*, 8–21. [CrossRef]
36. Chen, J.; Liu, M.J.; Wang, Q.; Du, H.Z.; Zhang, L.W. Deep Eutectic solvent-based microwave-assisted method for extraction of hydrophilic and hydrophobic components from radix salviae miltiorrhizae. *Molecules* **2016**, *21*, 1383. [CrossRef] [PubMed]
37. Dai, Y.T.; Witkamp, G.J.; Verpoorte, R.; Choi, Y.H. Tailoring properties of natural deep eutectic solvents with water to facilitate their applications. *Food Chem.* **2015**, *187*, 14–19. [CrossRef] [PubMed]
38. Skulcova, A.; Russ, A.; Jablonsky, M.; Sima, J. The pH behavior of seventeen deep eutectic solvents. *Bioresources* **2018**, *13*, 5042–5051. [CrossRef]
39. El Achkar, T.; Greige-Gerges, H.; Fourmentin, S. Basics and properties of deep eutectic solvents: A review. *Environ. Chem.* **2021**, *19*, 3397–3408. [CrossRef]
40. Liu, Y.; Friesen, J.B.; McAlpine, J.B.; Lankin, D.C.; Chen, S.N.; Pauli, G.F. Natural deep eutectic solvents: Properties, applications, and perspectives. *J. Nat. Prod.* **2018**, *81*, 679–690. [CrossRef]
41. Abolghasemi, M.M.; Piryaei, M.; Imani, R.M. Deep eutectic solvents as extraction phase in head-space single-drop microextraction for determination of pesticides in fruit juice and vegetable samples. *Microchem. J.* **2020**, *158*, 105041. [CrossRef]
42. Farajzadeh, M.A.; Abbaspour, M.; Kazemian, R. Synthesis of a green high density deep eutectic solvent and its application in microextraction of seven widely used pesticides from honey. *J. Chromatogr. A* **2019**, *1603*, 51–60. [CrossRef] [PubMed]

43. Farajzadeh, M.A.; Hojghan, A.S.; Mogaddam, M.R.A. Development of a new temperature-controlled liquid phase microextraction using deep eutectic solvent for extraction and preconcentration of diazinon, metalaxyl, bromopropylate, oxadiazon, and fenazaquin pesticides from fruit juice and vegetable samples followed by gas chromatography-flame ionization detection. *J. Food Compos. Anal.* **2018**, *66*, 90–97.
44. Florindo, C.; Branco, L.C.; Marrucho, I.M. Development of hydrophobic deep eutectic solvents for extraction of pesticides from aqueous environments. *Fluid Phase Equilib.* **2017**, *448*, 135–142. [CrossRef]
45. Jouyban, A.; Farajzadeh, M.A.; Mogaddam, M.R.A. In matrix formation of deep eutectic solvent used in liquid phase extraction coupled with solidification of organic droplets dispersive liquid-liquid microextraction; application in determination of some pesticides in milk samples. *Talanta* **2020**, *206*, 120169. [CrossRef]
46. Musarurwa, H.; Tavengwa, N.T. Deep eutectic solvent-based dispersive liquid-liquid micro-extraction of pesticides in food samples. *Food Chem.* **2021**, *342*, 127943. [CrossRef] [PubMed]
47. Nemati, M.; Farajzadeh, M.A.; Mohebbi, A.; Khodadadeian, F.; Mogaddam, M.R.A. Development of a stir bar sorptive extraction method coupled to solidification of floating droplets dispersive liquid-liquid microextraction based on deep eutectic solvents for the extraction of acidic pesticides from tomato samples. *J. Sep. Sci.* **2020**, *43*, 1119–1127. [CrossRef]
48. Nemati, M.; Tuzen, M.; Farazjdeh, M.A.; Kaya, S.; Mogaddam, M.R.A. Development of dispersive solid-liquid extraction method based on organic polymers followed by deep eutectic solvents elution; application in extraction of some pesticides from milk samples prior to their determination by HPLC-MS/MS. *Anal. Chim. Acta* **2022**, *1199*, 339570. [CrossRef]
49. Lin, Z.H.; Zhang, Y.H.; Zhao, Q.Y.; Chen, A.H.; Jiao, B.N. Ultrasound-assisted dispersive liquid-phase microextraction by solidifying L-menthol-decanoic acid hydrophobic deep eutectic solvents for detection of five fungicides in fruit juices and tea drinks. *J. Sep. Sci.* **2021**, *44*, 3870–3882. [CrossRef]
50. Torbati, M.; Farajzadeh, M.A.; Mogaddam, M.R.A. Deep eutectic solvent based homogeneous liquid-liquid extraction coupled with in-syringe dispersive liquid-liquid microextraction performed in narrow tube; application in extraction and preconcentration of some herbicides from tea. *J. Sep. Sci.* **2019**, *42*, 1768–1776. [CrossRef]
51. Skarpalezos, D.; Detsi, A. Deep eutectic solvents as extraction media for valuable flavonoids from natural sources. *Appl. Sci.* **2019**, *9*, 4169. [CrossRef]
52. Choi, Y.H.; Verpoorte, R. Green solvents for the extraction of bioactive compounds from natural products using ionic liquids and deep eutectic solvents. *Curr. Opin. Food Sci.* **2019**, *26*, 87–93. [CrossRef]
53. Ivanovic, M.; Razborsek, M.I.; Kolar, M. Innovative extraction techniques using deep eutectic solvents and analytical methods for the isolation and characterization of natural bioactive compounds from plant material. *Plants* **2020**, *9*, 1428. [CrossRef] [PubMed]
54. Socas-Rodriguez, B.; Torres-Cornejo, M.V.; Alvarez-Rivera, G.; Mendiola, J.A. Deep eutectic solvents for the extraction of bioactive compounds from natural sources and agricultural by-products. *Appl. Sci.* **2021**, *11*, 4897. [CrossRef]
55. Nakhle, L.; Kfoury, M.; Mallard, I.; Landy, D.; Greige-Gerges, H. Microextraction of bioactive compounds using deep eutectic solvents: A review. *Environ. Chem. Lett.* **2021**, *19*, 3747–3759. [CrossRef]
56. Bi, W.T.; Tian, M.L.; Row, K.H. Evaluation of alcohol-based deep eutectic solvent in extraction and determination of flavonoids with response surface methodology optimization. *J. Chromatogr. A* **2013**, *1285*, 22–30. [CrossRef] [PubMed]
57. Nam, M.W.; Zhao, J.; Lee, M.S.; Jeong, J.H.; Lee, J. Enhanced extraction of bioactive natural products using tailor-made deep eutectic solvents: Application to flavonoid extraction from Flos sophorae. *Green Chem.* **2015**, *17*, 1718–1727. [CrossRef]
58. Bajkacz, S.; Adamek, J. Evaluation of new natural deep eutectic solvents for the extraction of isoflavones from soy products. *Talanta* **2017**, *168*, 329–335. [CrossRef]
59. Meng, Z.R.; Zhao, J.; Duan, H.X.; Guan, Y.Y.; Zhao, L.S. Green and efficient extraction of four bioactive flavonoids from Pollen Typhae by ultrasound-assisted deep eutectic solvents extraction. *J. Pharm. Biomed.* **2018**, *161*, 246–253. [CrossRef]
60. Tang, W.Y.; Li, G.Z.; Chen, B.Q.; Zhu, T.; Row, K.H. Evaluating ternary deep eutectic solvents as novel media for extraction of flavonoids from Ginkgo biloba. *Sep. Sci. Technol.* **2017**, *52*, 91–99. [CrossRef]
61. Wan, Y.Y.; Wang, M.; Zhang, K.L.; Fu, Q.F.; Wang, L.J.; Gao, M.J.; Xia, Z.N.; Gao, D.E. Extraction and determination of bioactive flavonoids from *Abelmoschus manihot* (Linn.) Medicus flowers using deep eutectic solvents coupled with high-performance liquid chromatography. *J. Sep. Sci.* **2019**, *42*, 2044–2052. [CrossRef]
62. Dai, Y.; Row, K.H. Application of natural deep eutectic solvents in the extraction of quercetin from vegetables. *Molecules* **2019**, *24*, 2300. [CrossRef]
63. Duan, L.; Dou, L.L.; Guo, L.; Li, P.; Liu, E.H. Comprehensive evaluation of deep eutectic solvents in extraction of bioactive natural products. *ACS Sustain. Chem. Eng.* **2016**, *4*, 2405–2411. [CrossRef]
64. Radosevic, K.; Curko, N.; Srcek, V.G.; Bubalo, M.C.; Tomasevic, M.; Ganic, K.K.; Redovnikovic, I.R. Natural deep eutectic solvents as beneficial extractants for enhancement of plant extracts bioactivity. *LWT Food Sci. Technol.* **2016**, *73*, 45–51. [CrossRef]
65. Ma, W.; Row, K.H. Optimized extraction of bioactive compounds from Herba Artemisiae Scopariae with ionic liquids and deep eutectic solvents. *J. Liq. Chromatogr. Relat. Technol.* **2017**, *40*, 459–466. [CrossRef]
66. da Silva, D.T.; Pauletto, R.; Cavalheiro, S.D.; Bochi, V.C.; Rodrigues, E.; Weber, J.; da Silva, C.D.; Morisso, F.D.; Barcia, M.T.; Emanuelli, T. Natural deep eutectic solvents as a biocompatible tool for the extraction of blueberry anthocyanins. *J. Food Compos. Anal.* **2020**, *89*, 103470. [CrossRef]

67. Bosiljkov, T.; Dujmic, F.; Bubalo, M.C.; Hribar, J.; Vidrih, R.; Brncic, M.; Zlatic, E.; Redounikavic, I.R.; Jokic, S. Natural deep eutectic solvents and ultrasound-assisted extraction: Green approaches for extraction of wine lees anthocyanins. *Food Bioprod. Process.* **2017**, *102*, 195–203. [CrossRef]
68. Dai, Y.T.; Rozema, E.; Verpoorte, R.; Choi, Y.H. Application of natural deep eutectic solvents to the extraction of anthocyanins from Catharanthus roseus with high extractability and stability replacing conventional organic solvents. *J. Chromatogr. A* **2016**, *1434*, 50–56. [CrossRef]
69. Ruesgas-Ramon, M.; Figueroa-Espinoza, M.C.; Durand, E. Application of Deep Eutectic Solvents (DES) for phenolic compounds extraction: Overview, challenges, and opportunities. *J. Agric. Food Chem.* **2017**, *65*, 3591–3601. [CrossRef]
70. Redha, A.A. Review on extraction of phenolic compounds from natural sources using green deep eutectic solvents. *J. Agric. Food Chem.* **2021**, *69*, 878–912. [CrossRef]
71. Dai, Y.T.; Witkamp, G.J.; Verpoorte, R.; Choi, Y.H. Natural deep eutectic solvents as a new extraction media for phenolic metabolites in *carthamus tinctorius* L. *Anal. Chem.* **2013**, *85*, 6272–6278. [CrossRef]
72. Bubalo, M.C.; Curko, N.; Tomasevic, M.; Ganic, K.K.; Redovnikovic, I.R. Green extraction of grape skin phenolics by using deep eutectic solvents. *Food Chem.* **2016**, *200*, 159–166. [CrossRef] [PubMed]
73. Garcia, A.; Rodriguez-Juan, E.; Rodriguez-Gutierrez, G.; Rios, J.J.; Fernandez-Bolanos, J. Extraction of phenolic compounds from virgin olive oil by deep eutectic solvents (DESs). *Food Chem.* **2016**, *197*, 554–561. [CrossRef]
74. Chanioti, S.; Tzia, C. Extraction of phenolic compounds from olive pomace by using natural deep eutectic solvents and innovative extraction techniques. *Innov. Food Sci. Emerg. Technol.* **2018**, *48*, 228–239. [CrossRef]
75. Shishov, A.; Volodina, N.; Gagarionova, S.; Shilovskikh, V.; Bulatov, A. A rotating disk sorptive extraction based on hydrophilic deep eutectic solvent formation. *Anal. Chim. Acta* **2021**, *1141*, 163–172. [CrossRef] [PubMed]
76. Alam, M.A.; Muhammad, G.; Khan, M.N.; Mofijur, M.; Lv, Y.K.; Xiong, W.L.; Xu, J.L. Choline chloride-based deep eutectic solvents as green extractants for the isolation of phenolic compounds from biomass. *J. Clean. Prod.* **2021**, *309*, 127445. [CrossRef]
77. Ozturk, B.; Parkinson, C.; Gonzalez-Miquel, M. Extraction of polyphenolic antioxidants from orange peel waste using deep eutectic solvents. *Sep. Purif. Technol.* **2018**, *206*, 1–13. [CrossRef]
78. Fernandez, M.D.; Espino, M.; Gomez, F.J.V.; Silva, M.F. Novel approaches mediated by tailor-made green solvents for the extraction of phenolic compounds from agro-food industrial by-products. *Food Chem.* **2018**, *239*, 671–678. [CrossRef]
79. El Kantar, S.; Rajha, H.N.; Boussetta, N.; Vorobiev, E.; Maroun, R.G.; Louka, N. Green extraction of polyphenols from grapefruit peels using high voltage electrical discharges, deep eutectic solvents and aqueous glycerol. *Food Chem.* **2019**, *295*, 165–171. [CrossRef]
80. Fu, X.Z.; Wang, D.; Belwal, T.; Xu, Y.Q.; Li, L.; Luo, Z.S. Sonication-synergistic natural deep eutectic solvent as a green and efficient approach for extraction of phenolic compounds from peels of Carya cathayensis Sarg. *Food Chem.* **2021**, *355*, 129577. [CrossRef]
81. Cherif, M.M.; Grigorakis, S.; Halahlah, A.; Loupassaki, S.; Makris, D.P. High-efficiency extraction of phenolics from wheat waste biomass (Bran) by combining deep eutectic solvent, ultrasound-assisted pretreatment and thermal treatment. *Environ. Process.* **2020**, *7*, 845–859. [CrossRef]
82. Huang, Y.; Feng, F.; Jiang, J.; Qiao, Y.; Wu, T.; Voglmeir, J.; Chen, Z.G. Green and efficient extraction of rutin from tartary buckwheat hull by using natural deep eutectic solvents. *Food Chem.* **2017**, *221*, 1400–1405. [CrossRef] [PubMed]
83. Bubalo, M.C.; Vidovic, S.; Redovnikovic, I.R.; Jokic, S. New perspective in extraction of plant biologically active compounds by green solvents. *Food Bioprod. Process.* **2018**, *109*, 52–73. [CrossRef]
84. Bakirtzi, C.; Triantafyllidou, K.; Makris, D.P. Novel lactic acid-based natural deep eutectic solvents: Efficiency in the ultrasound-assisted extraction of antioxidant polyphenols from common native Greek medicinal plants. *J. Appl. Res. Med. Aromat. Plants* **2016**, *3*, 120–127. [CrossRef]
85. Peng, X.; Duan, M.H.; Yao, X.H.; Zhang, Y.H.; Zhao, C.J.; Zu, Y.G.; Fu, Y.J. Green extraction of five target phenolic acids from Lonicerae japonicae Flos with deep eutectic solvent. *Sep. Purif. Technol.* **2016**, *157*, 249–257. [CrossRef]
86. Gao, M.Z.; Cui, Q.; Wang, L.T.; Meng, Y.; Yu, L.; Li, Y.Y.; Fu, Y.J. A green and integrated strategy for enhanced phenolic compounds extraction from mulberry (*Morus alba* L.) leaves by deep eutectic solvent. *Microchem. J.* **2020**, *154*, 104598. [CrossRef]
87. Park, H.E.; Tang, B.; Row, K.H. Application of deep eutectic solvents as additives in ultrasonic extraction of two phenolic acids from herba artemisiae scopariae. *Anal. Lett.* **2014**, *47*, 1476–1484. [CrossRef]
88. Liu, Y.J.; Zhang, H.; Yu, H.M.; Guo, S.H.; Chen, D.W. Deep eutectic solvent as a green solvent for enhanced extraction of narirutin, naringin, hesperidin and neohesperidin from Aurantii Fructus. *Phytochem. Anal.* **2019**, *30*, 156–163. [CrossRef]
89. Oktaviyanti, N.D.; Kartini; Mun'im, A. Application and optimization of ultrasound-assisted deep eutectic solvent for the extraction of new skin-lightening cosmetic materials from Ixora javanica flower. *Heliyon* **2019**, *5*, e02950. [CrossRef]
90. Pontes, P.V.D.; Shiwaku, I.A.; Maximo, G.J.; Batista, E.A.C. Choline chloride-based deep eutectic solvents as potential solvent for extraction of phenolic compounds from olive leaves: Extraction optimization and solvent characterization. *Food Chem.* **2021**, *352*, 129346. [CrossRef]
91. Mogaddam, M.R.A.; Farajzadeh, M.A.; Tuzen, M.; Jouyban, A.; Khandaghi, J. Organic solvent-free elevated temperature liquid-liquid extraction combined with a new switchable deep eutectic solvent-based dispersive liquid-liquid microextraction of three phenolic antioxidants from oil samples. *Microchem. J.* **2021**, *168*, 106433. [CrossRef]
92. Li, M.Y.; Liu, Y.Z.; Hu, F.J.; Ren, H.W.; Duan, E.H. Amino Acid-based natural deep eutectic solvents for extraction of phenolic compounds from aqueous environments. *Processes* **2021**, *9*, 1716. [CrossRef]

93. Liu, L.L.; Tang, W.Y.; Tang, B.K.; Han, D.D.; Row, K.H.; Zhu, T. Pipette-tip solid-phase extraction based on deep eutectic solvent modified graphene for the determination of sulfamerazine in river water. *J. Sep. Sci.* **2017**, *40*, 1887–1895. [CrossRef]
94. Dogan, B.; Elik, A.; Altunay, N. Determination of paracetamol in synthetic urea and pharmaceutical samples by shaker-assisted deep eutectic solvent microextraction and spectrophotometry. *Microchem. J.* **2020**, *154*, 104645. [CrossRef]
95. Li, G.Z.; Zhu, T.; Row, K.H. Deep eutectic solvents for the purification of chloromycetin and thiamphenicol from milk. *J. Sep. Sci.* **2017**, *40*, 625–634. [CrossRef]
96. Jouyban, A.; Farajzadeh, M.A.; Khodadadeian, F.; Khoubnasabjafari, M.; Mogaddam, M.R.A. Development of a deep eutectic solvent-based ultrasound-assisted homogenous liquid-liquid microextraction method for simultaneous extraction of daclatasvir and sofosbuvir from urine samples. *J. Pharm. Biomed.* **2021**, *204*, 114254. [CrossRef] [PubMed]
97. Fan, Y.C.; Luo, H.; Zhu, C.Y.; Li, W.J.; Wu, D.; Wu, H.W. Hydrophobic natural alcohols based deep eutectic solvents: Effective solvents for the extraction of quinine. *Sep. Purif. Technol.* **2021**, *275*, 119112. [CrossRef]
98. Kurtulbas, E.; Pekel, A.G.; Toprakci, I.; Ozcelik, G.; Bilgin, M.; Sahin, S. Hydrophobic carboxylic acid based deep eutectic solvent for the removal of diclofenac. *Biomass Convers. Biorefin.* **2022**, *12*, 2219–2227. [CrossRef]
99. Shahi, M.; Javadi, A.; Mogaddam, M.R.A.; Mirzaei, H.; Nemati, M. Preparation of multiwall carbon nanotube/urea-formaldehyde nanocomposite as a new sorbent in solid-phase extraction and its combination with deep eutectic solvent-based dispersive liquid-liquid microextraction for extraction of antibiotic residues in honey. *J. Sep. Sci.* **2021**, *44*, 576–584. [CrossRef]
100. Hadi, N.A.; Ng, M.H.; Choo, Y.M.; Hashim, M.A.; Jayakumar, N.S. Performance of choline-based deep eutectic solvents in the extraction of tocols from crude palm oil. *J. Am. Oil Chem. Soc.* **2015**, *92*, 1709–1716. [CrossRef]
101. Mohammadi, B.; Shekaari, H.; Zafarani-Moattar, M.T. Selective separation of α-tocopherol using eco-friendly choline chloride—Based deep eutectic solvents (DESs) via liquid-liquid extraction. *Colloids Surf.* **2021**, *617*, 126317. [CrossRef]
102. Liu, W.; Fu, X.L.; Li, Z.Z. Extraction of tocopherol from soybean oil deodorizer distillate by deep eutectic solvents. *J. Oleo Sci.* **2019**, *68*, 951–958. [CrossRef] [PubMed]
103. Ge, D.D.; Wang, Y.; Jiang, Q.; Dai, E.R. A deep eutectic solvent as an extraction solvent to separate and preconcentrate parabens in water samples using in situ liquid-liquid microextraction. *J. Braz. Chem. Soc.* **2019**, *30*, 1203–1210. [CrossRef]
104. Hashemi, B.; Zohrabi, P.; Dehdashtian, S. Application of green solvents as sorbent modifiers in sorptive-based extraction techniques for extraction of environmental pollutants. *Trends Anal. Chem.* **2018**, *109*, 50–61. [CrossRef]
105. Shakirova, F.; Shishov, A.; Bulatov, A. Hydrolysis of triglycerides in milk to provide fatty acids as precursors in the formation of deep eutectic solvent for extraction of polycyclic aromatic hydrocarbons. *Talanta* **2022**, *237*, 122968. [CrossRef] [PubMed]
106. Nie, J.; Yu, G.W.; Song, Z.Y.; Wang, X.J.; Li, Z.G.; She, Y.B.; Lee, M. Microwave-assisted deep eutectic solvent extraction coupled with headspace solid-phase microextraction followed by GC-MS for the analysis of volatile compounds from tobacco. *Anal. Methods* **2017**, *9*, 856–863. [CrossRef]
107. Zhang, L.J.; Wang, M.S. Optimization of deep eutectic solvent-based ultrasound-assisted extraction of polysaccharides from Dioscorea opposita Thunb. *Int. J. Biol. Macromol.* **2017**, *95*, 675–681. [CrossRef]
108. Shang, X.C.; Chu, D.P.; Zhang, J.X.; Zheng, Y.F.; Li, Y.Q. Microwave-assisted extraction, partial purification and biological activity in vitro of polysaccharides from bladder-wrack (*Fucus vesiculosus*) by using deep eutectic solvents. *Sep. Purif. Technol.* **2021**, *259*, 118169. [CrossRef]
109. Nie, J.G.; Chen, D.T.; Lu, Y.B. Deep eutectic solvents based ultrasonic extraction of polysaccharides from edible brown seaweed sargassum horneri. *J. Mar. Sci. Eng.* **2020**, *8*, 440. [CrossRef]
110. Das, A.K.; Sharma, M.; Mondal, D.; Prasad, K. Deep eutectic solvents as efficient solvent system for the extraction of kappa-carrageenan from Kappaphycus alvarezii. *Carbohydr. Polym.* **2016**, *136*, 930–935. [CrossRef]
111. Liu, Y.H.; Li, J.; Fu, R.Z.; Zhang, L.L.; Wang, D.Z.; Wang, S. Enhanced extraction of natural pigments from *Curcuma longa* L. using natural deep eutectic solvents. *Ind. Crops Prod.* **2019**, *140*, 111620. [CrossRef]
112. Patil, S.S.; Pathak, A.; Rathod, V.K. Optimization and kinetic study of ultrasound assisted deep eutectic solvent based extraction: A greener route for extraction of curcuminoids from Curcuma longa. *Ultrason. Sonochem.* **2021**, *70*, 105267. [CrossRef] [PubMed]
113. Aydin, F.; Yilmaz, E.; Soylak, M. Vortex assisted deep eutectic solvent (DES)-emulsification liquid-liquid microextraction of trace curcumin in food and herbal tea samples. *Food Chem.* **2018**, *243*, 442–447. [CrossRef] [PubMed]
114. Zhang, H.; Tang, B.; Row, K.H. A green deep eutectic solvent-based ultrasound-assisted method to extract astaxanthin from shrimp byproducts. *Anal. Lett.* **2014**, *47*, 742–749. [CrossRef]
115. Wang, W.D.; Du, Y.G.; Xiao, Z.; Li, Y.; Li, B.F.; Yang, G.W. Determination of trace rhodamine B in chili oil by deep eutectic solvent extraction and an ultra high-performance liquid chromatograph equipped with a fluorescence detector. *Anal. Sci.* **2017**, *33*, 715–717. [CrossRef] [PubMed]
116. Su, E.Z.; Yang, M.; Cao, J.; Lu, C.; Wang, J.H.; Cao, F.L. Deep eutectic solvents as green media for efficient extraction of terpene trilactones from Ginkgo biloba leaves. *J. Liq. Chromatogr. Relat. Technol.* **2017**, *40*, 385–391. [CrossRef]
117. Ge, D.D.; Zhang, Y.; Dai, Y.X.; Yang, S.M. Air-assisted dispersive liquid-liquid microextraction based on a new hydrophobic deep eutectic solvent for the preconcentration of benzophenone-type UV filters from aqueous samples. *J. Sep. Sci.* **2018**, *41*, 1635–1643. [CrossRef] [PubMed]
118. Rabhi, F.; Di Pietro, T.; Mutelet, F.; Sifaoui, H. Extraction of butanol and acetonitrile from aqueous solution using carboxylic acid based deep eutectic solvents. *J. Mol. Liq.* **2021**, *325*, 115231. [CrossRef]

119. Wang, M.; Wang, J.Q.; Zhang, Y.; Xia, Q.; Bi, W.T.; Yang, X.D.; Chen, D.D.Y. Fast environment-friendly ball mill-assisted deep eutectic solvent-based extraction of natural products. *J. Chromatogr. A* **2016**, *1443*, 262–266. [CrossRef]
120. Zhang, H.; Tang, B.; Row, K. Extraction of catechin compounds from green tea with a new green solvent. *Chem. Res. Chin. Univ.* **2014**, *30*, 37–41. [CrossRef]
121. Zhao, Y.P.; Wang, P.; Zheng, W.; Yu, G.W.; Li, Z.G.; She, Y.B.; Lee, M. Three-stage microwave extraction of cumin (*Cuminum cyminum* L.) Seed essential oil with natural deep eutectic solvents. *Ind. Crops Prod.* **2019**, *140*, 111660. [CrossRef]
122. Xu, F.-X.; Zhang, J.-Y.; Jin, J.; Li, Z.-G.; She, Y.-B.; Lee, M.-R. Microwave-assisted natural deep eutectic solvents pretreatment followed by hydrodistillation coupled with GC-MS for analysis of essential oil from turmeric (*Curcuma longa* L.). *J. Oleo Sci.* **2021**, *70*, 1481–1494. [CrossRef] [PubMed]
123. Milani, G.; Vian, M.; Cavalluzzi, M.M.; Franchini, C.; Corbo, F.; Lentini, G.; Chemat, F. Ultrasound and deep eutectic solvents: An efficient combination to tune the mechanism of steviol glycosides extraction. *Ultrason. Sonochem.* **2020**, *69*, 105255. [CrossRef] [PubMed]

Article

Deep Eutectic Solvent Based Reversed-Phase Dispersive Liquid–Liquid Microextraction and High-Performance Liquid Chromatography for the Determination of Free Tryptophan in Cold-Pressed Oils

Slavica Ražić [1,*], Tamara Bakić [2], Aleksandra Topić [1], Jelena Lukić [2] and Antonije Onjia [3,*]

[1] Faculty of Pharmacy, University of Belgrade, Vojvode Stepe 450, 11221 Belgrade, Serbia
[2] Innovation Center of the Faculty of Technology and Metallurgy, Karnegijeva 4, 11120 Belgrade, Serbia
[3] Faculty of Technology and Metallurgy, University of Belgrade, Karnegijeva 4, 11120 Belgrade, Serbia
* Correspondence: slavica.razic@pharmacy.bg.ac.rs (S.R.); onjia@tmf.bg.ac.rs (A.O.)

Abstract: A fast and straightforward reversed-phase dispersive liquid–liquid microextraction (RP-DLLME) using a deep eutectic solvent (DES) procedure to determine free tryptophan in vegetable oils was developed. The influence of eight variables affecting the RP-DLLME efficiency has been studied by a multivariate approach. A Plackett–Burman design for screening the most influential variables followed by a central composite response surface methodology led to an optimum RP-DLLME setup for a 1 g oil sample: 9 mL hexane as the diluting solvent, vortex extraction with 0.45 mL of DES (choline chloride–urea) at 40 °C, without addition of salt, and centrifugation at 6000 rpm for 4.0 min. The reconstituted extract was directly injected into a high-performance liquid chromatography (HPLC) system working in the diode array mode. At the studied concentration levels, the obtained method detection limits (MDL) was 11 mg/kg, linearity in matrix-matched standards was $R^2 \geq 0.997$, relative standard deviations (RSD) was 7.8%, and average recovery was 93%. The combined use of the recently developed DES-based RP-DLLME and HPLC provides an innovative, efficient, cost-effective, and more sustainable method for the extraction and quantification of free tryptophan in oily food matrices. The method was employed to analyze cold-pressed oils from nine vegetables (Brazil nut, almond, cashew, hazelnut, peanut, pumpkin, sesame, sunflower, and walnut) for the first time. The results showed that free tryptophan was present in the range of 11–38 mg/100 g. This article is important for its contributions to the field of food analysis, and for its development of a new and efficient method for the determination of free tryptophan in complex matrices, which has the potential to be applied to other analytes and sample types.

Keywords: RP-DLLME; nuts; seeds; ionic liquids; factorial design; chemometric optimization; Plackett–Burman; HPLC

Citation: Ražić, S.; Bakić, T.; Topić, A.; Lukić, J.; Onjia, A. Deep Eutectic Solvent Based Reversed-Phase Dispersive Liquid–Liquid Microextraction and High-Performance Liquid Chromatography for the Determination of Free Tryptophan in Cold-Pressed Oils. *Molecules* 2023, 28, 2395. https://doi.org/10.3390/molecules28052395

Academic Editor: Alessandra Gentili

Received: 25 December 2022
Revised: 27 February 2023
Accepted: 3 March 2023
Published: 5 March 2023

Copyright: © 2023 by the authors. Licensee MDPI, Basel, Switzerland. This article is an open access article distributed under the terms and conditions of the Creative Commons Attribution (CC BY) license (https://creativecommons.org/licenses/by/4.0/).

1. Introduction

Tryptophan (Trp), an essential amino acid, is a precursor of many biologically active substances, including serotonin, melatonin, quinolinic acid, kynurenic acid, and tryptamine, as well as coenzymes important for electron transfer reactions (redox balance of metabolism), such as nicotinamide adenine dinucleotide (NAD+) [1]. A variety of pathological processes in humans are caused by the disorders of tryptophan metabolism, including neurologic disorders, inflammatory bowel disease, malignancies, and cardiovascular disease [2].

The human body cannot synthesize Trp and is dependent on dietary sources of Trp. It is found in foods that naturally contain protein, in dietetic and fortified food products, and in specific pharmaceutical formulations [3]. Nuts and seed oils are particularly rich in Trp [4,5]. Thus, detecting and quantifying this compound in vegetables is needed, but it is a significant analytical challenge [1].

Many analytical techniques have been recently investigated for Trp detection and quantification, including colorimetry [6–9], fluorimetry [10], voltammetry [11–15], capillary electrophoresis [16,17], gas chromatography–mass spectrometry (GC-MS) [18], high-performance liquid chromatography (HPLC) [19–28], and HPLC single [29] or tandem mass spectrometry (HPLC-MS/MS) [30–35]. Among these techniques, HPLC has been the most widely used due to its high selectivity and good sensitivity for quantifying Trp.

Since the complex matrix of food samples, which often contain large quantities of interfering substances, may influence the Trp content determination [36], a sample pretreatment is necessary. Acid hydrolysis is used in the analytical protocols of most amino acids [36]. However, when determining protein-bound Trp in food samples, it was shown that good results could only be obtained after the preparation of samples by alkaline hydrolysis [23] or enzymatic hydrolysis [25]. In any case, an additional extraction step prior to instrumental measurements is unavoidable. Whereas protein-bound Trp can be determined if only a sample hydrolysis step is included in the sample preparation procedure, the hydrolysis may be omitted if the content of free Trp is determined [37].

Time-consuming solid-phase extraction (SPE) and conventional liquid–liquid extraction (LLE), which need a large quantity of potentially toxic solvents [38,39], are recently being replaced with greener, miniaturized sample preparation techniques [40–42]. Numerous methods have been developed for this purpose, of which dispersive liquid-liquid microextraction (DLLME) has proven to be effective, simple, fast, economical, and environmentally friendly [43–46].

Reversed-phase DLLME (RP-DLLME) is a new modification of DLLME in which an extraction solvent compatible with the HPLC mobile phase is used [47,48]. In RP-DLLME, a small volume of extraction solvent is rapidly dispersed in the hydrophobic sample solution using a dispersive solvent to form a fine droplet phase. The analytes partition into the aqueous extraction phase, which is then collected and analyzed. The aqueous phase can often be injected directly into an HPLC system. In this way, the time required for solvent evaporation is saved. RP-DLLME has recently been used for the enrichment and extraction of a wide range of analytes from various lipophilic sample matrices, including biological samples [49], cosmetics [50], and vegetable oils [51–53].

The recently introduced DLLME using a deep eutectic solvent (DES) has attracted tremendous attention [54–58]. DES represents a mixture of a hydrogen bond donor (HBD) and hydrogen bond acceptor (HBA). Owing to the hydrogen bond interactions, this mixture is able to self-associate and form a eutectic mixture with a lower melting point than those of the individual constituents [57]. Unlike traditional ionic liquids with similar extraction properties, these compounds are more environmentally friendly and cheaper. In addition, hydrophilic DESs have the advantage of being compatible with the RP-HPLC mobile phase [59].

This work investigates a new analytical method for separating and determining free Trp in an oily matrix using DES–RP-DLLME followed by an HPLC–diode array detector (DAD). The DES–RP-DLLME variables were optimized using the design of experiments (DoE) [60,61]. The method was validated and applied to determine free Trp in cold-pressed vegetable oils.

The novelty of this study lies in the development of a method for the determination of free Trp in cold-pressed oils using a combination of DoE-optimized DES–RP-DLLME and HPLC. It offers several advantages over conventional extraction techniques, including improved selectivity and reduced solvent consumption. A small volume of DES, used as the extraction solvent, provides additional benefits, including its low toxicity, biodegradability, and cost effectiveness, making it an environmentally friendly and efficient choice for sample preparation.

2. Results and Discussion

2.1. Selection of DESs

Seven hydrophilic DESs based on ChCl receptor bonds have been evaluated as candidates for the RP-DLLME extractant. Two DESs, ChCl:DA and ChCl:CA, were not clear liquids at room temperature.

FTIR spectra of these DESs placed in pressed KBr tablets are illustrated in Figure 1. A comparison between the spectrums ChCl and its DES mixtures was made in order to identify the changes in the structure and the new interactions between the constituents in the synthesized DESs. It can be noted that ChCl has retained its structure, as some of its peaks were also observed in the DES spectrums. The presence of water had an insignificant effect on the vibration frequencies in the formed ChCl:water DES [62]. The bands related to pure ChCl compared to those related to the DESs showed a small frequency deviation and change in bandwidth [63]. Thus, the positions of some characteristic ChCl peaks, symmetrical C-H stretch at 2800 cm^{-1}, and asymmetric C-CH$_3$ stretch at 3000 cm^{-1} were changed in the DESs

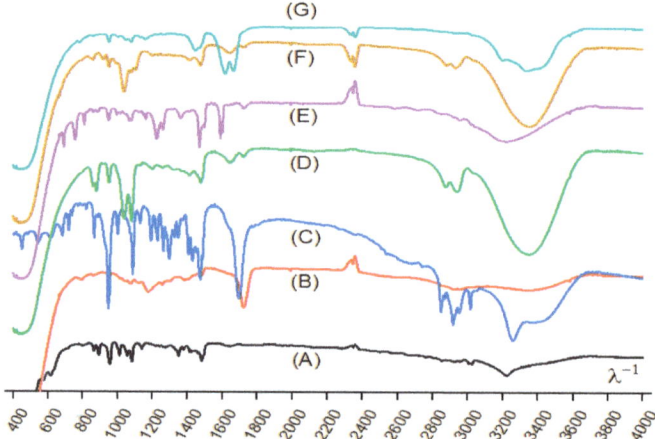

Figure 1. FTIR spectra of the synthesized deep eutectic solvents. (**A**) ChCl:W; (**B**) ChCl:CA; (**C**) ChCl:DA; (**D**) ChCl:EG; (**E**) ChCl:Ph; (**F**) ChCl:G; and (**G**) ChCl:U.

The OH vibration of the ChCl at 3200 cm^{-1} was shifted to 3400 cm^{-1}. This shift and the broadening of the O-H vibration bands indicate the presence of hydrogen bonds between ChCl and donor compounds when the DES is formed. This may be attributed to the transfer of oxygen atom cloud electrons to hydrogen bonding [64]. The peak at 1000 cm^{-1} is indicative of the C-N vibration. The carboxylic group at 1700 cm^{-1} can be observed in ChCl:CA and ChCl:DA. In the spectra of ChCl:DA, a polyunsaturated fatty acid chain is represented by the band at 3000 cm^{-1} [65,66], while the frequencies of 1600 cm^{-1} indicate the vibration of N-H and C-N in ChCl:U. At 1100 cm^{-1}, C-N bond vibration is shown.

These DESs were tested as the candidate solvent for RP-DLLME of Trp from oils. For this purpose, all experimental variables were set to the middle point in the RP-DLLME experimental domain (Table 1). Enrichment factor (EF), calculated as the ratio between the Trp concentration in the final DES solution (C_f) and the Trp concentration in the initial oil sample (C_i), was used to estimate the extraction recovery (ER) applying the following equation:

$$ER = 100 \times EF \times (V_f/V_i) \qquad (1)$$

where V_i and V_f are the initial sample volume and the volume of the final reconstituted DES extract, respectively.

Table 1. Variables and their ranges (−1, 0, +1) for the Plackett–Burman screening design.

No.	Variable	Symbol	Level −1	Level 0	Level +1
1.	Initial sample dilution ratio (:)	dil	1:1	1:5	1:9
2.	DES amount (μL)	DES	100	200	300
3.	Extraction time (min)	t_{ex}	1	3	5
4.	Extraction temperature (°C)	T	25	35	45
5.	Salt (NaCl) addition (%)	salt	0	5	10
6.	Stirring type (vortex or ultrasonic)	stir	Vor	-	Us
7.	Centrifuge speed (rpm)	w	2000	6000	10,000
8.	Centrifuge time (min)	t_{cfg}	2	6	10

Figure 2 shows the comparison of ERs in RP-DLLME of Trp from spiked samples for seven DESs. It is obvious that ChCl:U is capable of extracting the highest amount of Trp from an oily matrix (68%). In addition, this DES gives a clear solution and is straightforward for RP-DLME. In general, the results revealed that ChCl:U was the most suitable DES compared to other ones.

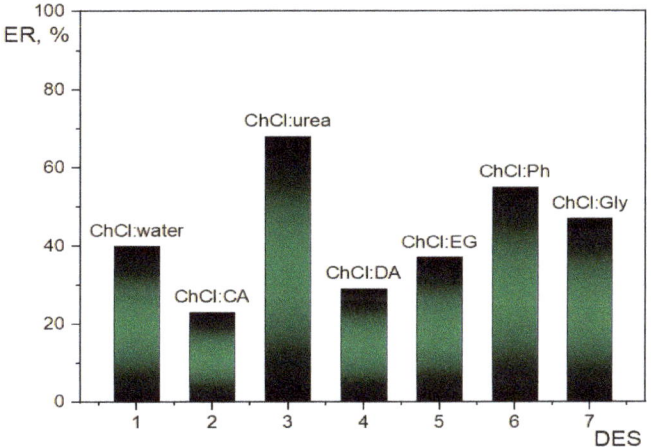

Figure 2. Extraction efficiency of different deep eutectic solvents for RP-DLLME of tryptophan from oil.

ChCl:urea was able to extract free Trp from an oily matrix by a combination of several intermolecular forces [50]. One of the most important forces is the hydrogen bonding that occurs between the hydrogen atom of the hydroxyl group of choline chloride and the nitrogen and oxygen atoms of the functional groups of Trp. This interaction facilitates the transfer of Trp from the sample matrix to the Chl:urea phase. Another important force is the hydrophobic interaction that occurs between the hydrophobic alkyl chains of the ChCl:urea DES and the nonpolar side chains of Trp. This interaction promotes the partitioning of Trp from the sample matrix into the ChCl:urea phase because Trp prefers to interact with nonpolar molecules rather than polar molecules. Furthermore, some additional intermolecular forces, such as electrostatic and van der Waals forces, contribute to the extraction mechanism. Unlike other DESs, both the ChCl and urea components of the DES have the ability to form hydrogen bonds, which increases the solubilizing power for Trp. Indeed, Trp is an aromatic amino acid with polar and nonpolar moieties, which has a relatively high polarity due to the presence of amino and carboxyl groups and can form hydrogen bonds with the polar functional groups of ChCl and urea.

2.2. Optimization of DLLME Procedure

Optimal microextraction of Trp was obtained by optimizing eight RP-DLLME variables in two steps. At first, the data obtained from the PBD experiments were analyzed using the ANOVA test, and the results are shown in Table 2. This helped in precisely choosing the variables with the largest influence using a minimal number of experiments. Here, the temperature and the DES amount appeared to be the two most influential variables, with the highest F-values (30.8 and 12.1). It was also observed that increasing the diluting factor, the extraction time, the centrifugation rate, and decreasing the amount of added salt and the centrifugation time resulted in an increase in ER. At the same time, vortexing was found to give better results than sonicating.

Table 2. ANOVA results from the Plackett–Burman (PBD) design.

Source	DF	Adj SS	Adj MS	F-Value	p-Value
Model	8	2222	277.7	8.97	0.049
Linear	8	2222	277.7	8.97	0.049
dil	1	184.0	184.1	5.94	0.093
DES	1	374.0	374.1	12.1	0.040
t_{ex}	1	24.08	24.08	0.78	0.443
T	1	954.0	954.0	30.8	0.012
salt	1	270.7	270.7	8.78	0.060
stir	1	310.1	310.1	10.01	0.051
W	1	90.75	90.75	2.93	0.185
t_{cfg}	1	14.08	14.08	0.45	0.548
Error	3	92.92	30.97		
Total	11	2314			

The next step in the optimization used RSM to find the best ER from RP-DLLM of Trp. In this case, the experimental domain was extended to 700 µL of DES. Figure 3 presents the response surface plot of ER as a function of the temperature and the DES amount. This method included quadratic and interaction terms in the model. Therefore, it was possible to account for the detailed effects of selected variables on each other and also on ER.

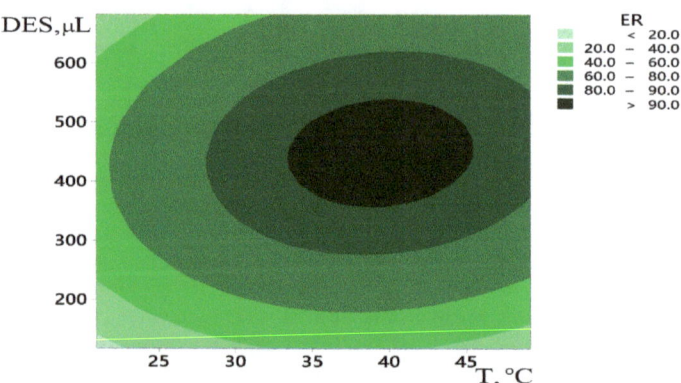

Figure 3. Response surface plot for optimization of RP-DLLME of tryptophan.

Thus, the experimental data were fitted by a second-order polynomial model (Equation (2)), which consisted of two main effect terms, two two-factor interaction effect terms and two curvature effect terms. The regression equation is

$$ER = A + B \cdot T + C \cdot DES + D \cdot T^2 + E \cdot DES^2 + F \cdot T \cdot DES \qquad (2)$$

where A (−144.2), B (7.92), C (0.3656), D (0.1087), E (−0.000472), and F (0.00147) terms were optimized iteratively to fit the model.

Finally, the optimized RP-DLLME variables may be summarized as follows: initial sample dilution ratio of 1:9, DES amount 450 µL, extraction time 3 min, extraction temperature 40 °C, no salt addition, vortex stirring type, centrifuge speed 6000 rpm, and centrifugation time 4 min.

Because samples containing a significant amount of tryptophan were analyzed in this study, an enrichment factor of 2.2 was sufficient for quantitative analysis.

2.3. Validation of RP-DLLME-HPLC Method

Analytical figures of the current method were determined using spiked samples. A typical chromatogram obtained under the optimum DES-RP–DLLME–HPLC–DAD conditions for Trp in a cold-pressed almond oil sample is shown in Figure 4.

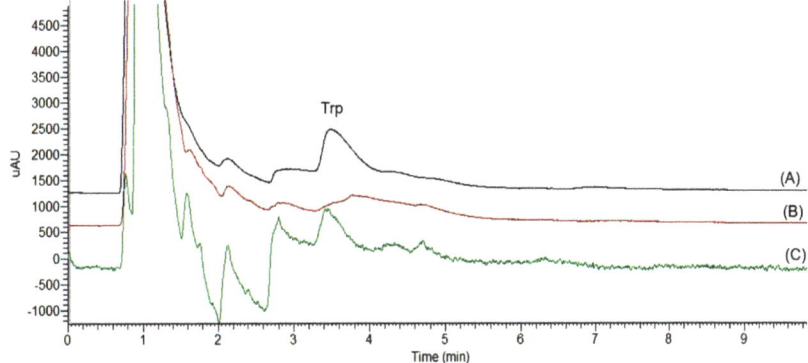

Figure 4. HPLC chromatograms of free tryptophan after RP−DLLME. (A) Matrix-matched standard (20 mg/kg 100 g oil); (B) blank; and (C) almond oil sample.

The method was validated by referring to US FDA official guidance [67]. The performance of the analytical method was evaluated by considering the recovery, precision, limit of detection (LOD), and matrix effect.

Six different calibration standards in the range of 10–400 mg/kg of Trp in methanol were analyzed by HPLC–DAD to obtain the linearity of the method. In addition, the matrix-matched standards at the same concentration levels were measured after RP-DLLME. The matrix effect (ME) was estimated by using the ratio between the slope of the calibration curve of the standards in methanol (b_1 = 923,541) and the slope of the matrix-matched calibration curve (b_2 = 775,774). In both cases, the linear correlation coefficients were r^2 > 0.996. However, the estimated ME was 26%, which indicates a matrix-induced suppression of the analytical signal.

According to the FDA guideline, limit of detection (LOD) values should be ≥ten times lower than quantified concentrations in the 0.1–10,000 mg/kg range. The LOD value of Trp was selected to be the concentration that gave the signal-to-noise ratio (S/N = 3) for the Trp peak which was 11 mg/kg. This LOD value allows the determination of Trp in cold-pressed vegetable oils.

Precision was evaluated by determining the RSD of six replicate spiked samples at three concentrations (50, 200, and 400 mg/kg), and it was determined to be 9%, 6%, and 5%, respectively. The predicted relative reproducibility standard deviation acceptable by the FDA for the concentration levels of 10, 100, and 1000 mg/kg are 11, 8, and 6%, respectively.

The average recovery for the same spikes was found to be 91%, 94%, and 93%, respectively. The FDA required recoveries for a quantitative method at the concentration level of 100 mg/L to be 90–107%.

2.4. Critical Analysis of the Method Performances

Although hydrophilic ChCl-based DESs have recently been used as an extraction solvent for RP-DLLME [58,67–69], no published method has been found for DLLME of Trp from foods The proposed method for the determination of Trp in vegetable oils using DoE-optimized DES–RP-DLLME and HPLC has a unique combination of features that make it a highly efficient and environmentally friendly technique. Comparing the developed RP-DLLME method for free Trp determination in oil samples with other reported methodologies reveals its pros and cons (Table 3). Electroanalytical techniques such as voltammetry cannot compete with chromatographic techniques in terms of separation power. For the analysis of real samples of complex matrices, only HPLC and LC-MS can be used. The latter has a much higher sensitivity. However, the detector MS is complex and expensive, requires extensive sample preparation, and is often not available in a small laboratory. It is likely that the current HPLC method requires a simpler and less expensive procedure that increases labor while reducing analytical costs.

Table 3. Comparison of the proposed method with other reported methods for the determination of tryptophan.

No.	Matrix	Concentration Range	Sample Preparation Method	Reagents Extractant	Analytical Technique	Limit of Detection	Linearity (R^2)	Recovery (%)	RSD (%)	Reference
1	Protein	10–100 μg	acid hydrolysis	HCl/ninhydrin	ViS	n.a.	n.a.	98.3	2.7	[6]
2	Yeast extract *	100–600 μM	enzymatic hydrolysis	hydroxylamine	ViS	100 μM	0.6969	86	n.a.	[7]
3	Solution *	10–100 mg/L	oxidation	NaOCl	ViS	10 mg/L	0.9996	90.5	1.19	[8]
4	Millets *	9–36 mg/L	biorecognition	MP@PDA-*E. coli*	ViS	5.6 μM	0.98	106	7.3	[9]
5	Beer *	0.02–0.12 mg/L	dSPE	graphene/clay/ Brij L23	FL	0.01 mg/L	0.9991	90	5.0	[10]
6	Dietary supplements *	1.0–7.0 μmol/L	dilution	GCE/p-ARG	SWV	0.30 μmol/L	0.990	97.6	2.1	[11]
7	Plasma *	0.08–20.0 μM	screen-printed electrode	PdCuCo/RGO	DPV	0.03 μM	0.997	103.7	2.8	[12]
8	Milk *	5.0–150 μM	electrochemical sensor	graphite electrode	DPV	5.78 μM	0.9841	99.3	8.6	[13]
9	Pharmaceutics *	1–350 μM	MIP electrochemical sensor	AuNPs@PVP@SiO2MIP	LSV	1 μM	0.995	105	4	[14]
10	Milk *	0.01–80 μM	MIP biosensor	MIP-AF	EIS	0.008 μM	0.99	98.2	1.8	[15]
11	Plasma *	0.005–0.1 mol/L	dilution	HBP/SSA	CE	5 μmol/L	0.998	101.9	5.4	[16]
12	Beer *	n.d.–40.7 mg/L	acid hydrolysis	HCl/HEC//BTP/EACA/AMPD	cITP	4.35 mg/L	0.9993	95.9	4.3	[17]
13	Leaf tissue *	n.a.	SPE	acetic anhydride deriv.	GC-MS	n.a.	n.a.	60	2	[18]
14	Soy sauces *	136–262 mg/L	precipitation	ethanol	HPLC	1 mg/L	0.995	108	4.9	[19]
15	Dietary supplements *	5.0–500 μg/m	HILIC	1-octane sulfonate	HPLC	1.2 mg/mL	0.979	96.5	2.3	[20]
16	Infant formula	0.018–30 mg/kg	enzymatic hydrolysis	pronase enzyme	HPLC	18 μg/kg	0.9999	93.8	6.9	[21]
17	Rapeseed	10–400 ng	alkaline hydrolysis	NaOH	HPLC	10 ng	0.998	98.6	1.6	[22]
18	Pig feed *	n.a.	dilution	HCl	HPLC	n.a.	n.a.	n.a.	5.0	[23]
19	Chicken feed	59–130 g/kg	alkaline hydrolysis	NaOH/o-phthalaldehyde	HPLC-FLD	n.a.	n.a.	86	4.0	[24]
20	White bread		alkaline hydrolysis	NaOH	HPLC	n.a.	n.a.	85	16.1	[25]
21	Wheat	1.3–14.8 g/kg	alkaline hydrolysis	NaOH/O-phthalaldehyde	HPLC-FLD	n.a.	n.a.	91.6	1.9	[26]
22	Yogurt	352–1220 mg/kg	alkaline hydrolysis	NaOH/5-methyl-l-tryptophan	HPLC-FLD	11 μg/kg	0.9995	93	1.1	[27]
23	Bee pollen *	0.069 mg/g	ultrasonic extraction	ACN	HPLC-FLD	0.003 mg/L	0.9998	93.8	3.82	[28]
24	Ryegrass shoot	0.5–40 μM	alkaline hydrolysis	NaOH	LC-MS	0.02 μM	0.99	89.9	8.5	[29]
25	Whole blood	0.1–25 ng/mL	VAMS	ACN/H_2O	LC-MS/MS	25 ng/mL	0.9987	85	9.6	[30]
26	Plasma	0–160 μM	acid hydrolysis	MeOH/$ZnSO_4$/TFA	LC-MS/MS	83 nM/L	0.995	88	11	[31]
27	Honey *	0.7–9.94 mg/kg	SPE	Oasis MCX 30 μm	LC-MS/MS	1.0 μg/kg	n.a.	60	4.3	[32]
28	Milk *	89.6–117	QuEChERS	CAN	LC-MS/MS	2 ng/mL	0.99	103.7	2.6	[33]
29	Plant material *	1–50 ng/mL	SPE	Hybrid SPE–phospholipids	LC-MS/MS	4 ng/mL	0.996	87.8	15	[34]

Table 3. Cont.

No.	Matrix	Concentration Range	Sample Preparation Method	Reagents Extractant	Analytical Technique	Limit of Detection	Linearity (R^2)	Recovery (%)	RSD (%)	Reference
30	Chicken feed	n.a.	microwave hydrolysis	AQC-derivatization	LC-MS/MS	1 fmol	n.a.	99	4.2	[35]
31	Hazelnut *	42–127 µg/g	water extraction	water	UPLC-MS/MS	n.a.	n.a.	117	30	[37]
32	Nuts and seed oils *	10–400 mg/kg	RP-DLLME	DES (ChCl:U)	HPLC	11 mg/kg	0.996	91	9.0	This study

*—free Trp was analyzed; ViS—visible spectrophotometry; DPV—differential pulse voltammetry; LSV—linear sweep voltammetry; SWV—square wave voltammetry; MIP—molecularly imprinted polymer; EIS—electrochemical impedance spectroscopy; VAMS—volumetric absorptive microsampling; FL—fluorescence; FLD—fluorescence detector; CE—capillary electrophoresis; cITP—capillary isotachophoresis; dSPE—dispersive solid phase extraction; SPE—solid-phase extraction; HILIC—hydrophilic interaction liquid chromatography; n.a.—not available; n.d.—not detected.

Unlike simple extraction or dilution in the analysis of hydrophilic samples, an oily matrix requires greater removal of interferences. An alternative technique to RP-DLLME, which can achieve lower LODs, is solid phase extraction (SPE). However, SPE consumes more organic solvents and requires an SPE manifold. Considering that sample cleanup is required for nuts and seed oils and the tryptophan concentration in the samples tested is high, the sensitivity of this method is quite acceptable.

In general, the green analytical chemistry (GAC) aspects of this method are its main advantages. All 12 GAC principles affecting the quality attributes of the analytical method were addressed here: green (G1. Toxicity of reagents, G2. Number and amount of reagents and waste, G3. Energy, G4. Direct impacts on the human), red (R1. Scope of application, R2. LOD and LOQ, R3. Precision, R4. Accuracy), and blue (B1. Cost-efficiency, B2. Time-efficiency, B3. Minimal requirements, B4. Operational simplicity).

Three greenness assessment approaches [70–72] were used to evaluate the environmental impact of this method: Analytical Eco-Scale, Green Analytical Procedure Index (GAPI), and Analytical Greenness metric (AGREE).

The Analytical Eco-Scale tool is used to assess the greenness level of analytical procedures in terms of the number of hazards. This method assigns penalty points for the amount and type of reagents used, hazards, energy consumed, and waste generated. The penalty points are then subtracted from a value of 100. It is considered an excellent green method if the method receives a score of ≥75. The method is considered inadequate if the Eco-scale value is less than 50. A high Eco-Scale score is primarily attributed to the amount and type of solvents consumed. The calculated penalty points (14) for our method resulted in an Eco-scale score of 86 out of 100, indicating that the developed method was excellent green (Supplementary Table S1).

GAPI evaluates the greenness level of an analytical method through five fields. Each field represents a different aspect of the developed method (Supplementary Table S2). Fields are colored green, yellow, or red depending on the ecological impact of each step. In this work, the GAPI pentagram had five green, seven yellow, and two red fields (Figure 5A). One of the red fields is because this method is not an in-, at-, or on-line method. Another pentagram is shaded red because sample preparation is required.

The AGREE method assesses the environmental impact of the method using a pictogram divided into twelve sections, which correspond to the twelve GAC principles. Each section and the middle zone of the AGREE pictogram is colored from green to red. According to the method greenness, the total score of the method is calculated and appears in the middle zone of the AGREE pictogram. The AGREE score in this study was 0.69 (Figure 5B), indicating that the method is environmentally friendly and has no negative impact on the environment.

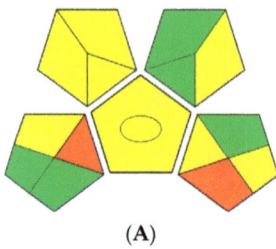 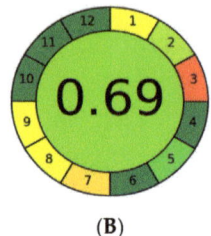

(A) (B)

Figure 5. Greenness assessment of the DES–RP-DLLME-HPLC method for tryptophan in oils. (**A**) GAPI pentagram; (**B**) AGREE pictogram.

2.5. Analysis of Real Samples

Five samples of each vegetable were purchased in different stories and analyzed (Supplementary Table S3). The analytical results of free Trp in different vegetable oils, using the DES–RP-DLLME–HPLC–DAD method, are presented in Table 4. The results, expressed in mg of Trp per 100 g of cold-pressed oils, are in the range between 11 and 38 mg/100 g oil.

Table 4. Trp (free) content in vegetable oils.

No.	Oils Made from	Trp Content (mg/100 g)	±	Variation of Trp Content between Samples (%)
1.	Almonds	16	±	16
2.	Brazil nuts	14	±	18
3.	Cashews	11	±	15
4.	Hazelnuts	17	±	14
5.	Peanuts	18	±	13
6.	Pumpkin seeds	32	±	26
7.	Sesame seeds	33	±	27
8.	Sunflower seeds	38	±	21
9.	Walnuts	12	±	14

Sunflower seed oil shows the highest average Trp concentration (38 mg/100 g), while cashew oil has the lowest (11 mg/100 g). Trp levels significantly differed among the nut and seed samples. In general, cold-pressed oils from seeds have higher Trp content than those from nuts.

3. Materials and Methods

3.1. Reagents and Chemicals

L-Tryptophan, reagent grade standard (≥98% purity), was purchased from Sigma-Aldrich Chemie GmbH (Taufkirchen, Germany). The stock solution (100 mg/kg) of Trp was prepared in methanol/deionized water (50:50 *v/v*). The working solutions were prepared by appropriately diluting the stock solution with methanol/deionized water (50:50 *v/v*). Methanol, acetonitrile (HPLC-grade), and sodium chloride (99%) were purchased from Merck KGaA, (Darmstadt, Germany). All standard solutions were stored at 4 °C and brought to an ambient temperature just prior to use.

Seven different types of DESs, at a molar ratio of 1:2, were prepared by mixing choline chloride:urea (ChCl:U), choline chloride:phenol (ChCl:Ph), choline chloride:citric acid (ChCl:CA), choline chloride:decanoic acid (ChCl:DA), choline chloride:glycerol (ChCl:G), choline chloride:ethylene glycol (ChCl:EG), and choline chloride:water (ChCl:W). The mixtures were stirred at 80 °C until DESs were formed. These DESs were used for RP-DLLME of Trp from vegetable oil samples.

3.2. Sample Preparation

Nine vegetable samples (sunflower seeds, sesame seeds, peanuts, cashews, Brazil nuts, pumpkin seeds, almonds, walnuts, and hazelnuts) underwent cold pressing using a press model VitaWAY OP650W (Gorenje d.o.o. Valjevo, Serbia), and the produced oils were stored at 4 °C until analysis. Trp was extracted from the oils using the following RP-DLLME procedure (Figure 6).

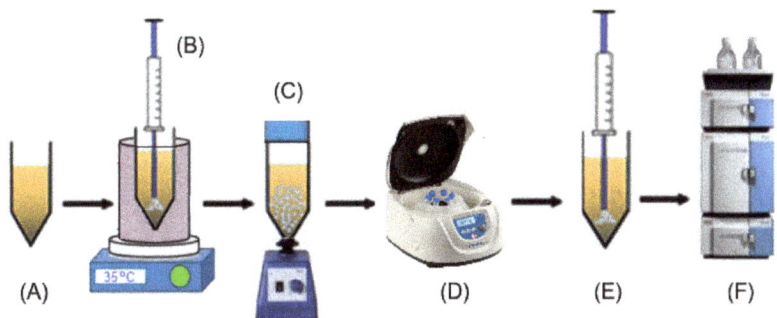

Figure 6. Scheme of the RP-DLLME-HPLC procedure. (**A**) Diluted oil sample; (**B**) injection of DES; (**C**) vortexing; (**D**) centrifugation; (**E**) retrieval of the DES phase; (**F**) HPLC measurement.

First, a 1.0 g sample measured in a 15 mL amber glass centrifuge tube was diluted to an appropriate ratio by adding n-hexane. Next, a volume of a DES, as the extractant, was rapidly added to the sample solution. The resulting mixture was vigorously shaken using a vortex agitator or ultrasonic bath. After centrifugation, two clear phases were observed, and the DES phase containing the extracted Trp was settled at the bottom of the tube. Next, an aliquot of the DES phase (lower phase) was withdrawn through a syringe and diluted with methanol. Finally, 5 µL of this solution was injected into an HPLC–DAD system for measurement. Spiked samples were prepared by adding an appropriate amount of Trp into a linoleic acid/oleic acid (1:1) mixture and processed in the same way.

A vortex (IKA model MS2 Mixer, IKA-Werke GmbH, Staufen, Germany) or ultrasonic bath (model Eumax 3L 100W, Skymen Cleaning Equipment Co. Ltd., Shenzhen, China) was used for mixing, whereas the phase separation was performed using a centrifuge (model Colo Lace16AS, Colo d.o.o., Rogatska Slatina, Slovenia). The extraction was performed in a temperature-controlled, shaking water bath (Memmert, model WNB 14, Memmert GmbH, Schwabach, Germany). Fourier-transform infrared (FTIR) spectra of the DESs were recorded in the wavelength range from 500 to 4000 cm^{-1} on a Thermo model Nicolet iS10 FTIR spectrometer equipped with the Omnic software (Thermo Electron Scientific Instruments LLC, Madison, WI, USA).

3.3. RP-DLLME Optimization

The RP-DLLME procedure has been optimized using the Plackett–Burman screening design (PBD) [46] for variable selection, followed by the central composite design (CCD) [45] used to find the optimal values for the variables. Table 1 shows the eight investigated RP-DLLME variables and their ranges.

After PBD screening experiments, thew variables that significantly influenced the RP-DLLME process were optimized following a response surface methodology using CCD.

3.4. HPLC Measurements

The Trp quantification was performed using an HPLC system comprising a pump, autosampler, and diode array detector (model Accela, Thermo Fisher Scientific Inc., Waltham, MA, USA). Isocratic elution at a 1.0 mL/min flow rate was used on a Thermo Scientific

Hypersil ODS (C18) Column (5 µm particle size, length 100 mm, 4.6 mm I.D.) at 35 °C. The mobile phase consisted of methanol:water (with 1% acetic acid) (40:60 v/v). The DAD spectrum was continuously recorded along with UV detection at a wavelength of 280 nm. The retention time for Trp was 3.54 min.

3.5. Method Validation

Analytical method validation was conducted using a mixture of linoleic:oleic (1:1) acids spiked with Trp at different levels. Linearity and the matrix effect were determined using calibration standards in methanol and matrix-matched standards. Recovery and relative standard deviation (RSD) values were determined using the spiked replicate samples.

3.6. Software

Statistical analysis for Plackett–Burman (PBD) design and ANOVA test was conducted using a software package of Minitab (release ver. 13.20).

4. Conclusions

RP-DLLME combined with HPLC–DAD was developed to determine the free Trp content of cold-pressed nuts and seed oils. A DES solvent comprised of choline chloride and urea was found as the most suitable. Plackett–Burman screening DoE, followed by the central composite RSM, was employed to estimate the optimum RP-DLLME conditions that yield the maximum extraction efficiency. Good analytical recovery and RSD for the method were obtained by analyzing spiked replicates. Applying this method, free Trp was determined in cold-pressed oils from cashews, walnuts, Brazil nuts, almonds, hazelnuts, peanuts, pumpkin seeds, sesame seeds, and sunflower seeds at the levels of 11, 12, 14, 16, 17, 18, 32, 33, and 38 mg/100 g, respectively. The findings of this study suggest that the use of DES-based methods for extraction and analysis of target analytes in complex food matrices can contribute to sustainable development in the analytical chemistry and food industry and may serve as a model for the development of greener and more sustainable food analytical methods in the future.

Supplementary Materials: The following supporting information can be downloaded at: https://www.mdpi.com/article/10.3390/molecules28052395/s1, Tables S1: Analytical eco-scale of the DES-RP-DLLME-HPLC method for tryptophan in oils. Table S2: GAPI of the DES-RP-DLLME-HPLC method for tryptophan in oils. Table S3: The content of tryptophan in nut vegetables from the following stores: A. JEZGRO (address: Ruzveltova 24, Belgrade); B. BIO ŠPAJZ (address: Bulevar Kralja Aleksandra 297, Belgrade); C. ZDRAVAC (address: Čumićevo sokače, Lokal 45, Belgrade); D. BIO MARKET (address: Svetogorska 18, Belgrade); E. DREN (address: Bulevar Despota Stefana 90, Belgrade); F. BIO SHOP (address: Braće Jerković 116, Belgrade); G. (address: EFEDRA, Prvomajska 8k, Zemun).

Author Contributions: Conceptualization, A.O.; methodology, J.L.; software, J.L.; validation, T.B.; investigation, T.B.; resources, A.T.; data curation, T.B.; writing—original draft preparation, A.O.; writing—review and editing, S.R.; visualization, A.T.; project administration, S.R.; funding acquisition, A.T. All authors have read and agreed to the published version of the manuscript.

Funding: This research was funded by the Ministry of Education, Science, and Technological Development, Republic of Serbia (Contract Nos. 451-03-47/2023-01/200161, 451-03-47/2023-01/200135 and 451-03-47/2023-01/200287).

Institutional Review Board Statement: Not applicable.

Informed Consent Statement: Not applicable.

Data Availability Statement: Data presented in this article are available upon request from the corresponding authors.

Acknowledgments: This article is supported by the Study Group *Sample Preparation* of the Division of Analytical Chemistry of the European Chemical Society (DAC-EuChemS).

Conflicts of Interest: The authors declare no conflict of interest. The funders had no role in the design of the study; in the collection, analyses, or interpretation of data; in the writing of the manuscript; or in the decision to publish the results.

References

1. Friedman, M. Analysis, Nutrition, and Health Benefits of Tryptophan. *Int. J. Tryptophan Res.* **2018**, *11*, 1178646918802282. [CrossRef] [PubMed]
2. Comai, S.; Bertazzo, A.; Brughera, M.; Crotti, S. Tryptophan in Health and Disease. In *Advances in Clinical Chemistry*; Elsevier: Amsterdam, The Netherlands, 2020; Volume 95, pp. 165–218. ISBN 978-0-12-821165-6.
3. Strasser, B.; Gostner, J.M.; Fuchs, D. Mood, Food, and Cognition: Role of Tryptophan and Serotonin. *Curr. Opin. Clin. Nutr. Metab. Care* **2016**, *19*, 55–61. [CrossRef] [PubMed]
4. Venkatachalam, M.; Sathe, S.K. Chemical Composition of Selected Edible Nut Seeds. *J. Agric. Food Chem.* **2006**, *54*, 4705–4714. [CrossRef] [PubMed]
5. Garcia-Aloy, M.; Hulshof, P.J.M.; Estruel-Amades, S.; Osté, M.C.J.; Lankinen, M.; Geleijnse, J.M.; de Goede, J.; Ulaszewska, M.; Mattivi, F.; Bakker, S.J.L.; et al. Biomarkers of Food Intake for Nuts and Vegetable Oils: An Extensive Literature Search. *Genes Nutr.* **2019**, *14*, 7. [CrossRef]
6. Pinter-Szakacs, M.; Molnar-Perl, I. Determination of Tryptophan in Unhydrolyzed Food and Feedstuffs by the Acid Ninhydrin Method. *J. Agric. Food Chem.* **1990**, *38*, 720–726. [CrossRef]
7. Wu, Y.; Wang, T.; Zhang, C.; Xing, X.-H. A Rapid and Specific Colorimetric Method for Free Tryptophan Quantification. *Talanta* **2018**, *176*, 604–609. [CrossRef]
8. Hosokawa, S.; Morinishi, T.; Ohara, K.; Yamaguchi, K.; Tada, S.; Tokuhara, Y. A Spectrophotometric Method for the Determination of Tryptophan Following Oxidation by the Addition of Sodium Hypochlorite Pentahydrate. *PLoS ONE* **2023**, *18*, e0279547. [CrossRef]
9. Li, L.; Luo, Y.; Jia, L. Genetically Engineered Bacterium-Modified Magnetic Particles Assisted Chiral Recognition and Colorimetric Determination of D/L-Tryptophan in Millets. *Food Chem.* **2023**, *407*, 135125. [CrossRef]
10. Fernández, L.G.; Vera-López, S.; Díez-Pascual, A.M.; San Andrés, M.P. Easy, Fast, and Clean Fluorescence Analysis of Tryptophan with Clays and Graphene/Clay Mixtures. *J. Food Compos. Anal.* **2022**, *114*, 104858. [CrossRef]
11. Lima, D.; Andrade Pessôa, C.; Wohnrath, K.; Humberto Marcolino-Junior, L.; Fernando Bergamini, M. A Feasible and Efficient Voltammetric Sensor Based on Electropolymerized L-Arginine for the Detection of L-Tryptophan in Dietary Supplements. *Microchem. J.* **2022**, *181*, 107709. [CrossRef]
12. Khoshsafar, H.; Bagheri, H.; Hashemi, P.; Bordbar, M.M.; Madrakian, T.; Afkhami, A. Combination of an Aptamer-Based Immunochromatography Assay with Nanocomposite-Modified Screen-Printed Electrodes for Discrimination and Simultaneous Determination of Tryptophan Enantiomers. *Talanta* **2023**, *253*, 124090. [CrossRef]
13. Tasić, Ž.Z.; Mihajlović, M.B.P.; Radovanović, M.B.; Simonović, A.T.; Medić, D.V.; Antonijević, M.M. Electrochemical Determination of L-Tryptophan in Food Samples on Graphite Electrode Prepared from Waste Batteries. *Sci. Rep.* **2022**, *12*, 5469. [CrossRef]
14. Rezaei, F.; Ashraf, N.; Zohuri, G.H. A Smart Electrochemical Sensor Based upon Hydrophilic Core–Shell Molecularly Imprinted Polymer for Determination of L-Tryptophan. *Microchem. J.* **2023**, *185*, 108260. [CrossRef]
15. Alam, I.; Lertanantawong, B.; Sutthibutpong, T.; Punnakitikashem, P.; Asanithi, P. Molecularly Imprinted Polymer-Amyloid Fibril-Based Electrochemical Biosensor for Ultrasensitive Detection of Tryptophan. *Biosensors* **2022**, *12*, 291. [CrossRef]
16. Forteschi, M.; Sotgia, S.; Assaretti, S.; Arru, D.; Cambedda, D.; Sotgiu, E.; Zinellu, A.; Carru, C. Simultaneous Determination of Aromatic Amino Acids in Human Blood Plasma by Capillary Electrophoresis with UV-Absorption Detection: Other Techniques. *J. Sep. Sci.* **2015**, *38*, 1794–1799. [CrossRef]
17. Jastrzębska, A.; Kowalska, S.; Szłyk, E. Determination of Free Tryptophan in Beer Samples by Capillary Isotachophoretic Method. *Food Anal. Methods* **2020**, *13*, 850–862. [CrossRef]
18. Michalczuk, L.; Bialek, K.; Cohen, J.D. Rapid Determination of Free Tryptophan in Plant Samples by Gas Chromatography-Selected Ion Monitoring Mass Spectrometry. *J. Chromatogr. A* **1992**, *596*, 294–298. [CrossRef]
19. Zhu, Y.; Yang, Y.; Zhou, Z.; Li, G.; Jiang, M.; Zhang, C.; Chen, S. Direct Determination of Free Tryptophan Contents in Soy Sauces and Its Application as an Index of Soy Sauce Adulteration. *Food Chem.* **2010**, *118*, 159–162. [CrossRef]
20. Lomenova, A.; Hroboňová, K. Application of Achiral–Chiral Two-dimensional HPLC for Separation of Phenylalanine and Tryptophan Enantiomers in Dietary Supplement. *Biomed. Chromatogr.* **2021**, *35*, e4972. [CrossRef]
21. Draher, J.; White, N. HPLC Determination of Total Tryptophan in Infant Formula and Adult/Pediatric Nutritional Formula Following Enzymatic Hydrolysis: Single-Laboratory Validation, First Action 2017.03. *J. AOAC Int.* **2018**, *101*, 824–830. [CrossRef]
22. Yust, M.M.; Pedroche, J.; Girón-Calle, J.; Vioque, J.; Millán, F.; Alaiz, M. Determination of Tryptophan by High-Performance Liquid Chromatography of Alkaline Hydrolysates with Spectrophotometric Detection. *Food Chem.* **2004**, *85*, 317–320. [CrossRef]
23. ISO 13904; Animal Feeding Stuffs—Determination of Tryptophan Content. 2005. Available online: https://www.iso.org/standard/37259.html (accessed on 2 March 2023).

24. Ravindran, G.; Ravindran, V.; Bryden, W.L. Total and Ileal Digestible Tryptophan Contents of Feedstuffs for Broiler Chickens. *J. Sci. Food Agric.* **2006**, *86*, 1132–1137. [CrossRef]
25. Allred, M.C.; Macdonald, J.L. Determination of Sulfur Amino Acids and Tryptophan in Foods and Food and Feed Ingredients: Collaborative Study. *J. AOAC Int.* **1988**, *71*, 603–606. [CrossRef]
26. Ravindran, G.; Bryden, W.L. Tryptophan Determination in Proteins and Feedstuffs by Ion Exchange Chromatography. *Food Chem.* **2005**, *89*, 309–314. [CrossRef]
27. Ritota, M.; Manzi, P. Rapid Determination of Total Tryptophan in Yoghurt by Ultra High Performance Liquid Chromatography with Fluorescence Detection. *Molecules* **2020**, *25*, 5025. [CrossRef]
28. Zhang, J.; Xue, X.; Zhou, J.; Chen, F.; Wu, L.; Li, Y.; Zhao, J. Determination of Tryptophan in Bee Pollen and Royal Jelly by High-Performance Liquid Chromatography with Fluorescence Detection. *Biomed. Chromatogr.* **2009**, *23*, 994–998. [CrossRef]
29. la Cour, R.; Jørgensen, H.; Schjoerring, J.K. Improvement of Tryptophan Analysis by Liquid Chromatography-Single Quadrupole Mass Spectrometry Through the Evaluation of Multiple Parameters. *Front. Chem.* **2019**, *7*, 797. [CrossRef]
30. Protti, M.; Cirrincione, M.; Mandrioli, R.; Rudge, J.; Regazzoni, L.; Valsecchi, V.; Volpi, C.; Mercolini, L. Volumetric Absorptive Microsampling (VAMS) for Targeted LC-MS/MS Determination of Tryptophan-Related Biomarkers. *Molecules* **2022**, *27*, 5652. [CrossRef]
31. Boulet, L.; Faure, P.; Flore, P.; Montérémal, J.; Ducros, V. Simultaneous Determination of Tryptophan and 8 Metabolites in Human Plasma by Liquid Chromatography/Tandem Mass Spectrometry. *J. Chromatogr. B* **2017**, *1054*, 36–43. [CrossRef]
32. Soto, M.E.; Ares, A.M.; Bernal, J.; Nozal, M.J.; Bernal, J.L. Simultaneous Determination of Tryptophan, Kynurenine, Kynurenic and Xanthurenic Acids in Honey by Liquid Chromatography with Diode Array, Fluorescence and Tandem Mass Spectrometry Detection. *J. Chromatogr. A* **2011**, *1218*, 7592–7600. [CrossRef]
33. Su, M.; Cheng, Y.; Zhang, C.; Zhu, D.; Jia, M.; Zhang, Q.; Wu, H.; Chen, G. Determination of the Levels of Tryptophan and 12 Metabolites in Milk by Liquid Chromatography-Tandem Mass Spectrometry with the QuEChERS Method. *J. Dairy Sci.* **2020**, *103*, 9851–9859. [CrossRef]
34. Vitalini, S.; Dei Cas, M.; Rubino, F.M.; Vigentini, I.; Foschino, R.; Iriti, M.; Paroni, R. LC-MS/MS-Based Profiling of Tryptophan-Related Metabolites in Healthy Plant Foods. *Molecules* **2020**, *25*, 311. [CrossRef]
35. Weber, P. Determination of Amino Acids in Food and Feed by Microwave Hydrolysis and UHPLC-MS/MS. *J. Chromatogr. B* **2022**, *1209*, 123429. [CrossRef]
36. Cooper, C.; Packer, N.; Williams, K. *Amino Acid Analysis Protocols*; Methods in Molecular Biology; Humana Press: Totowa, NJ, USA, 2000; Volume 159, ISBN 978-1-59259-047-6.
37. Taş, N.G.; Yılmaz, C.; Gökmen, V. Investigation of Serotonin, Free and Protein-Bound Tryptophan in Turkish Hazelnut Varieties and Effect of Roasting on Serotonin Content. *Food Res. Int.* **2019**, *120*, 865–871. [CrossRef]
38. Faraji, M.; Yamini, Y.; Gholami, M. Recent Advances and Trends in Applications of Solid-Phase Extraction Techniques in Food and Environmental Analysis. *Chromatographia* **2019**, *82*, 1207–1249. [CrossRef]
39. Lukić, J.; Radulović, J.; Lučić, M.; Đurkić, T.; Onjia, A. Chemometric Optimization of Solid-Phase Extraction Followed by Liquid Chromatography-Tandem Mass Spectrometry and Probabilistic Risk Assessment of Ultraviolet Filters in an Urban Recreational Lake. *Front. Environ. Sci.* **2022**, *10*, 916916. [CrossRef]
40. Rutkowska, M.; Płotka-Wasylka, J.; Sajid, M.; Andruch, V. Liquid–Phase Microextraction: A Review of Reviews. *Microchem. J.* **2019**, *149*, 103989. [CrossRef]
41. Lukić, J.; Đurkić, T.; Onjia, A. Dispersive Liquid–Liquid Microextraction and Monte Carlo Simulation of Margin of Safety for Octocrylene, EHMC, 2ES, and Homosalate in Sunscreens. *Biomed. Chromatogr.* **2023**, e5590. [CrossRef]
42. Mohammadi, A.; Barzegar, F.; Kamankesh, M.; Mousavi Khaneghah, A. Heterocyclic Aromatic Amines in Doner Kebab: Quantitation Using an Efficient Microextraction Technique Coupled with Reversed-phase High-performance Liquid Chromatography. *Food Sci. Nutr.* **2020**, *8*, 88–96. [CrossRef]
43. Wang, Y.; Li, J.; Ji, L.; Chen, L. Simultaneous Determination of Sulfonamides Antibiotics in Environmental Water and Seafood Samples Using Ultrasonic-Assisted Dispersive Liquid-Liquid Microextraction Coupled with High Performance Liquid Chromatography. *Molecules* **2022**, *27*, 2160. [CrossRef]
44. Han, Q.; Liu, Y.; Huo, Y.; Li, D.; Yang, X. Determination of Ultra-Trace Cobalt in Water Samples Using Dispersive Liquid-Liquid Microextraction Followed by Graphite Furnace Atomic Absorption Spectrometry. *Molecules* **2022**, *27*, 2694. [CrossRef] [PubMed]
45. Slavković-Beškoski, L.; Ignjatović, L.; Bolognesi, G.; Maksin, D.; Savić, A.; Vladisavljević, G.; Onjia, A. Dispersive Solid–Liquid Microextraction Based on the Poly(HDDA)/Graphene Sorbent Followed by ICP-MS for the Determination of Rare Earth Elements in Coal Fly Ash Leachate. *Metals* **2022**, *12*, 791. [CrossRef]
46. Tadić, T.; Marković, B.; Radulović, J.; Lukić, J.; Suručić, L.; Nastasović, A.; Onjia, A. A Core-Shell Amino-Functionalized Magnetic Molecularly Imprinted Polymer Based on Glycidyl Methacrylate for Dispersive Solid-Phase Microextraction of Aniline. *Sustainability* **2022**, *14*, 9322. [CrossRef]
47. Sereshti, H.; Karimi, M.; Samadi, S. Application of Response Surface Method for Optimization of Dispersive Liquid–Liquid Microextraction of Water-Soluble Components of *Rosa damascena* Mill. Essential Oil. *J. Chromatogr. A* **2009**, *1216*, 198–204. [CrossRef] [PubMed]

48. Hashemi, P.; Raeisi, F.; Ghiasvand, A.R.; Rahimi, A. Reversed-Phase Dispersive Liquid–Liquid Microextraction with Central Composite Design Optimization for Preconcentration and HPLC Determination of Oleuropein. *Talanta* **2010**, *80*, 1926–1931. [CrossRef]
49. Wang, Q.F.; Liang, L.J.; Sun, J.B.; Zhou, J. Application of a Reversed-Phase Ionic Liquid Dispersive Liquid-Liquid Microextraction Method for the Extraction and Preconcentration of Domoic Acid from Urine Samples. *Heliyon* **2022**, *8*, e10152. [CrossRef]
50. Schettino, L.; García-Juan, A.; Fernández-Lozano, L.; Benedé, J.L.; Chisvert, A. Trace Determination of Prohibited Acrylamide in Cosmetic Products by Vortex-Assisted Reversed-Phase Dispersive Liquid-Liquid Microextraction and Liquid Chromatography-Tandem Mass Spectrometry. *J. Chromatogr. A* **2023**, *1687*, 463651. [CrossRef]
51. Hassan, M.; Erbas, Z.; Alshana, U.; Soylak, M. Ligandless Reversed-Phase Switchable-Hydrophilicity Solvent Liquid–Liquid Microextraction Combined with Flame-Atomic Absorption Spectrometry for the Determination of Copper in Oil Samples. *Microchem. J.* **2020**, *156*, 104868. [CrossRef]
52. Shishov, A.; Volodina, N.; Semenova, E.; Navolotskaya, D.; Ermakov, S.; Bulatov, A. Reversed-Phase Dispersive Liquid-Liquid Microextraction Based on Decomposition of Deep Eutectic Solvent for the Determination of Lead and Cadmium in Vegetable Oil. *Food Chem.* **2022**, *373*, 131456. [CrossRef]
53. Ferreira, V.J.; Lemos, V.A.; Teixeira, L.S.G. Dynamic Reversed-Phase Liquid-Liquid Microextraction for the Determination of Cd, Cr, Mn, and Ni in Vegetable Oils by Energy Dispersive X-Ray Fluorescence Spectrometry. *J. Food Compos. Anal.* **2023**, *117*, 105098. [CrossRef]
54. Ma, S.; Li, F.; Liu, L.; Liao, L.; Chang, L.; Tan, Z. Deep-Eutectic Solvents Simultaneously Used as the Phase-Forming Components and Chiral Selectors for Enantioselective Liquid-Liquid Extraction of Tryptophan Enantiomers. *J. Mol. Liq.* **2020**, *319*, 114106. [CrossRef]
55. Santos, L.B.; Assis, R.S.; Barreto, J.A.; Bezerra, M.A.; Novaes, C.G.; Lemos, V.A. Deep Eutectic Solvents in Liquid-Phase Microextraction: Contribution to Green Chemistry. *TrAC Trends Anal. Chem.* **2022**, *146*, 116478. [CrossRef]
56. Wang, H.; Huang, X.; Qian, H.; Lu, R.; Zhang, S.; Zhou, W.; Gao, H.; Xu, D. Vortex-Assisted Deep Eutectic Solvent Reversed-Phase Liquid–Liquid Microextraction of Triazine Herbicides in Edible Vegetable Oils. *J. Chromatogr. A* **2019**, *1589*, 10–17. [CrossRef]
57. Hansen, B.B.; Spittle, S.; Chen, B.; Poe, D.; Zhang, Y.; Klein, J.M.; Horton, A.; Adhikari, L.; Zelovich, T.; Doherty, B.W.; et al. Deep Eutectic Solvents: A Review of Fundamentals and Applications. *Chem. Rev.* **2021**, *121*, 1232–1285. [CrossRef]
58. Cao, J.; Wang, C.; Shi, L.; Cheng, Y.; Hu, H.; Zeng, B.; Zhao, F. Water Based-Deep Eutectic Solvent for Ultrasound-Assisted Liquid–Liquid Microextraction of Parabens in Edible Oil. *Food Chem.* **2022**, *383*, 132586. [CrossRef]
59. Xie, Q.; Xia, M.; Lu, H.; Shi, H.; Sun, D.; Hou, B.; Jia, L.; Li, D. Deep Eutectic Solvent-Based Liquid-Liquid Microextraction for the HPLC-DAD Analysis of Bisphenol A in Edible Oils. *J. Mol. Liq.* **2020**, *306*, 112881. [CrossRef]
60. Heidari, H.; Ghanbari-Rad, S.; Habibi, E. Optimization Deep Eutectic Solvent-Based Ultrasound-Assisted Liquid-Liquid Microextraction by Using the Desirability Function Approach for Extraction and Preconcentration of Organophosphorus Pesticides from Fruit Juice Samples. *J. Food Compos. Anal.* **2020**, *87*, 103389. [CrossRef]
61. Lučić, M.; Sredović Ignjatović, I.; Lević, S.; Pećinar, I.; Antić, M.; Đurđić, S.; Onjia, A. Ultrasound-assisted Extraction of Essential and Toxic Elements from Pepper in Different Ripening Stages Using Box–Behnken Design. *Food Process. Preserv.* **2022**, *46*, e16493. [CrossRef]
62. Du, C.; Zhao, B.; Chen, X.-B.; Birbilis, N.; Yang, H. Effect of Water Presence on Choline Chloride-2urea Ionic Liquid and Coating Platings from the Hydrated Ionic Liquid. *Sci. Rep.* **2016**, *6*, 29225. [CrossRef]
63. Banjare, M.K.; Behera, K.; Satnami, M.L.; Banjare, R.K.; Ghosh, K.K. Self-Assembly of a Short-Chain Ionic Liquid within Deep Eutectic Solvents. *RSC Adv.* **2018**, *8*, 7969–7979. [CrossRef]
64. Khezeli, T.; Daneshfar, A.; Sahraei, R. A Green Ultrasonic-Assisted Liquid–Liquid Microextraction Based on Deep Eutectic Solvent for the HPLC-UV Determination of Ferulic, Caffeic and Cinnamic Acid from Olive, Almond, Sesame and Cinnamon Oil. *Talanta* **2016**, *150*, 577–585. [CrossRef] [PubMed]
65. dos Santos, C.; Padilha, C.; Damasceno, K.; Leite, P.; de Araújo, A.; Freitas, P.; Vieira, É.; Cordeiro, A.; de Sousa, F., Jr.; de Assis, C. Astaxanthin Recovery from Shrimp Residue by Solvent Ethanol Extraction Using Choline Chloride:Glycerol Deep Eutectic Solvent as Adjuvant. *J. Braz. Chem. Soc.* **2021**, *32*, 1030–1039. [CrossRef]
66. Ijardar, S.P.; Singh, V.; Gardas, R.L. Revisiting the Physicochemical Properties and Applications of Deep Eutectic Solvents. *Molecules* **2022**, *27*, 1368. [CrossRef] [PubMed]
67. FDA. *FDA Guidelines for the Validation of Chemical Methods for the FDA Foods Program*, 3rd ed.; FDA: Silver Spring, MD, USA, 2019.
68. Karimi, M.; Shabani, A.M.H.; Dadfarnia, S. Deep Eutectic Solvent-Mediated Extraction for Ligand-Less Preconcentration of Lead and Cadmium from Environmental Samples Using Magnetic Nanoparticles. *Microchim. Acta* **2016**, *183*, 563–571. [CrossRef]
69. Liu, W.; Zong, B.; Wang, X.; Yang, G.; Yu, J. Deep Eutectic Solvents as Switchable Solvents for Highly Efficient Liquid–Liquid Microextraction of Phenolic Antioxidant: Easily Tracking the Original TBHQ in Edible Oils. *Food Chem.* **2022**, *377*, 131946. [CrossRef]
70. Alañón, M.E.; Ivanović, M.; Gómez-Caravaca, A.M.; Arráez-Román, D.; Segura-Carretero, A. Choline Chloride Derivative-Based Deep Eutectic Liquids as Novel Green Alternative Solvents for Extraction of Phenolic Compounds from Olive Leaf. *Arab. J. Chem.* **2020**, *13*, 1685–1701. [CrossRef]

71. Zhang, K.; Guo, R.; Wang, Y.; Wang, J.; Nie, Q.; Li, B.; Zhu, G. Temperature-Controlled Air-Assisted Liquid–Liquid Microextraction Based on the Solidification of Floating Deep Eutectic Solvents for the Determination of Triclosan and Alkylphenols in Water Samples via HPLC. *Microchem. J.* **2022**, *182*, 107864. [CrossRef]
72. Moema, D.; Makwakwa, T.A.; Gebreyohannes, B.E.; Dube, S.; Nindi, M.M. Hollow Fiber Liquid Phase Microextraction of Fluoroquinolones in Chicken Livers Followed by High Pressure Liquid Chromatography: Greenness Assessment Using National Environmental Methods Index Label (NEMI), Green Analytical Procedure Index (GAPI), Analytical GREEnness Metric (AGREE), and Eco Scale. *J. Food Compos. Anal.* **2023**, *117*, 105131. [CrossRef]

Disclaimer/Publisher's Note: The statements, opinions and data contained in all publications are solely those of the individual author(s) and contributor(s) and not of MDPI and/or the editor(s). MDPI and/or the editor(s) disclaim responsibility for any injury to people or property resulting from any ideas, methods, instructions or products referred to in the content.

Article

Deep Eutectic Solvents Based on Carboxylic Acids and Glycerol or Propylene Glycol as Green Media for Extraction of Bioactive Substances from *Chamaenerion angustifolium* (L.) Scop.

Alena Koigerova [1], Alevtina Gosteva [2], Artemiy Samarov [3] and Nikita Tsvetov [1,2,*]

[1] Laboratory of Medical and Biological Technologies, Federal Research Centre "Kola Science Centre of the Russian Academy of Sciences", Fersmana Str. 14, Apatity 184209, Russia; a.koygerova@ksc.ru
[2] Tananaev Institute of Chemistry and Technology of Rare Elements and Mineral Raw Materials—Subdivision of the Federal Research Centre «Kola Science Centre of the Russian Academy of Sciences», Akademgorodok 26a, Apatity 184209, Russia; angosteva@list.ru
[3] Department of Chemical Thermodynamics and Kinetics, Institute of Chemistry, Saint Petersburg State University, Universitetskiy Prosp. 26, St. Petersburg 198504, Russia; samarov@yandex.ru
* Correspondence: tsvet.nik@mail.ru

Abstract: *Chamaenerion angustifolium* (L.) Scop. is one of the promising sources of biologically active compounds and a valuable industrial crop. Recently, green extraction methods have become more topical. One of them is the application of deep eutectic solvents (DESs). The aim of this work was the synthesis and characterization of DES consisting of glycerin or propylene glycol with malonic, malic, or citric acids, evaluation of their effectiveness for extracting useful substances from *C. angustifolium* during ultrasonic extraction, description of kinetics, and optimization of extraction conditions. DESs were obtained and characterized with FTIR. Their effectiveness in the process of ultrasound-assisted extraction of biologically active substances from *C. angustifolium* was estimated. Kinetic parameters describing the dependence of the total phenolic, flavonoids, and antioxidant content, free radical scavenging of DPPH, and concentration of flavonoid aglycons (myricetin, quercetin, and kaempferol) via time in the range of 5–60 min at 45 °C are obtained. Extraction conditions were optimized with the Box–Behnken design of experiment. The results of this work make it possible to expand the scope of DES applications and serve the development of *C. angustifolium* processing methods.

Keywords: *Chamaenerion angustifolium* (L.) Scop.; deep eutectic solvents; extraction

1. Introduction

In recent years, the attention of scientists has been increasingly focused on the study of the useful potential of wild plant raw materials characteristic of northern latitudes [1–3]. It is known that depending on the region of growth, the same plant species can accumulate different amounts of biologically active substances [4]. This is due to many factors, including the adaptive ability of plants to adapt to climate change. Thus, plants growing in northern latitudes can be promising raw materials for the production of biologically active compounds.

One of the most promising and valuable industrial crops growing in the northern latitudes is *Chamaenerion angustifolium* (L.) Scop., or fireweed. It is a perennial herbaceous plant of the Onagraceae family, growing in sparse woodlands, in clearings, in dry sandy places, and highly mineralized soils, with the absence of a fertile layer [5]. It is used in traditional medicine and herbal medicine [6]. Among the biologically active compounds contained in the aboveground parts of *C. angustifolium* are flavonoids, tannins, phenolic carboxylic acids, as well as a large amount of vitamin C. Extracts of this plant have anti-inflammatory, antifungal, antiproliferative, antiandrogenic, antioxidant, and anti-cancer properties [7]. In addition, it is a valuable summer honey plant. There are data on the high antimicrobial activity of honey from fireweed against *Streptococcus pneumoniae*, *S. pyogenes*, *Staphylococcus*

aureus, and methicillin-resistant *S. aureus* [8]. It is known that diluted decoctions in the form of tea or low doses of dry extracts of fermented leaves can be used in specialized nutrition to increase the antioxidant status and adaptive potential as well as enhance immunity and body resistance to stress and other damaging factors for people working in complicated environmental conditions, for example in the Arctic [9].

In recent years, the development of green methods of biologically active component extraction has become increasingly relevant—in particular, the use of deep eutectic solvents (DESs) [10–13]. They are binary or triple mixtures, the components of which are bound by strong hydrogen bonds and, as a result, have a reduced melting point below or close to room temperature [14]. A variety of combinations of DES and plant material compositions open up the widest prospects for research both from a fundamental and applied point of view. There are many applications of DESs as solvents for reactions [15], energy applications [16], applications in analytical chemistry [17], electrochemistry [18], and biomass treatment [19,20]. One of the most interesting applications of DES is extraction carried out with the use of additional exposure to ultrasound, microwave radiation, etc. [21]. This additionally expands the number of alternative directions of scientific and technological developments.

DESs may consist of "natural" organic components, such as acids and amino acids, sugars, vitamins, etc. [22], forming natural deep eutectic solvents (NADESs). DESs are divided into five main types, and NADESs in some cases can be classified as Type V (a combination of a donor (HBD) and a hydrogen bond acceptor (HBA)) [23]. It is often difficult to confirm experimentally the formation of a particular eutectic mixture (having a minimum melting point), and we can talk just about "low melting mixtures". However, following the majority of authors, in the future, the term "deep eutectic solutions", (D)ESs, will be used in the work. At the same time, the formation of specific hydrogen bonds should be confirmed using FTIR [24].

Mixtures of dicarboxylic acids with sugars can be attributed to such combinations [25]. The presence of hydroxyl and carboxyl groups allows the formation of sufficiently strong hydrogen bonds. The same type of interaction can be characteristic of mixtures of polybasic organic acids with simpler polyols: for example, ethylene glycol, propylene glycol, and glycerol. Some works reported carboxylic acids and polyols as components of DES of type V [26,27]. Glycerol and propylene glycol are used as solvents for plant extract production, and thus, they can be considered promising components of (D)ESs [28–31].

(D)ESs have rarely been used for the extraction of biologically active substances from *C. angustifolium*. Previously, it was shown that (D)ESs based on choline chloride and polybasic organic acids have greater extraction efficiency than traditional solvents such as water and ethanol [32].

The purpose of this work was the synthesis and characterization of (D)ES consisting of glycerol or propylene glycol with malonic, malic, or citric acids, evaluation of their effectiveness for extracting useful substances from *C. angustifolium* during ultrasonic extraction, description of kinetics and optimization of extraction conditions.

2. Results and Discussion

The (D)ES structure was confirmed using the ATR-FTIR method. IR spectra are shown in Figure 1a–f and in Table 1. The most intense vibration bands of the carboxyl group are in italics.

There is no OH-group in MA. Mal and CA contain an OH-group, the vibrations of which are observed at about 3443 and 3495 cm^{-1}, respectively. These peaks disappear when Mal and CA are mixed with alcohols and (D)ES is formed.

Mal and MA are anhydrous, while citric acid contains one water molecule. In the IR spectrum, the PG vibrational band at 3300 cm^{-1} refers to the O-H stretching of hydroxyl groups. A similar broad peak with a minimum of 3308 cm^{-1} is seen in the GL spectrum. The shift of these minima to the region of higher wavelengths occurs during the formation of the (D)ESs. This indicates the simultaneous interaction of all three components—carboxylic acid, polyhydric alcohol, and water. Therefore, the formation of new hydrogen bonds is indicative.

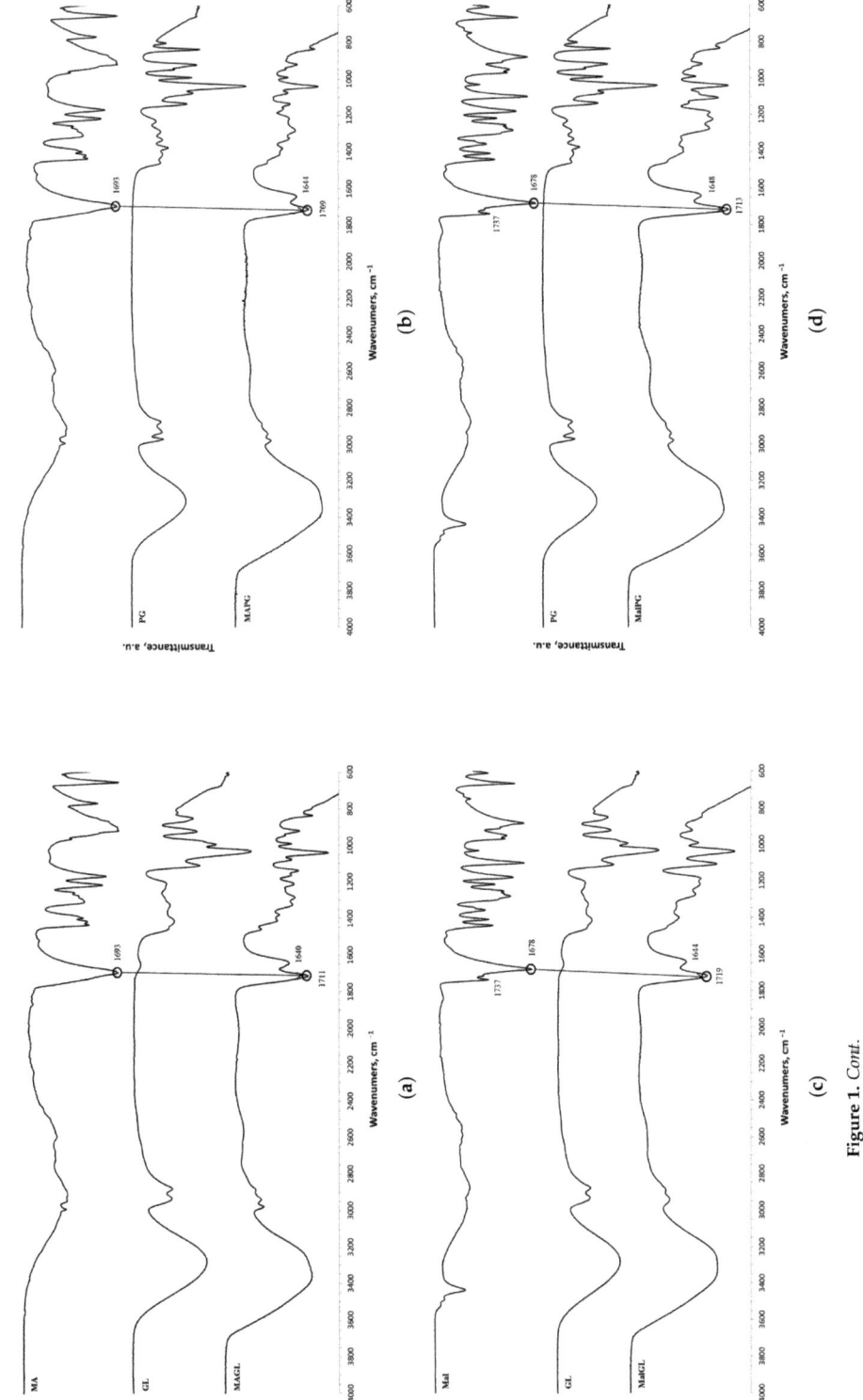

Figure 1. Cont.

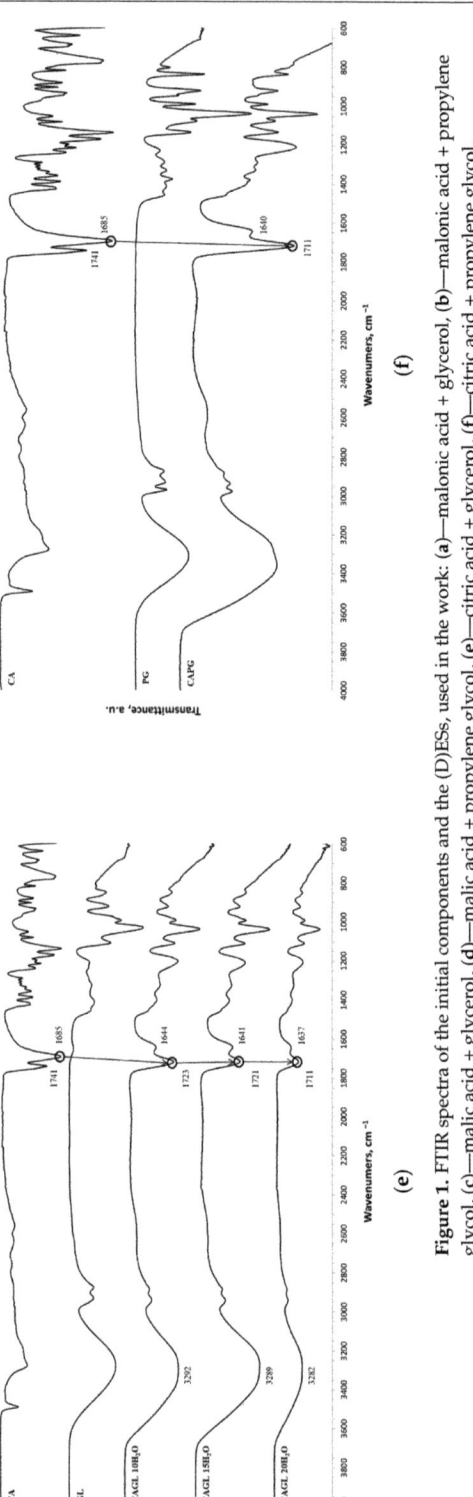

Figure 1. FTIR spectra of the initial components and the (D)ESs, used in the work: (**a**)—malonic acid + glycerol, (**b**)—malonic acid + propylene glycol, (**c**)—malic acid + glycerol, (**d**)—malic acid + propylene glycol, (**e**)—citric acid + glycerol, (**f**)—citric acid + propylene glycol.

Table 1. Characteristic bands of the initial components of the (D)ES and the (D)ES used in the work (the most intense peaks are highlighted in italics).

Component 1	Assignment, cm^{-1}	Component 2	Assignment, cm^{-1}	(D)ES + 10 H$_2$O
MA	ν(C=O) 1693	Gly PG	ν(O-H) 3278 ν(O-H) 3300	3358, *1711*, 1640 3357, *1709*, 1644
Mal	ν(C=O) 1737, *1678*	Gly PG	ν(O-H) 3278 ν(O-H) 3300	3297, *1719*, 1644 3357, *1713*, 1648
CA	ν(C=O) 1741, *1685*	Gly PG	ν(O-H) 3278 ν(O-H) 3300	3289, *1720*, 1644 3360, *1711*, 1640

No fluctuations are observed in the region of 1600–1800 cm^{-1} in the spectrum of PG. A weak peak is present at 1646 cm^{-1} for GL. In turn, one C=O stretching vibration at 1693 cm^{-1} is observed for MA. Two vibrations each are present in the spectra of Mal and CA: 1737 cm^{-1} and 1678 cm^{-1} for Mal and 1741 cm^{-1} and 1685 cm^{-1} for CA. Moreover, peaks that have a higher wavelength have a lower intensity. All the studied (D)ESs have two peaks in this region; the first one is more intense. When comparing the more intense peaks, it can be seen that there is also a shift to the region of higher wavelengths. The increased intensity of bands associated with carbonyl stretching may indicate the formation of hydrogen bonding interactions, given that the formation of (D)ES is characterized by intermolecular interactions such as hydrogen bonding, Van der Waals forces, and electrostatic interactions [24].

Figure 1e demonstrates the effect of the amount of additional water on the structure of the resulting (D)ESs. The general patterns discussed above for three water contents (10, 15, and 20 molar parts) are preserved. However, the greatest difference in the intensities of the first and second peaks of the carboxyl group is observed for (D)ESs with 10 molar parts of water (1600–1750 cm^{-1}). There is also a slight shift in the frequency of these peaks in the direction of decreasing wavelengths as the water content increases. A similar pattern is observed for the minimum of a wide peak in the region of 3000–3700 cm^{-1}—the absorption region of OH groups. Consequently, there is a weakening of hydrogen bonds.

The results of these experimentally measured properties of density and viscosity of (D)ESs under study (with 10 molar parts of water) at 30 °C are presented in Supplementary Materials S1.

It can be seen that the highest density and viscosity are observed for CAGL, while MAPG has the lowest density and MAGL has the lowest viscosity. In the pair of (D)ESs with the same acid, a higher density was obtained for glycerol. For the viscosity, there is no dependence on polyol type. The highest viscosities were obtained for (D)ESs based on citric acid. The viscosities of (D)ESs can affect the extraction rate due to diffusion limitations, so it is important to compare the kinetics of the processes for each (D)ES.

The kinetics of extraction of polyphenols and flavonoids in a medium of various (D)ES is well described by second-order reaction equations, and the 1/Yt vs. t dependences for all (D)ES are described by linear equations with $R^2 > 0.99$ (Supplementary Materials S2 and S3). The obtained parameters of the kinetic equations are presented in Supplementary Materials S4. It can be noted that the fastest polyphenols are extracted using MAGL (k = 17.0 × 10^{-3} g/mg×min), while the slowest extraction process occurs in the case of MalGL (k = 0.6 × 10^{-3} g/mg×min). Flavonoids are extracted faster by MAPG (k = 65.6 × 10^{-3} g/mg×min) and slower when using CAGL (k = 4.4 × 10^{-3} g/mg×min), which may be due to the high density and viscosity. The lack of correlation between the kinetic parameters for TPC and TFC can be explained by the difference in the behavior of different groups of substances in different (D)ES and, accordingly, the different rates of their extraction. In our previous study [32], the highest rate constant for polyphenols was observed for ethanol, but its value was just 2.5 × 10^{-3} g/mg×min. Thus, the extraction of polyphenols is faster for MAGL than ethanol. The extraction rate for flavonoids in this work is also higher than that for ethanol in the work [32].

The highest yield of polyphenols is obtained for (D)ES CAGL, and then relatively close TPC values are obtained for MAGL and CAPG. Flavonoids are extracted better when using MAGL, while CAGL and CAPG have a statistically insignificant difference in efficiency (Figure 2a,b). It should be mentioned that in this work, the TPC and TFC values (300–350 mg GAE/g and 100–120 mg RE/g, respectively) are greater than the ones reached in previous work (250–300 mg GAE/g and 60–80 mg RE/g, respectively) [32].

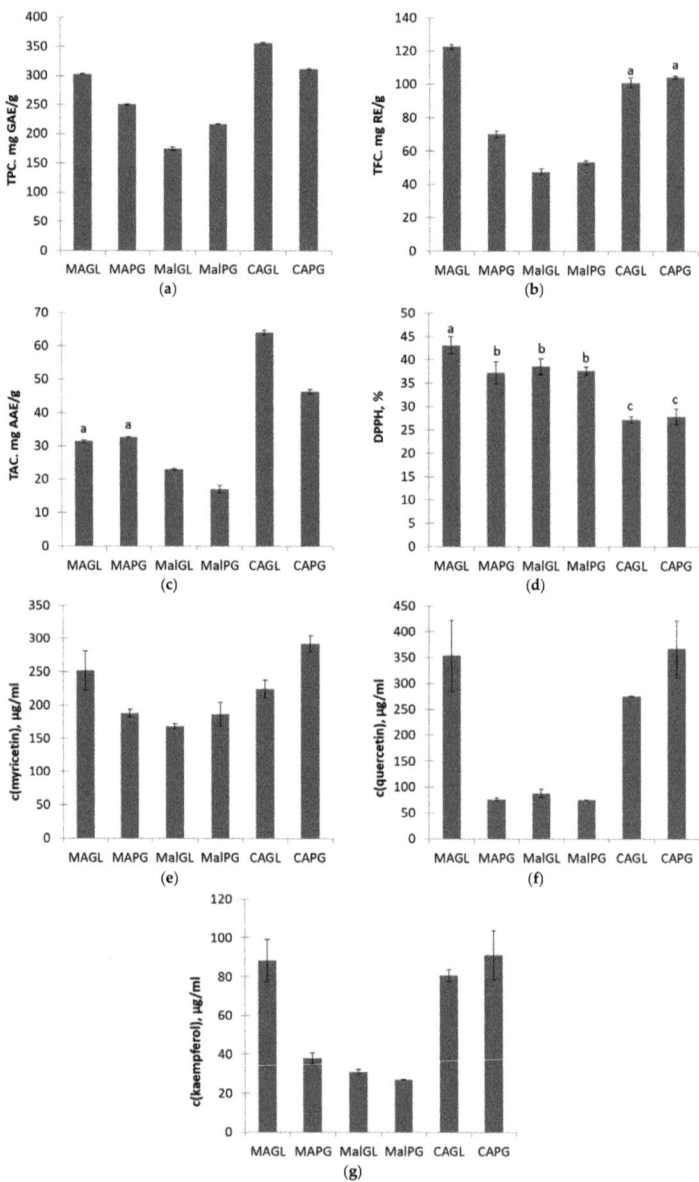

Figure 2. Comparison of total polyphenols (**a**) and total flavonoids (**b**) content, total antioxidant activity (**c**), free radical scavenging (**d**) of extracts, and content of glycosides of myricetin (**e**), quercetin (**f**), and kaempferol (**g**) in extracts based on various (D)ESs. The same letters denote the values, the difference between which is not statistically significant at $p < 0.05$.

The dependence of these characteristics of the extracts obtained on time, in general, is satisfactorily described by the kinetic equation, and for most (D)ESs, the linear approximation of $1/Yt$ vs. t is characterized by $R^2 > 0.99$ (Supplementary Materials S5 and S6). The kinetic parameters of TAC and DPPH are presented in Supplementary Material S6. However, in the case of CAGL, it was not possible to describe the change in DPPH over time using the equation used, since there is a decrease in the indicator after 20 min of extraction. It is also impossible to talk about a satisfactory description of TAC and DPPH for MalPG. This indicates more complex processes associated with the simultaneous extraction, decomposition, and/or evaporation from the liquid phase of substances with antiradical and antioxidant activity in these cases. In general, DPPH is characterized by kinetic dependencies that differ from other parameters of extracts [32,33]. The highest rate of change in TAC is observed for CAPG, and DPPH is observed for MalGL.

Extracts based on CAGL have the greatest antioxidant activity. However, according to the DPPH parameter, both (D)ES compositions with citric acid exhibit the lowest activity, while MAGL has the greatest free radical scavenging (Figure 2c,d).

The dependence of the yield of flavonoids, which are glycosides of myricetin, quercetin, and kaempferol, on time is described fairly accurately by the second-order reaction equation (Supplementary Material S11); however, for quercetin, strictly speaking, a different picture is observed: a gradual decrease in concentration during extraction in the first 20 min is obtained (Supplementary Materials S8–S10). After this time, the concentration of quercetin glycosides reaches a plateau. Kaempferol glycosides are completely extracted within 20–40 min, while the concentration of myricetin glycosides does not reach the equilibrium value within 60 min of extraction. However, in general, based on the data on the kinetics of the extraction of the sum of flavonoids, it can be assumed that the main process is completed in 40 min.

All glycosides of myricetin, quercetin, and kaempferol are best extracted using MAGL and CAPG. The extraction is worst for (D)ESs with malic acid (Figure 2e–g). It should be noted that in a pair of MalGL and MalPG, quercetin and kaempferol are better extracted to MalGL, while myricetin is better extracted to MalPG.

Based on the obtained data on extraction kinetics and comparison of extraction efficiency, (D)ES CAGL was selected for further work on the optimization of extraction conditions. (D)ESs based on citric acid also show higher effectiveness in the previous work [32].

The results of BBD optimization (Figure 3) are described quite well by polynomials

$$Y = a_0 + a_1 A + a_2 B + a_3 C + a_4 AB + a_5 AC + a_6 BC + a_7 A^2 + a_8 B^2 + a_9 C^2 + a_{10} A^2 B + a_{11} A^2 C + a_{12} AB^2$$

For all parameters, the most suitable models and their parameters were selected; in each case, lack of fit is not significant: the R^2 for TPC was 0.914, TFC—0.995, TAC—0.959, concentration of myricetin—0.973, quercetin—0.988, kaempferol—0.992. The numerical values of model coefficients are given in Table 2. From the shape of the surfaces and the p-values for specific terms of the model, it can be seen that the water content in (D)ESs has a very weak effect on the yield of the final product, while the influence of temperature and volume/mass ratio is significant.

From the obtained equations, the optimal extraction conditions were estimated as follows: for TPC, TFC, and TAC, the optimal temperature is 60 °C, volume/mass ratio—24, and moles of water in (D)ES—20. For the aglycones of flavonoids, the optimal temperature is 55 °C, volume/mass ratio—10, and moles of water in (D)ES—20. In these conditions, the following yields may be obtained: TPC—212 mg GAE/g, TFC—74 mg RE/g, TAC—33 mg AAE/g, c(myricetin)—157 µg/mL, c(quercetin)—143 µg/mL, c(kaempferol)—53 µg/mL.

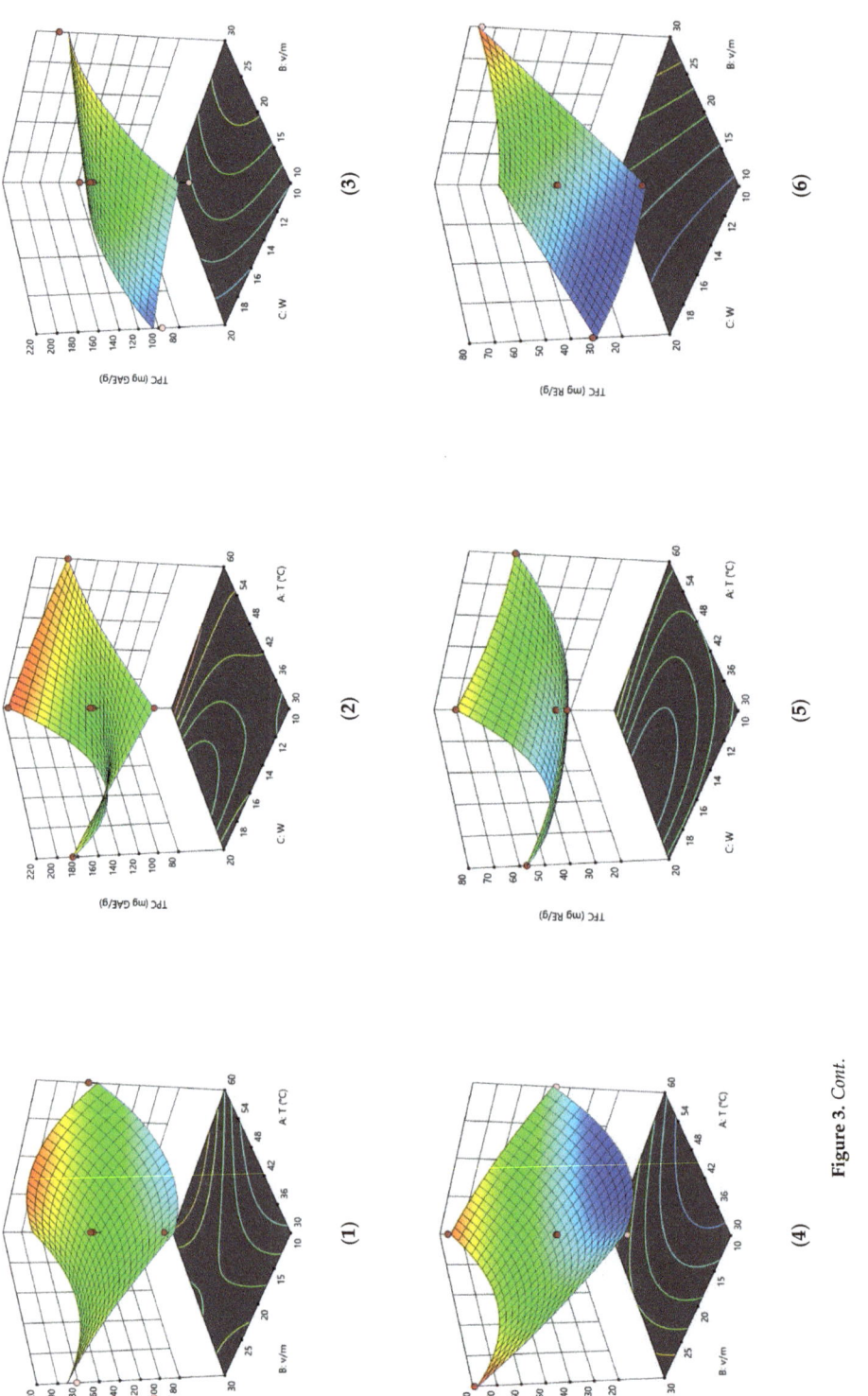

Figure 3. *Cont.*

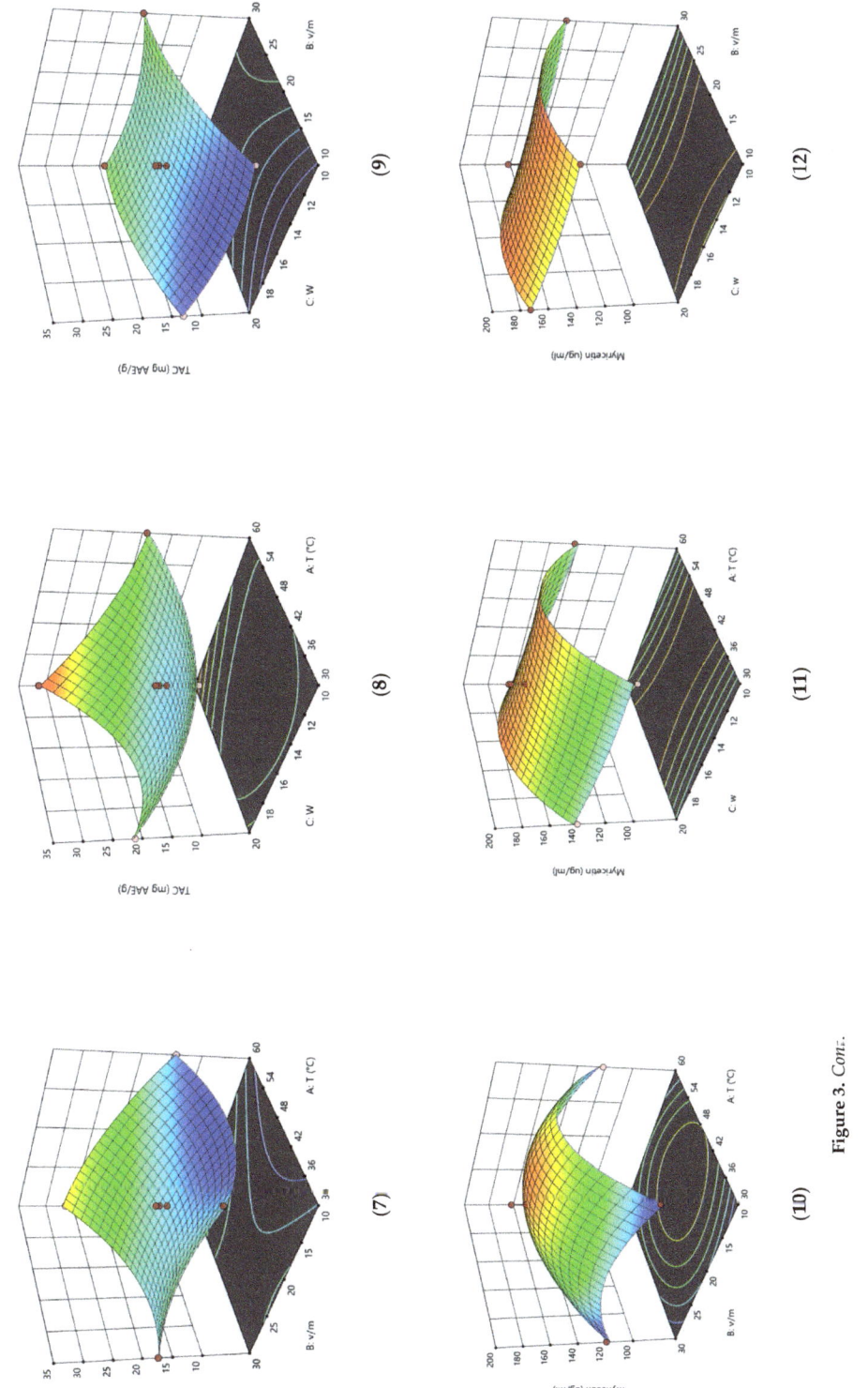

Figure 3. *Cont.*

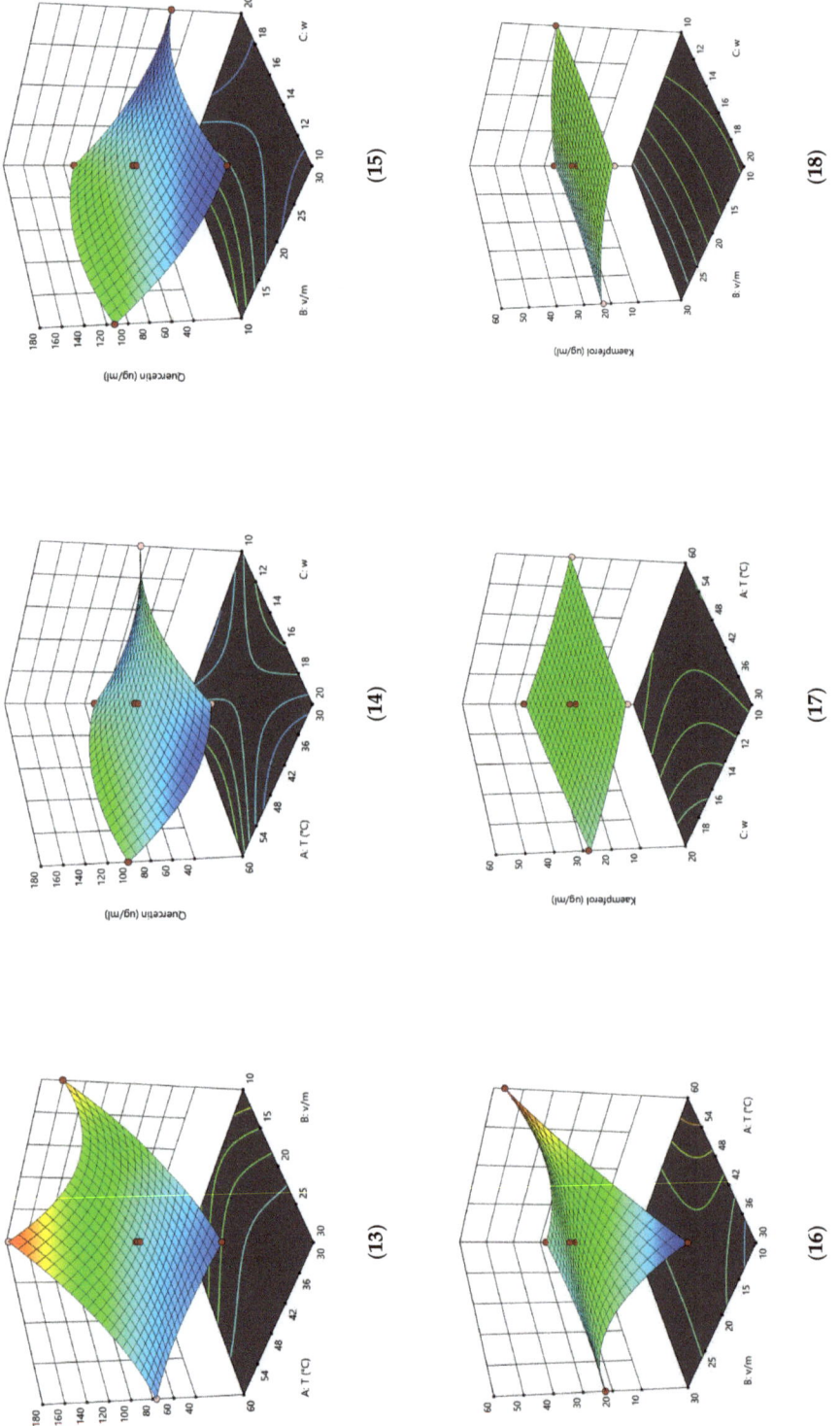

Figure 3. Response surfaces displaying the effect of extraction temperature (A), extraction time (B), and water content (C) on the extraction yield of TPC (1–3), TFC (4–6), TAC (7–9), myricetin (10–12), quercetin (13–15), and kaempferol (16–18).

Table 2. Coefficients of models for BBD results fitting.

	Intercept	A	B	C	AB	AC	BC	A^2	B^2	C^2	A^2B	A^2C	AB^2
TPC	160.140	15.700	19.750	−22.125	−6.525	−1.400	−7.675	28.405	−17.970	−1.745		35.725	−10.575
p-values		0.090	0.314	0.032	0.423	0.859	0.352	0.011	0.057	0.820		0.020	0.363
TFC	45.800	3.225	15.925	−7.225	−2.275	3.225	−5.025	14.688	0.138	2.788		8.200	−2.400
p-values		0.016	<0.0001	0.001	0.053	0.016	0.003	<0.0001	0.882	0.025		0.001	0.119
TAC	16.100	2.013	3.463	0.200	2.775	2.650	−0.650	4.938	−2.213	2.113		3.100	
p-values		0.013	0.001	0.814	0.014	0.017	0.455	0.001	0.032	0.037		0.036	
Myricetin	178.420	4.708	−12.700	0.017	1.042	3.935	−0.747	−35.169	−21.229	2.908	13.489	6.482	
p-values		0.113	0.015	0.996	0.776	0.308	0.838	0.0001	0.002	0.429	0.040	0.244	
Quercetin	91.060	5.633	−24.525	−2.389	−2.954	5.350	−0.685	17.285	14.206	−16.995	−22.855	2.857	
p-values		0.047	0.001	0.468	0.377	0.139	0.831	0.002	0.005	0.002	0.003	0.537	
Kaempferol	32.991	2.330	−7.112	−1.141	−8.938	1.878	0.081	−0.404	−1.846	0.850	1.955		9.326
p-values		0.020	0.0002	0.068	<0.0001	0.043	0.912	0.577	0.042	0.265	0.104		0.0002

Here, we offer a workflow for a qualitative assessment of biologically active substances obtained by the extraction of (D)ESs based on glycerol and propylene glycol from a promising plant. In addition, this workflow can be extended to all types of plant extracts from fruits, flowers, leaves and roots. Glycols are a viable alternative to other reference solvents, such as ethanol and methanol, for the production of plant extracts with wide application in the medical, food, cosmetic and agricultural industries [28].

Glycerol is a widely used, non-toxic ingredient in cosmetic products. However, due to its high viscosity, it is not possible to use it undiluted for the preparation of plant extracts [30].

3. Materials and Methods

3.1. Plant Material

Leaves of *C. angustifolium* were collected on the experimental site of the Polar Alpine Botanical Garden Institute (67°34' N 33°24' E) during the flowering vegetation period in mid-August 2020. An average sample of aerial parts of plants was obtained from an area of 5 × 5 m. The drying and storage of plant material was following the recommendations from [34]. For extraction, plant material was milled with a household grinder and sieved through a sieve with 0.5 mm holes. The milled plant material was additionally dried at 45 °C until a constant mass was reached. The humidity of the dry sample was 0.134%.

3.2. Chemicals and Reagents

Malonic, malic, tartaric and citric acids, glycerol and propylene glycol (>99%, Vekton, St. Petersburg, Russia) were used for (D)ES preparation. 2,2-diphenyl-1-picrylhydrazyl (99%, Sigma-Aldrich, Burlington, MA, USA), Folin–Ciocalteu reagent (2M, Sigma-Aldrich), ammonium molybdate, potassium dihydrogenphosphate, aluminum chloride (>99%, Vekton, St. Petersburg, Russia), concentrated sulfuric and hydrochloric acids (>94%, Nevareactiv, St. Petersburg, Russia), gallic acid (98% Sigma-Aldrich, Burlington, MA, USA), rutin (≥94%, Sigma-Aldrich, Burlington, MA, USA), ascorbic acid (>99.7%, Hugestone, Nanjing, China), and ethanol (96%, RFK Company, St. Petersburg, Russia) were used for chemical analysis. Myricetin, quercetin and kaempferol analytical standards (>98%, Sigma-Aldrich), acetonitrile HPLC grade (Component Reactive, Moscow, Russia), glacial acetic acid (Vekton, St.Petersburg, Russia), and deionized water were obtained with the water purification system "Millipore Element" (Millipore, Burlington, MA, USA) and used for LC-UV analysis.

3.3. Deep Eutectic Solvents Preparation

The preparation of (D)ESs was carried out by mixing acid mixtures with glycols, and then the mixtures were heated at 60 °C for 3 h. Various combinations of components were tested, and ratios were selected that gave a homogeneous mixture (Table 3). To lower the (D)ES viscosity, water was added to them. To compare the efficiency of extraction using different (D)ESs, a water additive of 10 molar parts was used. To optimize extraction conditions using the Box–Behnken design of experiment, the amount of water varied.

In all experiments, freshly prepared (D)ESs were used to minimize the effect of the possible formation of esters, which is well known for (D)ES based on choline chloride and carboxylic acids [35].

3.4. Characterization of (D)ESs

IR spectra were recorded on a Nicolet 6700 FT-IR spectrophotometer (Thermo Fisher Scientific Inc., Hillsboro, OR, USA, 2010) in the 4000–550 cm^{-1} region (diamond ATR, 16 scans, resolution 4).

Table 3. Composition of (D)ES used in the work and their abbreviations.

Component 1	Component 2	Abbreviation	Molar Ratio Component 1: Component 2: Water
Malonic acid (COOH-CH2-COOH)	Glycerol	MAGL	1:1:10
	Propylene glycol	MAPG	1:2:10
Malic acid (HOOC-CH(OH)-CH2-COOH)	Glycerol	MalGL	1:2:10
	Propylene glycol	MalPG	1:2:10
Citric acid (HOOC-CH2-C(OH)(COOH)-CH2-COOH)	Glycerol	CAGL 10 H_2O CAGL 15 H_2O CAGL 20 H_2O	1:4:10 1:4:15 1:4:20
	Propylene glycol	CAPG	1:4:10

There are plenty of experimental data for (D)ESs formed by choline chloride and glycerol or ethylene glycol, and choline chloride and organic acids. However, data on (D)ESs formed by organic acids and glycerol or propylene glycol are not found in the literature. In this work, we measured the density of the (D)ESs under study using a DMA 5000 M density meter (Anton Paar GmbH, Graz, Austria); the measurement uncertainty is 0.00001 g cm^{-3}. Also for the (D)ESs, the viscosity was determined using a Modular Compact Rheometer MCR 702 (Anton Paar GmbH, Austria); the measurement uncertainty is 0.08 mPa·s.

3.5. Ultrasound-Assisted Extraction, Kinetics and Box–Behnken Design Optimization

Extraction was performed in the thermostated ultrasound bath Vilitek VBS 3-DP (Vilitek, Moscow, Russia, 2018) with an ultrasound power of 120 W and a frequency of 40 kHz.

To determine the optimal extraction time and select the most suitable extractant, extraction was carried out for 5–60 min at 45 °C with a volume/mass ratio of 20. The change in concentrations of target groups of components or flavonoids in the form of their aglycones was described by the second-order reaction kinetic equation, following the work [32]. From the coefficients of the linear equation for dependence, $1/Y_t$ vs. t (where Y_t—yield of a target compound or group of compounds at time t), equilibrium yield (Y^{eq}), and the rate constant (k) were calculated.

After selecting the most suitable (D)ES composition and extraction time, the extraction conditions, such as temperature, volume/mass ratio and water content in (D)ES were optimized. A Box–Behnken design of experiment (BBD) combined with response surface methodology (RSM) was applied for the optimal condition calculation. In this design of experiment, three parameters of extraction conditions have three levels. The temperature was in the range of 30–60 °C, the volume/mass ratio was in the range of 10–30, and the water content in (D)ESs was in the range of 10–30 molar parts. The combinations of parameters used in this work are presented in Table 4.

Table 4. Parameters and their levels used in Box–Behnken design of experiment.

Temperature °C	Volume to Mass Ratio	Molar Parts of H_2O
30 (−1)	20 (0)	5 (−1)
30 (−1)	10 (−1)	10 (0)
30 (−1)	30 (+1)	10 (0)
30 (−1)	20 (0)	15 (+1)
45 (0)	10 (−1)	5 (−1)
45 (0)	30 (+1)	5 (−1)
45 (0)	20 (0)	10 (0)
45 (0)	20 (0)	10 (0)
45 (0)	20 (0)	10 (0)
45 (0)	20 (0)	10 (0)
45 (0)	10 (−1)	15 (+1)
45 (0)	30 (+1)	15 (+1)
60 (+1)	20 (0)	5 (−1)
60 (+1)	10 (−1)	10 (0)
60 (+1)	30 (+1)	10 (0)
60 (+1)	20 (0)	15 (+1)

3.6. Chemical Analysis of Extracts

The total phenolic content (TPC) was measured using the Folin–Ciocalteu method, and the total flavonoid content (TFC) was measured using the reaction of complexation with aluminum chloride in a 90% (v/v) ethanol–water mixture. TPC is expressed as mg/g of gallic acid equivalent (GAE) per one gram of plant weight. TFC is expressed as mg/g of rutin equivalent (RE) per one gram of plant weight. All the methods are described in detail in [33].

3.7. Antioxidant Activity Measuring

The total antioxidant content (TAC) was measured using the phosphomolybdate method and was expressed as mg/g of ascorbic acid equivalent (AAE) per one gram of plant weight. Free radical scavenging was measured using the DPPH method and was expressed as % of inhibition in comparison with blank solution. Here, 95% ethanol was used as a solvent for the DPPH solution. The incubation was performed for 30 min at 25 °C in the dark. All these methods are also described in detail in [33].

3.8. LC-UV Analysis for Quantification of Aglycons of Extracted Flavonoids

For the quantitative determination of flavonoid aglycones, the acid hydrolysis described in [24] was performed with modifications. First, 100 µL of freshly prepared ascorbic acid solution in 4 M hydrochloric acid with a concentration of 1 mg/mL was mixed with 100 µL of row plant extract. The obtained mixture was incubated at 70 °C for 1 h in the plastic test tubes with screw caps. Then, the mixture was dissolved 5 times with 95% ethanol and was centrifugated for 10 min at 4 rpm in the laboratory centrifuge MiniSpin (Eppendorf, Hamburg, Germany, 2018).

LC-UV analysis was performed with a liquid chromatograph Milichrom A-02 with a ProntoSil-120-5-C18 AQ column, 75 × 2 mm, with a particle size of 5 µm (Econova, Novosibirsk, Russia, 2022). The gradient elution was performed using 1% (v/v) acetic acid in water (A) and acetonitrile (B) with the following program: 0–5 min—0% B, 5–30 min 0–100% B,

30–32 min 100% B, 32–35 min 100–0% B, flow rate—100 µL/min, sample volume—2 µL. The wavelength of detection was 254 nm. Solutions of myricetin, quercetin, and kaempferol in the concentration range of 6.25–100 µg/mL were used for calibration.

3.9. Statistical Analysis

All the measurements were made three times for each analysis. Statistical comparison was performed using factorial analysis of variance (ANOVA) and post hoc Tukey's HSD test at $p \leq 0.05$. The calculations were performed using MS Excel 2010 (Microsoft, Redmond, WA, USA). DesignExpert 11 (Stat-Ease, Minneapolis, MN, USA) software was used for the Box–Behnken design and the UAE condition optimization using response surface methodology.

4. Conclusions

In the present work, for the first time, (D)ESs based on carboxylic acids (malonic, malic, and citric) with glycerol or propylene glycol were applied for the ultrasound-assisted extraction of biologically active substances from *Chamaenerion angustifolium* (L.) Scop.

The (D)ESs used in the work are characterized in detail using FTIR; data on density and viscosity are obtained. (D)ESs based on glycerol have relatively higher densities in combination with the same acid. Dependences of the total phenolic, flavonoids, and antioxidant content (TPC, TFC, and TAC), free radical scavenging of DPPH, and concentration of flavonoid aglycons (myricetin, quercetin, and kaempferol) via time in the range of 5–60 min at 45 °C were approximated using second-order reaction equations. The highest yield of the target components was achieved when using (D)ES—citric acid + glycerol. For this (D)ES, extraction conditions were optimized with the Box–Behnken design of experiment, and the optimal conditions are the following: for the TPC, TFC, and TAC, the optimal temperature is 60 °C and the volume/mass ratio is 24, while for flavonoid aglycons, the optimal temperature is 55 °C and the volume/mass ratio is 10. The optimal molar content of water was 20. In these conditions, the following yields may be obtained: TPC—212 mg GAE/g, TFC—74 mg RE/g, TAC—33 mg AAE/g, c(myricetin)—157 µg/mL, c(quercetin)—143 µg/mL, and c(kaempferol)—53 µg/mL.

The results of this work make it possible to expand the scope of (D)ES application and serve the development of *C. angustifolium* processing methods within the framework of green chemistry technologies for the needs of the production of cosmetics, biologically active additives, and pharmaceuticals. In addition, a relevant issue remains regarding (D)ESs recycling, and this should be considered a further direction of research in (D)ESs application.

Supplementary Materials: The following supporting information can be downloaded at: https://www.mdpi.com/article/10.3390/molecules28196978/s1, S1. Experimental values of density and dynamic viscosity of suited DESs with 10 molar part of water at 30 °C; S2. Kinetical curves and linearization for TPC. 1–2—MAGL, 3–4—MAPG, 5–6—MalGL, 7–8—MalPG, 9–10—CAGL, 11–12—CAPG; S3. Kinetical curves and linearization for TFC. 1–2—MAGL, 3–4—MAPG, 5–6—MalGL, 7–8—MalPG, 9–10—CAGL, 11–12—CAPG; S4. Calculated equilibrium yields (Yeq) and rate constants (k) for TPC and TFC; S5. Kinetical curves and linearization for TAC. 1–2—MAGL, 3–4—MAPG, 5–6—MalGL, 7–8—MalPG, 9–10—CAGL, 11–12—CAPG; S6. Kinetical curves and linearization for DPPH. 1–2—MAGL, 3–4—MAPG, 5–6—MalGL, 7–8—MalPG, 9–10—CAGL, 11–12—CAPG; S7. Calculated equilibrium yields (Yeq) and rate constants (k) for TAC and DPPH; S8. Kinetical curves and linearization for myricetin. 1–2—MAGL, 3–4—MAPG, 5–6—MalGL, 7–8—MalPG, 9–10—CAGL, 11–12—CAPG; S9. Kinetical curves and linearization for quercetin. 1–2—MAGL, 3–4—MAPG, 5–6—MalGL, 7–8—MalPG, 9–10—CAGL, 11–12—CAPG; S10. Kinetical curves and linearization for kaempferol. 1–2—MAGL, 3–4—MAPG, 5–6—MalGL, 7–8—MalPG, 9–10—CAGL, 11–12—CAPG; S11. Calculated equilibrium yields (Yeq) and rate constants (k) for describing of the experimental data on the kinetics of myricetin, quercetin, and kaempferol glycosides extraction.

Author Contributions: Conceptualization, N.T. and A.K.; methodology, N.T. and A.K.; software, N.T.; validation, A.K.; formal analysis, N.T. and A.G.; investigation, A.K., A.G. and A.S.; resources, N.T.; data curation, N.T.; writing—original draft preparation, A.K. and N.T.; writing—review and editing, N.T. and A.K.; visualization, N.T.; supervision, N.T.; project administration, N.T.; funding acquisition, N.T. All authors have read and agreed to the published version of the manuscript.

Funding: This research was funded by the Russian Science Foundation, grant number 22-26-20114.

Institutional Review Board Statement: Not applicable.

Informed Consent Statement: Not applicable.

Data Availability Statement: Not applicable.

Acknowledgments: The experimental work was facilitated by the equipment of Centre for Diagnostics of Functional Materials for Medicine, Pharmacology and Nanoelectronics, St. Petersburg State University Research Park.

Conflicts of Interest: The authors declare no conflict of interest.

Sample Availability: Not applicable.

References

1. Stepanova, E.M.; Lugovaya, E.A. Mineral Composition of Wild Berry Fruits from the Forest Zone of the City Of Magadan. *Chem. Plant Raw Mater.* **2022**, 343–350. [CrossRef]
2. Kaminskii, I.P.; Kadyrova, T.V.; Kalinkina, G.I.; Larkina, M.S.; Ermilova, E.V.; Belousov, M.V. Comparative Pharmacognostic Research of Centaurea Scabiosa L. Wild-Growing and Culti-Vated in the Conditions of Tomsk. *Chem. Plant Raw Mater.* **2020**, 119–126. [CrossRef]
3. Velikorodov, A.V.; Pilipenko, V.N.; Pilipenko, T.A.; Malyi, S.V. Studying the Chemical Composition of Essential Oil Received From Fruits of Prangos Odon-Talgica Wild-Growing in Astrakhan Region. *Chem. Plant Raw Mater.* **2019**, 95–101. [CrossRef]
4. Sergeeva, I.; Zaushintsena, A.; Bryukhachev, E. Photosynthetic Pigments and Phenolic Potential of Rhodiola Rosea L. from Plant Communities of Different Ecology and Geography. *Food Process. Tech. Technol.* **2020**, *50*, 393–403. [CrossRef]
5. Pasichnik, E.A.; Paukshta, O.I.; Nikolaev, V.G.; Tsvetov, N.S. Extraction of Bioactive Components from *Chamaenerion angustifolium* (L.) Herb Growing in Kola Peninsula Using Deep Eutectic Solvents. *IOP Conf. Ser. Earth Environ. Sci.* **2022**, *981*, 032080. [CrossRef]
6. Tsarev, V.N.; Bazarnova, N.G.; Dubenskiy, M.M. Narrow-Leaved Cypress (*Chamerion angustifolium* L.) Chemical Composition, Biological Activity (Review). *Chem. Plant Raw Mater.* **2016**, 15–26. (In Russian) [CrossRef]
7. Irinina, O.I.; Eliseeva, S.A. Study of the Biochemical Composition and Medicinal Properties of the Narrow-Leaved Cypress Plant (Ivan-Tea). *Polzunovskiy Vestn.* **2021**, *2*, 44–54. (In Russian) [CrossRef]
8. Huttunen, S.; Riihinen, K.; Kauhanen, J.; Tikkanen-Kaukanen, C. Antimicrobial Activity of Different Finnish Monofloral Honeys against Human Pathogenic Bacteria. *APMIS* **2013**, *121*, 827–834. [CrossRef]
9. Volodina, S.O.; Volodin, V.V.; Nekrasova, E.V.; Syrov, V.N.; Khushbaktova, Z.A. Stress-Protective Effect of Aqueous Infusion of Fermented Leaves Chamaenerion An-Gustifolium (L.) Scop. *Chem. Plant Raw Mater.* **2020**, 267–272. [CrossRef]
10. Zainal-Abidin, M.H.; Hayyan, M.; Hayyan, A.; Jayakumar, N.S. New Horizons in the Extraction of Bioactive Compounds Using Deep Eutectic Solvents: A Review. *Anal. Chim. Acta* **2017**, *979*, 1–23. [CrossRef]
11. Dai, Y.; Witkamp, G.J.; Verpoorte, R.; Choi, Y.H. Tailoring Properties of Natural Deep Eutectic Solvents with Water to Facilitate Their Applications. *Food Chem.* **2015**, *187*, 14–19. [CrossRef] [PubMed]
12. Jurić, T.; Mićić, N.; Potkonjak, A.; Milanov, D.; Dodić, J.; Trivunović, Z.; Popović, B.M. The Evaluation of Phenolic Content, in Vitro Antioxidant and Antibacterial Activity of Mentha Piperita Extracts Obtained by Natural Deep Eutectic Solvents. *Food Chem.* **2021**, *362*, 130226. [CrossRef] [PubMed]
13. Shishov, A.; Gagarionova, S.; Bulatov, A. Deep Eutectic Mixture Membrane-Based Microextraction: HPLC-FLD Determination of Phenols in Smoked Food Samples. *Food Chem.* **2020**, *314*, 126097. [CrossRef] [PubMed]
14. Abbott, A.P.; Boothby, D.; Capper, G.; Davies, D.L.; Rasheed, R. Deep Eutectic Solvents Formed Between Choline Chloride and Carboxylic Acids. *J. Am. Chem. Soc.* **2004**, *126*, 9142.
15. Piemontese, L.; Sergio, R.; Rinaldo, F.; Brunetti, L.; Perna, F.M.; Santos, M.A.; Capriati, V. Deep Eutectic Solvents as Effective Reaction Media for the Synthesis of 2-Hydroxyphenylbenzimidazole-Based Scaffolds En Route to Donepezil-Like Compounds. *Molecules* **2020**, *25*, 574. [CrossRef]
16. Boldrini, C.L.; Quivelli, A.F.; Manfredi, N.; Capriati, V.; Abbotto, A. Deep Eutectic Solvents in Solar Energy Technologies. *Molecules* **2022**, *27*, 709. [CrossRef]
17. Santana, A.P.R.; Mora-Vargas, J.A.; Guimarães, T.G.S.; Amaral, C.D.B.; Oliveira, A.; Gonzalez, M.H. Sustainable Synthesis of Natural Deep Eutectic Solvents (NADES) by Different Methods. *J. Mol. Liq.* **2019**, *293*, 17–20. [CrossRef]

18. Arnaboldi, S.; Mezzetta, A.; Grecchi, S.; Longhi, M.; Emanuele, E.; Rizzo, S.; Arduini, F.; Micheli, L.; Guazzelli, L.; Mussini, P.R. Natural-Based Chiral Task-Specific Deep Eutectic Solvents: A Novel, Effective Tool for Enantiodiscrimination in Electroanalysis. *Electrochim. Acta* **2021**, *380*, 138169. [CrossRef]
19. Afonso, J.; Mezzetta, A.; Marrucho, I.M.; Guazzelli, L. History Repeats Itself Again: Will the Mistakes of the Past for ILs Be Repeated for DESs? From Being Considered Ionic Liquids to Becoming Their Alternative: The Unbalanced Turn of Deep Eutectic Solvents. *Green Chem.* **2023**, *25*, 59–105. [CrossRef]
20. Morais, E.S.; Lopes, A.M.d.C.; Freire, M.G.; Freire, C.S.R.; Coutinho, J.A.P.; Silvestre, A.J.D. Use of Ionic Liquids and Deep Eutectic Solvents in Polysaccharides Dissolution and Extraction Processes towards Sustainable Biomass Valorization. *Molecules* **2020**, *25*, 3652. [CrossRef]
21. Cunha, S.C.; Fernandes, J.O. Extraction Techniques with Deep Eutectic Solvents. *TrAC Trends Anal. Chem.* **2018**, *105*, 225–239. [CrossRef]
22. Choi, Y.H.; van Spronsen, J.; Dai, Y.; Verberne, M.; Hollmann, F.; Arends, I.W.C.E.; Witkamp, G.J.; Verpoorte, R. Are Natural Deep Eutectic Solvents the Missing Link in Understanding Cellular Metabolism and Physiology? *Plant Physiol.* **2011**, *156*, 1701–1705. [CrossRef] [PubMed]
23. Ijardar, S.P.; Singh, V.; Gardas, R.L. Revisiting the Physicochemical Properties and Applications of Deep Eutectic Solvents. *Molecules* **2022**, *27*, 1368. [CrossRef] [PubMed]
24. Santana, A.P.R.; Andrade, D.F.; Guimarães, T.G.S.; Amaral, C.D.B.; Oliveira, A.; Gonzalez, M.H. Synthesis of Natural Deep Eutectic Solvents Using a Mixture Design for Extraction of Animal and Plant Samples Prior to ICP-MS Analysis. *Talanta* **2020**, *216*, 120956. [CrossRef]
25. Dai, Y.; van Spronsen, J.; Witkamp, G.-J.; Verpoorte, R.; Choi, Y.H. Natural Deep Eutectic Solvents as New Potential Media for Green Technology. *Anal. Chim. Acta* **2013**, *766*, 61–68. [CrossRef]
26. Zhang, X.; Su, J.; Chu, X.; Wang, X. A Green Method of Extracting and Recovering Flavonoids from Acanthopanax Senticosus Using Deep Eutectic Solvents. *Molecules* **2022**, *27*, 923. [CrossRef]
27. Chen, W.; Jiang, J.; Lan, X.; Zhao, X.; Mou, H.; Mu, T. A Strategy for the Dissolution and Separation of Rare Earth Oxides by Novel Brønsted Acidic Deep Eutectic Solvents. *Green Chem.* **2019**, *21*, 4748–4756. [CrossRef]
28. Nastasi, J.R.; Daygon, V.D.; Kontogiorgos, V.; Fitzgerald, M.A. Qualitative Analysis of Polyphenols in Glycerol Plant Extracts Using Untargeted Metabolomics. *Metabolites* **2023**, *13*, 566. [CrossRef]
29. Ciganović, P.; Jakupović, L.; Momchev, P.; Nižić Nodilo, L.; Hafner, A.; Zovko Končić, M. Extraction Optimization, Antioxidant, Cosmeceutical and Wound Healing Potential of Echinacea Purpurea Glycerolic Extracts. *Molecules* **2023**, *28*, 1177. [CrossRef]
30. Juszczak, A.M.; Marijan, M.; Jakupović, L.; Tomczykowa, M.; Tomczyk, M.; Zovko Končić, M. Glycerol and Natural Deep Eutectic Solvents Extraction for Preparation of Luteolin-Rich Jasione Montana Extracts with Cosmeceutical Activity. *Metabolites* **2022**, *13*, 32. [CrossRef]
31. de Brito, V.P.; de Souza Ribeiro, M.M.; Viganó, J.; de Moraes, M.A.; Veggi, P.C. Silk Fibroin Hydrogels Incorporated with the Antioxidant Extract of Stryphnodendron Adstringens Bark. *Polymers* **2022**, *14*, 4806. [CrossRef] [PubMed]
32. Tsvetov, N.; Pasichnik, E.; Korovkina, A.; Gosteva, A. Extraction of Bioactive Components from *Chamaenerion angustifolium* (L.) Scop. with Choline Chloride and Organic Acids Natural Deep Eutectic Solvents. *Molecules* **2022**, *27*, 4216. [CrossRef]
33. Tsvetov, N.; Sereda, L.; Korovkina, A.; Artemkina, N.; Kozerozhets, I.; Samarov, A. Ultrasound-Assisted Extraction of Phytochemicals from Empetrum Hermafroditum Hager. Using Acid-Based Deep Eutectic Solvent: Kinetics and Optimization. *Biomass Convers. Biorefinery* **2022**, *12*, 145–156. [CrossRef]
34. World Health Organization. *Pharmaceuticals Unit Quality Control Methods for Medicinal Plant Materials*; WHO: Geneva, Switzerland, 1998; p. 122.
35. Rodriguez Rodriguez, N.; Van Den Bruinhorst, A.; Kollau, L.J.B.M.; Kroon, M.C.; Binnemans, K. Degradation of Deep-Eutectic Solvents Based on Choline Chloride and Carboxylic Acids. *ACS Sustain. Chem. Eng.* **2019**, *7*, 11521–11528. [CrossRef]

Disclaimer/Publisher's Note: The statements, opinions and data contained in all publications are solely those of the individual author(s) and contributor(s) and not of MDPI and/or the editor(s). MDPI and/or the editor(s) disclaim responsibility for any injury to people or property resulting from any ideas, methods, instructions or products referred to in the content.

Article

Folic Acid Ionic-Liquids-Based Separation: Extraction and Modelling

Alexandra Cristina Blaga *, Elena Niculina Dragoi, Alexandra Tucaliuc, Lenuta Kloetzer and Dan Cascaval

"Cristofor Simionescu" Faculty of Chemical Engineering and Environmental Protection, "Gheorghe Asachi" Technical University of Iasi, D. Mangeron 73, 700050 Iasi, Romania; elena.dragoi@tuiasi.ro (E.N.D.); alexandra.tucaliuc@academic.tuiasi.ro (A.T.); lenuta.kloetzer@academic.tuiasi.ro (L.K.); dan.cascaval@academic.tuiasi.ro (D.C.)
* Correspondence: acblaga@tuiasi.ro

Abstract: Folic acid (vitamin B9) is an essential micronutrient for human health. It can be obtained using different biological pathways as a competitive option for chemical synthesis, but the price of its separation is the key obstacle preventing the implementation of biological methods on a broad scale. Published studies have confirmed that ionic liquids can be used to separate organic compounds. In this article, we investigated folic acid separation by analyzing 5 ionic liquids (CYPHOS IL103, CYPHOS IL104, [HMIM][PF_6], [BMIM][PF_6], [OMIM][PF_6]) and 3 organic solvents (heptane, chloroform, and octanol) as the extraction medium. The best obtained results indicated that ionic liquids are potentially valuable for the recovery of vitamin B9 from diluted aqueous solutions as fermentation broths; the efficiency of the process reached 99.56% for 120 g/L CYPHOS IL103 dissolved in heptane and pH 4 of the aqueous folic acid solution. Artificial Neural Networks (ANNs) were combined with Grey Wolf Optimizer (GWO) for modelling the process, considering its characteristics.

Keywords: CYPHOS IL103; vitamin B9; ionic liquid; extraction; modelling

1. Introduction

Vitamins are chemical substances whose derivatives are engaged in the vital metabolic pathways of all living organisms [1]. Since only bacteria, yeast and plants have endogenous routes for vitamin production, humans must obtain most of these crucial nutrients from food [2]. A form of B vitamin called folic acid (folate in its natural form [3])—Figure 1 aids in the maintenance and production of new cells in the body and prevents nucleic acid alterations [2]. Several physiological processes in humans, including the biosynthesis of nucleotides, cell division, and gene expression, as well as the prevention of vascular diseases, megaloblastic anemia, and neural tube defects in developing children, depend on a proper supply of folic acid (FA) or folates.

Figure 1. Folic acid (FA) chemical structure.

FA has numerous applications in the pharmaceutical, nutraceutical, food and beverages industries [4], but its stability decreases when exposed to light, moisture, strong acidic or alkaline media, oxygen and high temperatures. It can be found in foods, including oranges, whole-wheat products, dry beans, peas, lentils, oranges, liver, asparagus, beets, broccoli, brussels sprouts and spinach [5].

The estimated value of the global FA market in 2022 was USD 166.8 million. By 2028, it is anticipated to grow to USD 220.9 million, the leading producers being BASF, DSM, Nantong Changhai Food Additive Co., Ltd., Niutang, Zhejiang Shengda, Parchem Fine & Specialty Chemicals, Xinjiang Wujiaqu Xingnong Cycle Chemical Co., Ltd., Xinfa Pharmaceutical Co., Medicamen Biotech Ltd., Jiangxi Tianxin Pharmaceutical Co., Ltd., and Zydus Pharmaceuticals Ltd. [6].

FA is used as a dietary supplement and is currently produced through chemical synthesis, but this process requires the reduction of ecologically harmful effects. Significant research has been employed to develop microbial strains (*Ashbya gossypii* ATCC 10895, *Lactococcus lactis* NZ9000, *Bacillus subtilis* 168) in order to manufacture FA [7–10]. The highest production value to date has been recorded for *A. gossypii*, which can synthesize 6.59 mg/L of folic acid after metabolic engineering (from its natural ability to produce 0.04 mg/L) [11]. For this process to be successfully employed, efficient separation methods must be developed. More research is required to create a method that is both economical and environmentally benign for producing high-purity FA, because its industrial separation involves many expensive downstream steps and requires mild conditions due to acid's instability.

Various technologies can be used to separate carboxylic acids (ion exchange, electrodialysis, ultrafiltration, solvent extraction, and membrane processes) [12]. However, reactive extraction using particular extractants has been proven to be an excellent alternative to classical methods, due to its many advantages [13]. Reactive extraction is based on a reaction between an extractant (dissolved in the organic solvent) and the target solute (e.g., carboxylic acids dissolved in the aqueous phase). Several carboxylic acids (gallic acid [14,15], keto-gluconic acid [16], pseudo-monic acid [17], lactic acid [18]), and vitamins (vitamin C [19], vitamin B5 [20]) have been successfully separated through this method at laboratory scale. For sustainability of this process, finding a selective, affordable, and effective extractant and diluent system based on maximal efficiency and minimal toxicity and determining the ideal implementation circumstances are the key challenges in using reactive extraction for the recovery of organic acids. Several characteristics must be considered for the choice of the organic phase, such as selectivity, solubility, cost and operational safety, hydrophobicity, density, polarity, viscosity, recoverability and environmental effects (the use of volatile organic solvents harms the environment). Thus, based on their superior characteristics, ionic liquids (tunable organic salts obtained as a combination of an organic cation and either an organic or a polyatomic inorganic anion in a liquid state below 100 °C) are effective alternatives to classical solvents. The use of most ionic liquids has several advantages over organic solvents, and they play crucial roles in the extraction processes; high thermal stability, negligible vapor pressure, and biocompatibility make them environmentally friendly substances with excellent solvation ability. Based on their properties, most ionic liquids can be employed in green chemistry concepts [21,22]. Lactic acid, citric acid, mevalonic acid [23], and butyric acid [24] have all been successfully extracted using ionic liquids. For the micro-solid phase extraction (for preconcentration and analysis) of pyridoxine and folic acid from biological samples, Zare et al. (2015) investigated a sorbent obtained through the synthesis of gold nanoparticles (Au NPs) and their subsequent transfer to aqueous solution by the application of the ionic liquid: 1-hexyl-3-methylimidazolium bis(trifluoro-methyl-sulfonyl)imide [25].

For the scale-up application of reactive extraction, rigorous modelling and optimization of laboratory-scale studies are essential and different models can be applied [15–17]. ANNs are inspired by the biological brain, and GWO is inspired by the grey wolf's social hierarchy and hunting mechanism [26]. In this work, ANN represents the process model, while GWO is used for model optimization. This combination was considered based on the difficulties in identifying the optimal characteristics of an ANN for a given problem. From the multitude of ANN types, the fully connected feed-forward multilayer perceptron was selected because it is well suited to the complexity and characteristics of the studied process. Moreover, this type of network was successfully applied to solve different problems, including modeling of phytocompounds extraction from dragon fruit peel [27], pectinase extraction from cashew apple juice [28], and separation of pseudo-monic acids [16]. As a bio-inspired metaheuristic, GWO was efficiently applied to optimize the extraction of essential oil from Cleome Coluteiodes Boiss [29], or biodiesel production from waste oils [30,31].

To the authors' knowledge, this is the first study regarding folic acid's separation using ionic liquids (IL) and a liquid–liquid approach. For this research, five ionic liquids (Trihexyl-tetradecyl-phosphonium decanoate, Trihexyl-tetradecyl-phosphonium bis(2,4,4-trimethylpentyl)phosphinate, 1-Butyl-3-methylimidazolium hexafluorophosphate, 1-Octyl-3-methylimidazolium hexafluorophosphate, 1-Hexyl-3-methylimidazolium hexafluorophosphate—Figure 2) were analyzed for FA extraction in order to find an optimal system for the separation process.

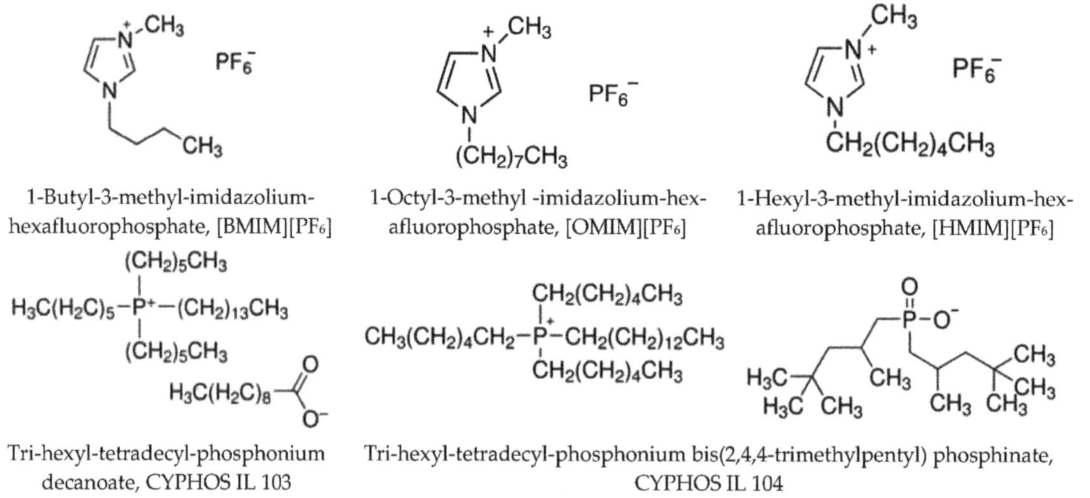

Figure 2. Ionic Liquids (IL) chemical structure.

Considering the physical characteristics of the ionic liquid (such as high density, viscosity and surface tension) and their high price, three solvents were analyzed as diluents (heptane, chloroform and octanol). The results were discussed from the viewpoint of the extraction mechanism, separation yield and distribution coefficient for different extraction conditions (aqueous phase pH and ionic liquids concentration in the organic phase). Supplementary Artificial Neural Networks (ANNs) were combined with Grey Wolf Optimizer (GWO) to model the considered process.

2. Results and Discussions

2.1. Extraction Process

Liquid–liquid extraction is a low-energy separation process with simple technical requirements and gentle operating conditions. Its effectiveness is influenced by several variables, including the type of solute and solvent, the utilized diluent and its physicochemical properties, and the solution pH. For FA extraction 5 ionic liquids with different chemical structures (Figure 2) and properties (Table 1) and 3 organic solvents with different dielectric constants (heptane—1.92, chloroform—4.81 and octanol—10.3 [32]) were investigated. The extraction system was chosen based on its low environmental impact. The ionic liquids chosen are highly hydrophobic [33] and were successfully used for the separation of other carboxylic acids (e.g., lactic acid) [34]. Phosphonium ILs offer, in specific cases and applications, several advantages over other types of ILs, including higher thermal stability, lower viscosity and higher stability in strongly basic or strongly reducing conditions [35].

Table 1. Main ionic liquids' physical properties used in biosynthetic compounds extraction [21,36,37].

Ionic Liquid	Molecular Formula	mol. wt., g/mol	Viscosity, cP, 25 °C	log P	Toxicity
[BMIM][PF_6]	$C_8H_{15}F_6N_2P$	284.18	274	4.49	log EC_{50} * = 3.32 μM
[OMIM][PF_6]	$C_{12}H_{23}F_6N_2P$	340.29	682	6.05	log EC_{50} * = 2.24 μM
[HMIM][PF_6]	$C_{10}H_{19}F_6N_2P$	312.24	585	5.27	log EC_{50} * = 1.25 μM
CYPHOS IL 103	$C_{42}H_{87}O_2P$	665.11	319	14.32	Inhib. ** = 1.5 cm
CYPHOS IL 104	$C_{48}H_{102}O_2P_2$	773.27	805.8	18.28	Inhib. ** = 2.6 cm

* Determined against *A. fischeri*. ** Determined against *Shewanella* sp. (inhibition zone, ±0.2).

The results obtained are presented in Figure 3. It can be observed that, without regarding the used organic solvent, the separation yield is very low, proving that physical extraction in classical organic solvents (based only on diffusion and solubilization) is practically impossible for FA. Simultaneously, the ionic liquids can effectively remove FA from the aqueous phase. In general, various variables, including the hydrophobic effect, hydrogen bonding, steric hindrance, and π-π interaction, affect how well ILs can extract carboxylic acids.

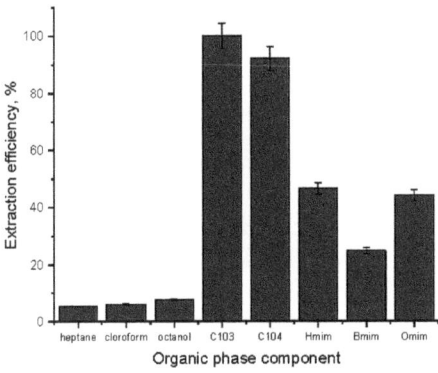

Figure 3. The organic phase composition influence on FA extraction (pH = 4, pure substances).

The highest efficiency was obtained for quaternary phosphonium salts: CYPHOS IL103 (99.98%) and CYPHOS IL104 (92.85%), the anions decanoate and bis(2,4,4-trimethylpentyl) phosphinate providing significantly superior yields than the hexafluorophosphate $\left[PF_6^-\right]$ anion, due to stronger hydrogen bonds established between the anion and FA, and to the hydrophobic behavior of the trihexyl(tetradecyl)phosphonium cation (phosphonium IL possesses the highest hydrophobicity among ILs [34]), present in both CYPHOS IL103 and CYPHOS IL104 structure). The superior efficiency that was obtained using CYPHOS IL103 can be explained by the effect of both the interference of sterically hindrance in CYPHOS IL104 case during the chemical reaction (CYPHOS IL104 has a larger structure compared to CYPHOS IL103 due to the presence of bis(2,4,4-trimethylpentyl)phosphinate ion compared to decanoate anion) and the superior viscosity of CYPHOS IL104 (Table 1)—according to the Wilke-Chang equation, diffusivity varies inversely with viscosity [34]. Similar results (superior values for CYPHOS IL103 compared with other ionic liquids) were obtained by Schlosser et al., 2018, for lactic and butyric acids [38]. These results proved that FA could be successfully separated using ionic liquids; however, their high viscosity and high price are vital points that require more research on this matter.

Due to mass transfer at the liquid–liquid interface, which influences the time required to set the equilibration stage between the aqueous and IL phases, viscosity is a crucial element that affects the kinetics of IL-based extraction systems. Conventional extraction systems using organic solvents may typically reach equilibrium in a short contact time (minutes), whereas IL-based systems require a longer contact time (minutes to hours) due to the high viscosity value, which is in this case (Table 1), between 274 and 805 cP (viscosity of water is 0.89 cP at 25 °C). Because of the influence on the Coulombic interaction between ions, adding an inert solvent could reduce IL viscosity [39]. In this context, the use of heptane, a non-polar solvent, as a diluent to decrease the viscosity and surface tension of the very viscous ionic liquids used, was analyzed as an alternative solution to pure ionic liquids for both CYPHOS IL. For [HMIM][PF$_6$], the solvent considered was chloroform, since this ionic liquid and heptane are not miscible. The FA from the aqueous phase can react with the strong hydrophobic ionic liquid dissolved in heptane to generate complexes that are only soluble in the organic phase:

$$R(COOH)COOH(aq) + [C_{14}C_6C_6C_6P][BTMPP] \rightarrow [C_{14}C_6C_6C_6P][R(COOH)COO] + BTMPPH$$

$$R(COOH)COOH(aq) + [C_{14}C_6C_6C_6P][CH_3(CH_2)_8COO] \rightarrow [C_{14}C_6C_6C_6P][R(COOH)COO] + CH_3(CH_2)_8COOH$$

The extraction mechanism could be characterized in terms of displacement reaction since ionic liquids are organic salts and contain tri-hexyl(tetradecyl) phosphonium ($[C_{14}C_6C_6C_6P]$) as a cation and bis(2,4,4-trimethylpentyl) phosphinate ([BTMPP]—CYPHOS IL104 and decanoate -CYPHOS IL103) as an anion. The anionic component of the acid ($C_{13}H_{11}N_6O\text{-}CH(CH_2\text{-}CH_2\text{-}COOH)\text{-}COO$) can displace the anionic species of IL in this reaction (pKa of decanoic acid is 5.7 while folic acid pKa are 4.69; 6.80). The extraction mechanism can also imply hydrogen bonding, the values for hydrogen-bonding interaction energy in the equimolar cation-anion mixture (E_{HB}/(kJ/mol) are -38.64 for decanoate $[C_9H_{20}CO_2]^-$ and -38.45 for Bis(2,4,4-trimethylpentyl)phosphinate, $[C_{16}H_{34}O_2P]^-$ [39].

FA is a weak acid in aqueous solutions, stable between pH 2–10, without heating [40], but its maximum stability is in the pH range of 4–10. The aqueous phase pH has a significant effect on extraction efficiency as it controls acid dissociation:

$$R(COOH)COOH_{(aq)} \leftrightarrow R(COOH)COO^-_{(aq)} + H^+, pKa_1 = 4.69$$
$$R(COOH)COO^-_{(aq)} \leftrightarrow R(COO^-)COO^-_{(aq)} + H^+, pKa_2 = 6.8$$

The extraction efficiency using the purposed extraction system (ionic liquids mixed with an organic diluent) decreases with the increase of aqueous phase pH, as highlighted by the experimental results depicted in Figure 4. Better results were obtained for CYPHOS (dissolved in heptane) compared to [HMIM][PF$_6$] (dissolved in chloroform), similar to the results obtained for protocatechuic acid or adipic acid [21]. Due to this fact (extraction efficiency much lower for [HMIM][PF$_6$]), the ionic liquid concentration influence was only analyzed for CYPHOS IL103 and CYPHOS IL104.

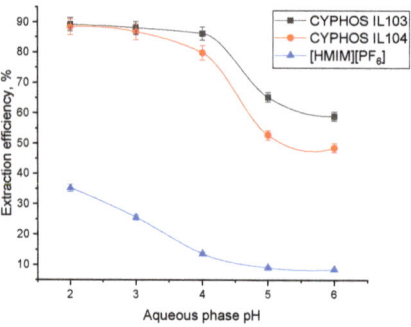

Figure 4. Aqueous phase pH influence on extraction efficiency (ionic liquid concentration 40 g/L).

The experimental results proved that FA could only be extracted by CYPHOS IL103 and CYPHOS IL104 in its undissociated state through H-bond coordination. FA is present in the aqueous solution in an undissociated form at pH lower than 4.5, as determined by the pKa values of 4.69 for the first carboxylic group and 6.80 for the second carboxylic group. This supports the idea that FA will be reactively extracted utilizing a coordination mechanism similar to lactic acid extraction, using CYPHOS IL104 [41]. Furthermore, the results prove that satisfactory separation efficiency can be achieved even at pH equal to 4, at which FA stability is considered maximum.

The extraction efficiency of FA increases with IL (CYPHOS IL103 and CYPHOS IL104) concentrations in heptane, as seen in Figure 5, proving that this parameter has a critical impact on extraction efficiency. This variation is due to the increase of one reactant concentration at the reaction interface; a higher concentration of IL is more likely to extract a higher concentration of the targeted folic acid from the aqueous phase. No third phase formation was observed in the experiments.

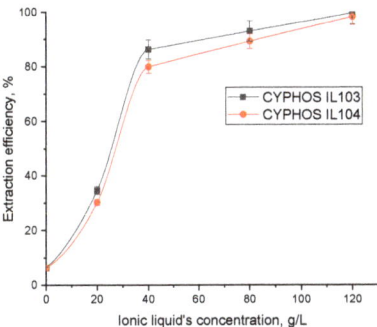

Figure 5. Ionic liquid concentration influence on extraction efficiency (initial phase pH = 4).

The influence of ionic liquids concentration on extraction efficiency was analyzed at pH 4 to avoid denaturation [40]. In order to establish how many molecules of acid and ionic liquid are involved in the formation of the interfacial complex, the loading factor, Z, defined as the ratio between the concentrations of FA and ionic liquid in the organic phase: [FA]org/[IL]org, was calculated (Table 2).

Table 2. Loading factor and distribution coefficient values obtained for C103 and C104 in heptane.

	CYPHOS IL103 conc., M	D	Loading Factor	CYPHOS IL104 conc., M	D	Loading Factor
1	0.03	0.52	0.43	0.02	0.43	0.43
2	0.06	6.24	0.53	0.05	3.95	0.57
3	0.12	12.99	0.28	0.10	8.19	0.32
4	0.18	231.18	0.20	0.15	53.51	0.23

Two cases can be considered when the variation of the Z parameters is assessed [42]. The first implies that, as the concentration of the ionic liquid increases, so do the loading factor values. This phenomenon, known as overloading (loading larger than unity), shows the formation of complexes with more than one acid per ionic liquid molecule. The maximum value obtained for the loading factor was 0.53 and 0.57 for both ionic liquids, so no overloading was noted in the system. The second case implies that complexes include more than one ionic liquid molecule if the loading factor values decrease as the IL concentration rises. The obtained results, presented in Table 2, showed that for ionic concentrations below the value equal to 40 g/L (0.05 M and 0.06 M for CYPHOS IL104 and CYPHOS IL103, respectively), the loading ratio slowly increases (from 0.43 to 0.53/0.57) between the concentrations 0.03 M and 0.06 M (20–40 g/L), followed by a significant decrease of Z with the increase in ionic liquid concentration. Thus, two types of complexes are formed in direct connection with the ionic liquid concentration:

- For concentrations lower than 40 g/L, echi-molecular complexes (involving only one molecule of both FA and ionic liquid) for both cases: [FA][CYPHOS IL103] and [FA][CYPHOS IL104] are formed, while
- For concentrations higher than 40 g/L, the decrease of Z with increasing ionic liquid concentration in the organic phase indicates the formation of complexes involving two molecules of ionic liquid per folic acid molecule: $[FA][CYPHOS\ IL103]_2$ and $[FA][CYPHOS\ IL104]_2$. This suggests that it may be more cost-effective to enhance the extractant concentration when the process efficiency is below the optimum level than to raise the process efficiency while maintaining the extractant concentration [42].

Schlosser [37] described two methods for water coextraction in the organic phase using Cyphos IL-104 dissolved in dodecane: the production of reverse micelles, and the incorporation of water into hydrated complexes including lactic acid and IL, complexes that include two molecules of water. In this study, no modifications of the two phases volume were recorded after the extraction using heptane as an organic solvent. This could be due to the short extraction time—10 min, or to the large structure of folic acid and its lower concentrations than the ionic liquid.

2.2. Modeling

The proposed ANN-GWO approach was applied to model the process. To reach this objective, the extraction yield for FA was determined as a function of the type of solvent used, aqueous phase pH, type of extractant and its concentration. Since the type of solvent and type of extractant are categorical values, they were coded with numerical integer values. This strategy allowed the development of a single model for all possible combinations regarding the considered parameters. This represents one significant advantage over the classical regression methods usually used to determine a process model.

Since the experimental work aimed to perform a reduced number of experiments and since ANNs work better with large datasets, in this work the experimental data were supplemented with a series of additional points by performing an interpolation procedure for simple, individual cases of combinations of parameters. For these cases, based on 2D plot representation, the trendlines that best fit the experimental points were determined using the R^2 metric, and the identified equation was then used to generate additional data. In this manner, the available dataset was extended from 27 experimental points to 103, thus allowing the ANN model to better capture the dynamic and influence of all process parameters on the extraction yield. Next, following the standard procedures regarding the application of ANNs, the data was normalized and randomly split into two groups (training and testing). The normalization procedure scales all the parameters to the $[-1,1]$ interval and reduces the impact of inputs with higher orders of magnitude. The type of normalization considered in this work is the Min-Max approach [43]. Afterward, the entire available dataset was randomly attributed: 75% to training and 25% to testing. The statistical indicators for these two subsets are presented in Table 3.

In the next step, the limits for the maximum ANN topology and the settings for GWO were set. Regarding topology, to reduce the number of ANN parameters that need to be identified based on the available data, the maximum number of hidden layers was set to 1, with 20 neurons in the hidden layer. This limitation was based on a set of preliminary runs that indicated that, for the current process, an ANN with a single hidden layer could efficiently capture the system's dynamic. Concerning the GWO parameters, the population size was set to 50 individuals, and the number of runs was set to 500. Next, 50 simulations were performed to determine the best ANN model for the process. The statistics of these runs are presented in Table 4, where Fitness measures model efficiency and is determined based on the Mean Squared Error (MSE) obtained in the training phase. The ANN with the highest fitness, referred to as ANN (4:05:01), indicates the best model for the process. In Table 4, topology is represented using an Input:HiddenLayer:Output notation, where Input represents the number of inputs corresponding to the process parameters (type of solvent used, aqueous phase pH, type of extractant and its concentration), HiddenLayer indicates the number of neurons in the hidden layer and Output indicates the number of process outputs (which for the current problem is FA).

Table 3. Statistics descriptors for the subsets used for ANN training and testing.

Subset	Indicator	Type of Solvent	Type of Extractant	Aqueous Phase pH	Extractant Concentration	FA
Training	Mean	1.155844	1.727273	4.084416	48.7013	63.0724
	Median	1	2	4	40	79.8336
	Standard Deviation	0.365086	0.718851	0.874164	28.2191	32.09486
	Sample Variance	0.133288	0.516746	0.764162	796.3175	1030.08
	Kurtosis	1.792505	−0.94144	0.632868	0.556316	−1.11127
	Skewness	1.935617	0.461642	−0.01714	0.909524	−0.65705
	Minimum	1	1	2	0	6.229
	Maximum	2	3	6	120	98.33493
	Count	77	77	77	77	77
Testing	Mean	1.269231	1.923077	3.75	51.15385	62.5145
	Median	1	2	4	40	81.54103
	Standard Deviation	0.452344	0.796145	0.806226	26.08861	32.47939
	Sample Variance	0.204615	0.633846	0.65	680.6154	1054.911
	Kurtosis	−0.84995	−1.37721	1.646948	0.895838	−1.59354
	Skewness	1.105353	0.143288	−0.10853	1.188646	−0.49504
	Minimum	1	1	2	15	8.340475
	Maximum	2	3	5.75	120	99.5693
	Count	26	26	26	26	26

Table 4. Statistic indicators for 50 runs for the ANN-GWO approach.

	Fitness	MSE Training	MSE Testing	Topology
Best	647.7774	0.001544	0.000624	4:05:01
Worst	137.7863	0.007258	0.006111	4:08:01
Confidence interval	318.34 ± 41.06	0.004 ± 0.0005	0.003 ± 0.0004	0.985 ± 0.0018

The average absolute error computed in the training phase for ANN (4:05:01) presented in Figure 6 was 6.8%, and in the testing phase was 3.8%. In Figure 6, the hidden layer contains in total 5 neurons, numbered Neuron1 through Neuron5.

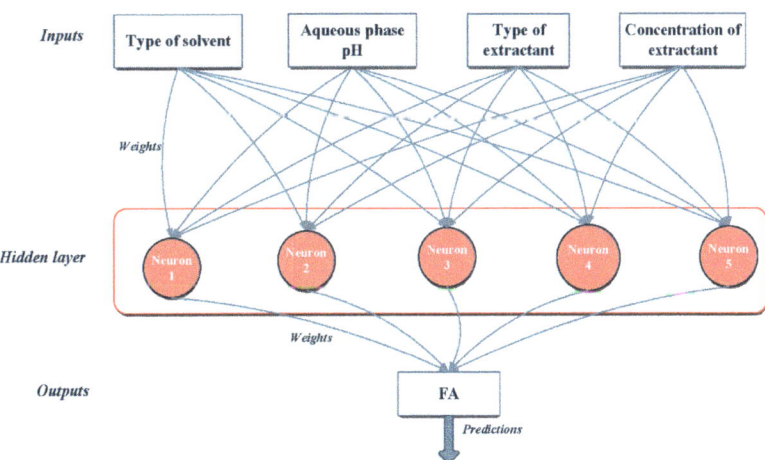

Figure 6. Topology of the best ANN model obtained.

These results and the low values for MSE (Table 4) indicate the selected model's performance. One explanation for the MSE in the training phase being higher than in the testing phase is related to the fact that it contains the only two examples in the entire dataset with a 0 value for the extractant concentration. Only for these two cases is the absolute error high (~50%), with an experimental value of 6.2 and predicted values of ~9.2.

An analysis of the data obtained in laboratory settings showed that the identified ANN captures well the system's behavior in different combinations of solvent-extractant (Figure 7a–c). The relations describing the model are presented in the Supplementary Material.

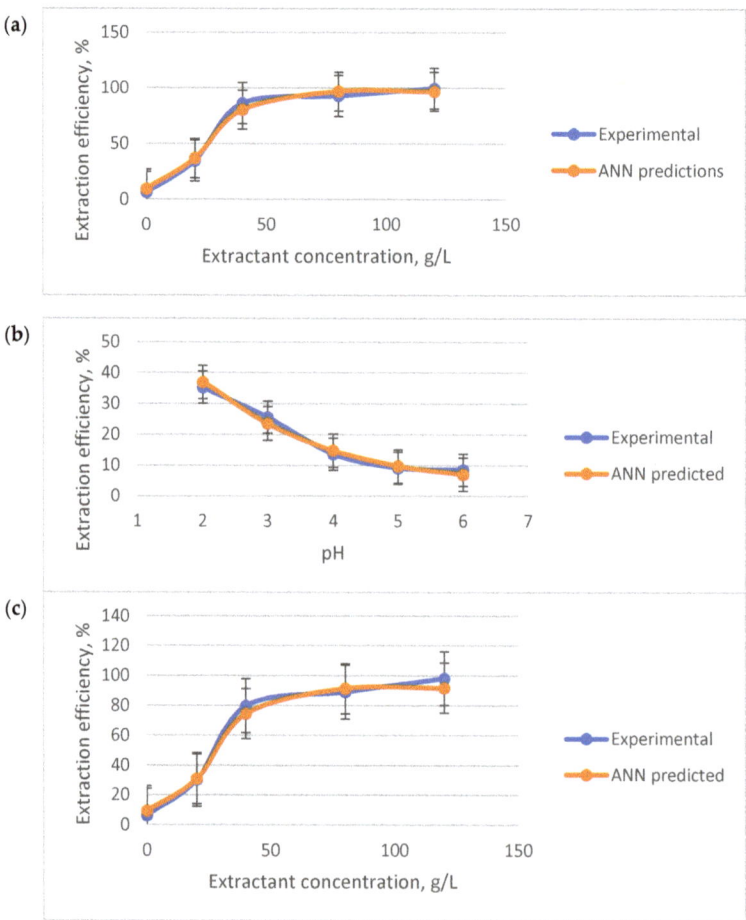

Figure 7. Comparison between the experimental data and the ANN predictions for (**a**) Heptane and CYPHOS IL103 when pH = 4; (**b**) chloroform and [HMIM][PF$_6$] when extractant concentration is 40; (**c**) Heptane and CYPHOS IL104 when pH = 4.

3. Materials and Methods

3.1. Extraction Process

The experiments performed for FA extraction were carried out using a vibration shaker (WIZARD IR Infrared Vortex Mixer, VELP Scientifica Srl, Usmate (MB), Italy) that ensured a stirring speed of 1200 rpm (extraction time—10 min and temperature 22 °C), using equal volumes (2 mL) of FA solution, and the organic phase using a glass cell. FA was extracted from aqueous solutions whose initial concentration was 0.04 g/L (due to limited solubility in water). The extraction was carried out either using only pure ionic liquids (preliminary studies): [BMIM][PF_6]: 1-Butyl-3-methylimidazolium hexafluorophosphate, [HMIM][PF_6]: 1-Hexyl-3-methylimidazolium hexafluorophosphate, [OMIM][PF_6]: 1-Octyl-3-methylimidazolium hexafluorophosphate, CYPHOS IL103—Trihexyl-tetra-decyl-phosphonium decanoate, and CYPHOS IL104—Tri-hexyl-tetra-decyl-phosphonium bis(2,4,4-trimethylpentyl)phosphinate, or organic solvents with different dielectric constants (heptane, chloroform and octanol) and a mixture of ionic liquid and organic solvent (CYPHOS IL103 and heptane, CYPHOS IL104 and heptane and [HMIM][PF_6] and chloroform). All reagents were procured from Sigma-Aldrich [Merck KGaA, Darmstadt, Germany]. The ionic liquid concentration in the organic phase varied between 0 and 120 g/L. The pH of the initial aqueous phase was corrected to the predetermined value, using 4% sulfuric acid and sodium hydroxide solutions, based on the indications of a Hanna Instruments pH 213 digital pH meter (Woonsocket, Rhode Island). After extraction, the samples were separated by centrifugation at 4000 rpm for 5 min, using a DLAB centrifuge (Beijing, China).

The analysis of the acid extraction process was carried out using the separation yield. It was calculated by determining the FA concentration from the initial solution and the raffinate solution, using a Dionex Ultimate HPLC system (Thermo Fisher Scientific Inc., Waltham, MA, USA) equipped with an Acclaim PA2 column, the mobile phase being a mixture of acetonitrile and 30 mM KH_2PO_4 solution with a flow rate of 0.5 mL/min, detection at 270 nm. All experiments were performed in triplicate (n = 3, error between 1.5 and 4.5%).

3.2. Modeling

Combining mathematical modeling with experimental research can offer a tool for anticipating extraction performance. In order to extend experimental knowledge, the FA separation process was modelled. The methodology applied in this work combined ANNs with the GWO algorithm to model the considered process. The role of GWO is to optimize the ANN characteristics. This combination of bio-inspired metaheuristic-ANN belongs to the neuro-evolutive procedures. Neuro-evolution can be applied at different levels: (a) topology; (b) connection weights; (c) learning rule; (d) node behavior [44]. In this work, GWO was applied simultaneously for topology (the number of hidden layers and neurons in each hidden layer), connections' weights (parameters usually set in the training procedure with algorithms such as Backpropagation) and node behavior (the selection of transfer function and its associated properties) and, to incorporate all these parameters into the structure of real numbers used by GWO, a direct encoding procedure is used. Once this structure is established, the GWO optimizes a population of randomly generated networks until a stop criterion is reached. The stop function is represented by the number of iterations set at the start of the run.

Distinctively from other bio-inspired metaheuristics such as Genetic Algorithms or Differential Evolution, in GWO the population is divided into four groups that simulate the hierarchical structure of wolfs: alpha (the leaders), beta (that supports the alpha), delta (that executes the commands of alpha and beta), omega (that are managed by the delta and are at the bottom of the hierarchy) [45]. The hunting mechanisms that GWO mathematically simulates include: (a) encircling (where the entire pack works together to chase and direct the prey in a strategy to increase the chance of catching it); (b) hunting (movement of wolfs around the prey guided by the alpha); and (c) attacking (attacking the prey when it stopped moving). These principles were implemented in Visual Studio C#, following the mathematical relations described in the work of Mirjalili et al., 2014 [26].

4. Conclusions

This study presents the extraction of FA from aqueous solutions using reactive extraction with hydrophobic IL diluted with heptane (a greener alternative to classical solvents [46]) as an efficient technological alternative for this vitamin's separation. Almost

100% (99.56%) extraction yield was obtained for 120 g/L CYPHOS IL103 and aqueous phase pH equal to 4, the reactive extraction mechanism being based on hydrogen bonding between FA and IL. The analysis of the loading factor indicated no overloading in the extraction systems in optimum conditions. The process was modelled using a combined ANNs with the GWO algorithm; the data obtained showed good accordance between predicted and experimental results. According to the findings of this study, it is possible to use efficiently phosphonium-based ionic liquids for folic acid extraction or reactive extraction. Moreover, the separation procedures were straightforward and free of volatile organic solvents.

Supplementary Materials: The following supporting information can be downloaded at: https://www.mdpi.com/article/10.3390/molecules28083339/s1, Figure S1: FTIR spectra for folic acid, Cyphos IL103 ionic liquid in heptane and extract; Figure S2: UV-VIS spectra for Cyphos IL103 ionic liquid in heptane and Chyphos Il103 loaded with FA.

Author Contributions: Conceptualization, A.C.B.; methodology, A.C.B., A.T. and L.K.; software and modelling, E.N.D.; validation, D.C.; investigation, A.C.B.; resources, A.C.B.; writing—original draft preparation, A.C.B.; writing—review and editing, D.C.; visualization, A.T. and L.K.; project administration, A.C.B.; funding acquisition, A.C.B. All authors have read and agreed to the published version of the manuscript.

Funding: This work was supported by a grant of the Ministry of Research, Innovation and Digitization, CNCS—UEFISCDI, project number PN-III-P1-1.1-TE-2021-0153, within PNCDI III.

Institutional Review Board Statement: Not applicable.

Informed Consent Statement: Not applicable.

Data Availability Statement: Data is contained in the article and Supplementary Material.

Conflicts of Interest: The authors declare no conflict of interest.

References

1. Yang, H.; Zhang, X.; Liu, Y.; Liu, L.; Li, J.; Du, G.; Chen, J. Synthetic Biology-Driven Microbial Production of Folates: Advances and Perspectives. *Bioresour. Technol.* **2021**, *324*, 124624. [CrossRef]
2. Menezo, Y.; Elder, K.; Clement, A.; Clement, P. Folic Acid, Folinic Acid, 5 Methyl TetraHydroFolate Supplementation for Mutations That Affect Epigenesis through the Folate and One-Carbon Cycles. *Biomolecules* **2022**, *12*, 197. [CrossRef] [PubMed]
3. Park, S.Y.; Do, J.R.; Kim, Y.J.; Kim, K.S.; Lim, S.D. Physiological Characteristics and Production of Folic Acid of Lactobacillus plantarum JA71 Isolated from Jeotgal, a Traditional Korean Fermented Seafood. *Korean J. Food Sci. Anim. Resour.* **2014**, *34*, 106–114. [CrossRef] [PubMed]
4. San, H.H.M.; Alcantara, K.P.; Bulatao, B.P.I.; Sorasitthiyanukarn, F.N.; Nalinratana, N.; Suksamrarn, A.; Vajragupta, O.; Rojsitthisak, P.; Rojsitthisak, P. Folic Acid-Grafted Chitosan-Alginate Nanocapsules as Effective Targeted Nanocarriers for Delivery of Turmeric Oil for Breast Cancer Therapy. *Pharmaceutics* **2023**, *15*, 110. [CrossRef] [PubMed]
5. Crider, K.S.; Bailey, L.B.; Berry, R.J. Folic Acid Food Fortification—Its History, Effect, Concerns, and Future Directions. *Nutrients* **2011**, *3*, 370–384. [CrossRef]
6. Quadintel. *Global Folic Acid Market Size study, by Application (Dietary Supplements, Cosmetics, Pharmaceuticals, and Other Applications), and Regional Forecasts 2022–2028*; Report ID: QI037; Bizwit Research & Consulting: Indore, India, 2022.
7. Hamzehlou, P.; Akhavan Sepahy, A.; Mehrabian, S.; Hosseini, F. Production of Vitamins B3, B6 and B9 by Lactobacillus isolated from Traditional Yogurt Samples from 3 Cities in Iran, Winter 2016. *Appl. Food Biotechnol.* **2018**, *5*, 107–120.
8. Laiño, J.E.; Levit, R.; de LeBlanc, A.d.M.; Savoy de Giori, G.; LeBlanc, J.G. Characterization of folate production and probiotic potential of *Streptococcus gallolyticus* subsp. macedonicus CRL415. *Food Microbiol.* **2019**, *79*, 20–26. [CrossRef]
9. Hugenschmidt, S.; Miescher Schwenninger, S.; Lacroix, C. Concurrent high production of natural folate and vitamin B12 using a co-culture process with *Lactobacillus plantarum* SM39 and *Propionibacterium freudenreichii* DF13. *Process Biochem.* **2011**, *46*, 1063–1070. [CrossRef]
10. Serrano-Amatriain, C.; Ledesma-Amaro, R.; Lopez-Nicolas, R.; Ros, G.; Jimenez, A.; Revuelta, J.L. Folic acid production by engineered *Ashbya gossypii*. *Metab. Eng.* **2016**, *38*, 473–482. [CrossRef]
11. Wang, Y.; Liu, L.; Jin, Z.; Zhang, D. Microbial Cell Factories for Green Production of Vitamins. *Front. Bioeng. Biotechnol.* **2021**, *9*, 661562. [CrossRef]
12. Roy, S.; Olokede, O.; Wu, H.; Holtzapple, M. In-situ carboxylic acid separation from mixed-acid fermentation of cellulosic substrates in batch culture. *Biomass Bioenergy* **2021**, *151*, 106165. [CrossRef]

13. Galaction, A.I.; Blaga, A.C.; Cascaval, D. The influence of pH and solvent polarity on the mechanism and efficiency of folic acid extraction with Amberlite LA2. *Chem. Ind. Chem. Eng. Q.* **2005**, *11*, 63–68. [CrossRef]
14. Aras, S.; Demir, Ö.; Gök, A. Reactive extraction of gallic acid by trioctylphosphine oxide in different kinds of solvents: Equilibrium modeling and thermodynamic study. *Braz. J. Chem. Eng.* **2022**. [CrossRef]
15. Blaga, A.C.; Dragoi, E.N.; Munteanu, R.E.; Cascaval, D.; Galaction, A.I. Gallic Acid Reactive Extraction with and without 1-Octanol as Phase Modifier: Experimental and Modeling. *Fermentation* **2022**, *8*, 633. [CrossRef]
16. Lazar, R.G.; Blaga, A.C.; Dragoi, E.N.; Galaction, A.I.; Cascaval, D. Application of reactive extraction for the separation of pseudomonic acids: Influencing factors, interfacial mechanism, and process modelling. *Can. J. Chem. Eng.* **2021**, *100*, S246–S257. [CrossRef]
17. Georgiana, L.R.; Cristina, B.A.; Niculina, D.E.; Irina, G.A.; Dan, C. Mechanism, influencing factors exploration and modelling on the reactive extraction of 2-ketogluconic acid in presence of a phase modifier. *Sep. Purif. Technol.* **2020**, *255*, 117740. [CrossRef]
18. Demmelmayer, P.; Kienberger, M. Reactive extraction of lactic acid from sweet sorghum silage press juice. *Sep. Purif. Technol.* **2022**, *282*, 120090. [CrossRef]
19. Blaga, A.C.; Malutan, T. Selective Separation of Vitamin C by Reactive Extraction. *J. Chem. Eng. Data* **2012**, *57*, 431–435. [CrossRef]
20. Poştaru, M.; Bompa, A.S.; Galaction, A.I.; Blaga, A.C.; Caşcaval, D. Comparative Study on Pantothenic Acid Separation by Reactive Extraction with Tri-n-octylamine and Di-(2-ethylhexyl) Phosphoric Acid. *Chem. Biochem. Eng. Q.* **2016**, *30*, 81–92. [CrossRef]
21. Blaga, A.C.; Tucaliuc, A.; Kloetzer, L. Applications of Ionic Liquids in Carboxylic Acids Separation. *Membranes* **2022**, *12*, 771. [CrossRef]
22. de Jesus, S.S.; Filho, R.M. Are ionic liquids eco-friendly? *Renew. Sustain. Energy Rev.* **2022**, *157*, 112039–112061. [CrossRef]
23. Li, Q.; Jiang, X.; Zou, H.; Cao, Z.; Zhang, H.; Xian, M. Extraction of short-chain organic acids using imidazolium-based ionic liquids from aqueous media. *J. Chem. Pharm. Res.* **2014**, *6*, 374–381.
24. Blahusiak, M.; Schlosser, S.; Martak, J. Extraction of butyric acid with a solvent containing ammonium ionic liquid. *Sep. Purif. Technol.* **2013**, *119*, 102–111. [CrossRef]
25. Zare, F.; Ghaedi, M.; Daneshfar, A. Application of an ionic-liquid combined with ultrasonic-assisted dispersion of gold nanoparticles for micro-solid phase extraction of unmetabolized pyridoxine and folic acid in biological fluids prior to high-performance liquid chromatography. *RSC Adv.* **2015**, *5*, 70064–70072. [CrossRef]
26. Mirjalili, S.; Mirjalili, S.M.; Lewis, A. Grey Wolf Optimizer. *Adv. Eng. Softw.* **2014**, *69*, 46–61. [CrossRef]
27. Raj, G.B.; Dash, K.K. Ultrasound-assisted extraction of phytocompounds from dragon fruit peel: Optimization, kinetics and thermodynamic studies. *Ultrason. Sonochem.* **2020**, *68*, 105180. [CrossRef]
28. Abdullah, S.; Pradhan, R.C.; Pradhan, D.; Mishra, S. Modeling and optimization of pectinase-assisted low-temperature extraction of cashew apple juice using artificial neural network coupled with genetic algorithm. *Food Chem.* **2021**, *339*, 127862. [CrossRef]
29. Sodeifian, G.; Ardestani, N.S.; Sajadian, S.A.; Ghorbandoost, S. Application of supercritical carbon dioxide to extract essential oil from Cleome coluteoides Boiss: Experimental, response surface and grey wolf optimization methodology. *J. Supercrit. Fluids* **2016**, *114*, 55–63. [CrossRef]
30. Sukpancharoen, S.; Srinophakun, T.R.; Aungkulanon, P. Grey wolf optimizer (GWO) with multi-objective optimization for biodiesel production from waste cooking oil using central composite design (CCD). *Int. J. Mech. Eng.* **2020**, *9*, 1219–1225. [CrossRef]
31. Samuel, O.D.; Okwu, M.O.; Oyejide, O.J.; Taghinezhad, E.; Afzal, A.; Kaveh, M. Optimizing biodiesel production from abundant waste oils through empirical method and grey wolf optimizer. *Fuel* **2020**, *281*, 118701. [CrossRef]
32. Weast, R.C. *Handbook of Chemistry and Physics*, 54th ed.; CRC Press: Cleveland, OH, USA, 1974.
33. Nosrati, S.; Jayakumar, N.; Hashim, M. Performance evaluation of supported ionic liquid membrane for removal of phenol. *J. Hazard. Mater.* **2011**, *192*, 1283–1290. [CrossRef] [PubMed]
34. Matsumoto, M.; Panigrahi, A.; Murakami, Y.; Kondo, K. Effect of Ammonium- and Phosphonium-Based Ionic Liquids on the Separation of Lactic Acid by Supported Ionic Liquid Membranes (SILMs). *Membranes* **2011**, *1*, 98–108. [CrossRef]
35. Fraser, K.J.; MacFarlane, D.R. Phosphonium-Based Ionic Liquids: An Overview. *Aust. J. Chem.* **2009**, *62*, 309–321. [CrossRef]
36. Gonçalves, A.R.P.; Paredes, X.; Cristino, A.F.; Santos, F.J.V.; Queirós, C.S.G.P. Ionic Liquids—A Review of Their Toxicity to Living Organisms. *Int. J. Mol. Sci.* **2021**, *22*, 5612. [CrossRef] [PubMed]
37. Kebaili, H.; Pérez de los Ríos, A.; José Salar-García, M.; Ortiz-Martínez, V.M.; Kameche, M.; Hernández-Fernández, J.; Hernández-Fernández, F.J. Evaluating the Toxicity of Ionic Liquids on *Shewanella* sp. for Designing Sustainable Bioprocesses. *Front. Mater.* **2020**, *7*, 578411. [CrossRef]
38. Schlosser, Š.; Marták, J.; Blahušiak, M. Specific phenomena in carboxylic acids extraction by selected types of hydrophobic ionic liquids. *Chem. Pap.* **2018**, *72*, 567–584. [CrossRef]
39. Kurnia, K.A.; Lima, F.; Cláudio, A.F.; Coutinho, J.A.; Freire, M.G. Hydrogen-bond acidity of ionic liquids: An extended scale. *Phys. Chem. Chem. Phys.* **2015**, *17*, 18980–18990. [CrossRef]
40. Gazzali, A.M.; Lobry, M.; Colombeau, L.; Acherar, S.; Azaïs, H.; Mordon, S.; Arnoux, P.; Baros, F.; Vanderesse, R.; Frochot, C. Stability of folic acid under several parameters. *Eur. J. Pharm. Sci.* **2016**, *10*, 419–430. [CrossRef]
41. Marták, J.; Schlosser, S. Extraction of lactic acid by phosphonium ionic liquids. *Sep. Purif. Technol.* **2007**, *57*, 483–494. [CrossRef]

42. DeSimone, D.; Counce, R.; Watson, J. Predicting the loading ratio and optimum extractant concentration for solvent extraction. *Chem. Eng. Res. Des.* **2020**, *161*, 271–278. [CrossRef]
43. Priddy, K.; Keller, P. *Artificial Neural Networks: An introduction*; SPIE Press: Washington, DC, USA, 2005.
44. Islam, M.; Yao, X. Evolving Artificial Neural Network Ensembles. In *Computational Intelligence: A Compendium*; Fulcher, J., Jain, L., Eds.; Springer: Berlin/Heidelberg, Germany, 2008.
45. Torabi, S.; Safi-Esfahani, F. Improved Raven Roosting Optimization algorithm (IRRO). *Swarm Evol. Comput.* **2018**, *40*, 144–154. [CrossRef]
46. Jessop, P.G. Searching for green solvents. *Green Chem.* **2011**, *13*, 1391–1398. [CrossRef]

Disclaimer/Publisher's Note: The statements, opinions and data contained in all publications are solely those of the individual author(s) and contributor(s) and not of MDPI and/or the editor(s). MDPI and/or the editor(s) disclaim responsibility for any injury to people or property resulting from any ideas, methods, instructions or products referred to in the content.

Article

Extraction of Bioactive Compounds from *C. vulgaris* Biomass Using Deep Eutectic Solvents

Maria Myrto Dardavila [1,*], Sofia Pappou [1,2], Maria G. Savvidou [1], Vasiliki Louli [1], Petros Katapodis [3], Haralambos Stamatis [3], Kostis Magoulas [1] and Epaminondas Voutsas [1,*]

[1] Laboratory of Thermodynamics and Transport Phenomena, Zografou Campus, School of Chemical Engineering, National Technical University of Athens, 9 Iroon Polytechniou Str., 15780 Athens, Greece
[2] Department of Marine Sciences, University of Aegean, University Hill, Lesvos Island, 81100 Mytilene, Greece
[3] Laboratory of Biotechnology, Department of Biological Applications and Technologies, University of Ioannina, 45110 Ioannina, Greece
* Correspondence: mirtodar@mail.ntua.gr (M.M.D.); evoutsas@chemeng.ntua.gr (E.V.); Tel.: +302107723971 (M.M.D. & E.V.)

Abstract: *C. vulgaris* microalgae biomass was employed for the extraction of valuable bioactive compounds with deep eutectic-based solvents (DESs). Particularly, the Choline Chloride (ChCl) based DESs, ChCl:1,2 butanediol (1:4), ChCl:ethylene glycol (1:2), and ChCl:glycerol (1:2) mixed with water at 70/30 w/w ratio were used for that purpose. The extracts' total carotenoid (TCC) and phenolic contents (TPC), as well as their antioxidant activity (IC50), were determined within the process of identification of the most efficient solvent. This screening procedure revealed ChCl:1,2 butanediol (1:4)/H$_2$O 70/30 w/w as the most compelling solvent; thus, it was employed thereafter for the extraction process optimization. Three extraction parameters, i.e., solvent-to-biomass ratio, temperature, and time were studied regarding their impact on the extract's TCC, TPC, and IC50. For the experimental design and process optimization, the statistical tool Response Surface Methodology was used. The resulting models' predictive capacity was confirmed experimentally by carrying out two additional extractions under conditions different from the experimental design.

Keywords: DES; *C. vulgaris*; carotenoid content; phenolic content; antioxidant activity; RSM; analysis of variance

1. Introduction

Phenolic and carotenoid compounds possess important pharmacological activities such as antioxidant, antibacterial, and anti-inflammatory [1–3]. The recovery of high-value biomolecules from several natural sources is largely realized with the conventional extraction process using traditional organic solvents [4]. Nevertheless, these solvents are often related to low yield efficiency and increased energy consumption, as well as toxicity, volatility, flammability, non-biodegradability, and non-renewability [5,6]. Given these drawbacks and considering the principles of Green Chemistry, new alternative solvents have been introduced for the extraction of essential biocompounds, such as switchable hydrophilicity solvents, SHS [7]; switchable ionic liquids, S-IL [8]; and deep eutectic solvents, DESs [8,9].

DESs, since their initial introduction by Abbott et al., 2003 [10], have dynamically emerged as means of surpassing the above-referred limitations. They can be produced by naturally occurring, biodegradable, and low-cost components via simple synthesis routes [6]. Usually, DESs consist of two components, a hydrogen bond acceptor (HBA) and a hydrogen bond donor (HBD). Choline chloride (ChCl) is the most widely studied HBA, while different polyalcohols, organic acids, sugars, and amino acids have been employed as HBDs [4,6,11,12]. The possible combinations of HBAs and HBDs are almost unlimited, giving the ability to design task-specific DESs [6,13,14]. Some of their most important characteristics are low volatility at high temperatures, selectivity, strong dissolving ability, adjustable polarity, biocompatibility, pharmaceutical acceptance, and high viscosity. These characteristics are severely influenced by the type and molar ratio of the HBAs and HBDs [13,15,16]. DESs with polyols, as HBDs in particular, exhibit lower freezing points and can even exist in a liquid state below room temperature [17].

DESs have been used in many scientific fields ranging from chemical synthesis to bio-catalysis and nanomaterials fabrication, as greener, eco-friendlier, and more efficient alternatives to traditional solvents [18–24]. The exploitation of DESs for the extraction of various non-polar and polar bioactive compounds from different natural sources has also shown an admirable trend [9,11,25–28]. It has been reported that DESs can provide higher extraction yields and stabilization capacity of the targeted biomolecules in comparison to conventional solvents [14,29,30].

However, high viscosity is one of the most important drawbacks in the exploitation of DESs as extractants at an industrial scale. To reduce the viscosity and enhance the mass transport phenomena of the biomolecules from the solid to the liquid, water, an abundant natural substance, is frequently added in various ratios, also influencing the polarity of the DESs and affecting the dissolution of the compounds of interest [12,31,32].

The extraction efficiency of the bioactive compounds is also dependent on several extraction parameters. Some of the most influential are the biomass-to-solvent ratio, the mode of agitation, the time and temperature of the extraction, the types and ratio of HBA and HBD, the viscosity of the DES [12,26,33], etc. Hence, the investigation of the most efficient extraction parameters for a given DES and natural source combination is important.

Chlorella vulgaris is a microalgae strain which is consisted of 4% phenolic compounds, 2% carotenoids, 16% lipids, 10% carbohydrates, and other valuable components [34]. *Chlorella* species are verified to have one of the highest percentages of phenolic compounds and carotenoids compared to other microalgae strains [35]. The recovery of compounds with antioxidant activity, such as polyphenols and carotenoids, from *C. vulgaris* using conventional solvents is well documented [36–39]. On the contrary, the use of DESs for the same purpose has hardly been addressed. According to Mahmood et al., 2019 [40], polyol-based DESs have been found to outperform conventional solvents in terms of polyphenolic extraction efficiency, the antioxidant activity of the extracts, and the selectivity of target antioxidants from *C. vulgaris*.

In this study, *C. vulgaris* biomass extractions were conducted using deep eutectic-based solvents. The DESs ChCl:1,2 butanediol (1:4), ChCl:ethylene glycol (1:2), and ChCl:glycerol (1:2) were synthesized, and water was added to a 70/30 w/w ratio. The resulting mixtures were employed for the extractions of *C. vulgaris* under given conditions. Subsequently, the extracts' total carotenoid and phenolic contents, as well as their antioxidant activity, were determined. The screening procedure of the DES/water mixtures resulted in the determination of the most efficient one, namely ChCl:1,2 butanediol (1:4)/H_2O 70/30 w/w, which was employed thereafter for optimizing the extraction process. For that purpose, the influence of three important extraction parameters, namely biomass-to-solvent ratio, temperature, and time was studied regarding their impact on the recovery of carotenoids and phenolics and on the extracts' antioxidant activity. An experimental design was implemented, and the Response Surface Methodology was employed for the process optimization. The influence of the independent parameters on each dependent one was determined through Analysis of Variance, and the resulting models were evaluated and confirmed experimentally by carrying out two additional extractions under conditions different from the experimental design.

2. Results and Discussion

2.1. Physical Properties of DESs

The resulting viscosities and densities of the DESs and their mixtures with water (70/30 w/w) at 60 °C are included in Table 1. The reported values are the means of three measurements. The DES1/w, DES2/w, and DES3/w abbreviations correspond to the ChCl:1,2 butanediol (1:4)/water, ChCl:glycerol (1:2)/water, and ChCl:ethylene glycol (1:2)/water, respectively. It is observed that by far the most viscus DES was the ChCl:glycerol (1:2) (DES2), ChCl:1,2 butanediol (1:4) (DES1) follows, and ChCl: Ethylene glycol (1:2) (DES3) shows the lowest viscosity. In the case of polyol-based DESs, the hydrogen bonds that are formed between the HBAs and HBDs are proportional to the hydroxyl groups present in the molecules of the HBDs. The presence of more hydroxyl groups increases the intermolecular forces resulting in higher η values [15]. Indeed, the glycerol molecule ($HOCH_2CHOHCH_2OH$) has one more hydroxyl group than the diols 1,2 butanediol ($HOCH_2CHOHCH_2CH_3$) and ethylene glycol ($HOCH_2CH_2OH$), justifying the higher η value of DES2. Moreover, the viscosity of DESs is also affected by the molecular structure (molecular weight and size) of the HBD [41]. Between the diols serving as the HBDs of DES1 and DES3, 1,2 butanediol is a larger molecule than ethylene glycol; hence, DES1 is more viscus than DES3.

Table 1. Viscosities and densities of DESs and their 70/30 w/w water mixtures at 60 °C.

Solvent	Viscosity, η [cP]	Density, ϱ [g cm^{-3}]
DES1	15.01 ± 0.04	1.0120 ± 0.0001
DES2	53.07 ± 0.22	1.1725 ± 0.0001
DES3	13.49 ± 0.11	1.0965 ± 0.0001
DES1/w	3.54 ± 0.01	1.0145 ± 0.0001
DES2/w	4.34 ± 0.02	1.1179 ± 0.0001
DES3/w	2.78 ± 0.05	1.0678 ± 0.0001

As expected, water addition to the DESs led to a large reduction of their viscosities, which is due to the weakening of the hydrogen bonding between their constituents [42]. The viscosities of the DES/water mixtures followed the order that the pure DESs exhibited as well, i.e., DES2/w > DES1/w > DES3/w. The values of the viscosities of the three DES/water mixtures indicate that these can be used for industrial applications as solvents [43].

The measured densities of the pure DESs and their water mixtures employed in the present study are also shown in Table 1. It is observed that the density values of the pure DESs diminished according to the order: DES2 > DES1 > DES3, following the same order as for the viscosity. The DESs densities depend on the hydrogen bonds developed between the HBA and the HBD. In particular, when a larger number of hydrogen bonds are formed, the available free space in the DESs is reduced, resulting in increased density [44,45]. Therefore, the higher density value measured for DES2 is attributed to the surplus of hydroxyl functional groups found in the molecule of glycerol as compared to 1,2 butanediol and ethylene glycol. Moreover, DESs' density is affected by the length of the alkyl chain of the HBD molecule. According to the literature, a longer alkyl chain results in lower densities [46,47]. This conclusion is confirmed by our research, too, since DES1 has a longer alkyl chain than DES3.

DES2/w and DES3/w demonstrated lower densities in comparison to pure DES2 and DES3 due to the weakening of the hydrogen bond network caused by water addition [46]. However, the level of the reduction is low; thus, it can be claimed that the specific physical property of these two DESs is not significantly affected by water, at least for the given water ratio and temperature. Florindo et al. [46] came to the same conclusion for five different choline chloride-based DESs.

The density of DES1/w was found to be marginally higher (0.0025 g cm^{-3}) than that of DES1. This implies a positive excess molar volume (V^E) upon mixing of ChCl:1,2 butanediol (1:4) with water at a 70/30 w/w ratio and 60 °C, pointing to volume compression, hence density increase. Such a phenomenon hints at stronger intramolecular interactions (i.e., among DES1 molecules or among water molecules) than interspecies interactions (i.e., between water and DES1 molecules) for the specific composition of the DES/water mixture and temperature [48]. Further investigation of the ChCl:1,2 butanediol (1:4)/H$_2$O mixtures within the whole compositional range and also within a vast temperature range should be performed to obtain a clear view of the key physical property of density for this mixture. Such a study exceeds the scope of the present work and is planned for the near future.

2.2. Solvent Screening

The three DES-based solvents, i.e., DES1/w, DES2/w, and DES3/w, were compared for their capacity to extract bioactive compounds from *C. vulgaris*, while the EtOH/w mixture served as the control solvent. The measured TCC, TPC, and IC50 values of the obtained extracts are included in Table 2.

Table 2. TCC, TPC, and IC50 values measured in *C. vulgaris* extracts obtained at T = 60 °C, r = 20:1 g$_{SW}$ g^{-1}$_{DW}$, and t = 3 h using the DES-based solvents and EtOH/w. The acronym SW stands for solvent weight, and DW for dry weight biomass.

Solvent	TCC [mg g^{-1}$_{DW}$]	TPC [mg$_{GAE}$ g^{-1}$_{DW}$]	IC50 [g$_{DW}$ mL^{-1}$_{sol}$]
DES1/w	3.462 ± 0.121	8.553 ± 0.213	0.180 ± 0.011
DES2/w	0.218 ± 0.011	4.407 ± 0.128	0.260 ± 0.014
DES3/w	0.293 ± 0.014	4.687 ± 0.131	0.360 ± 0.018
EtOH/w	8.436 ± 0.211	7.686 ± 0.219	0.139 ± 0.010

According to the results of Table 2, the DES1/w solvent outperformed the two other DES-based ones in extracting carotenoid compounds from *C. vulgaris* biomass. In fact, the TCC value measured for the DES1/w extract was 93.7% and 91.5% greater than the corresponding values measured for the DES2/w and DES3/w extracts, respectively. The control solvent is the best among the four tested for carotenoid extraction. This finding is in accordance with the literature since ethanol/water mixtures are known to be particularly efficient solvents in extracting carotenoids from micro and macro algae [37,49,50].

The superiority of the DES1/w solvent regarding the extraction of phenolics from *C. vulgaris* biomass can be observed in Table 2. The TPC value that was determined for the DES1/w extracts was almost double the TPC values found at the DES2/w and DES3/w *C. vulgaris* extracts. In comparison to the extract obtained from the control solvent, the DES1/w delivered a total phenolics content greater by about 10%. It is also noticed that all the DES-based solvents used in the present study performed significantly better in extracting phenolic compounds than carotenoids. Several DES/water mixtures have been acknowledged for their efficiency in extracting phenolic compounds from various natural sources [5,16,35]. According to the generally accepted concept known as "like–dissolve–like", it is suggested that the polar DES-based solvents perform better in extracting polar species, such as phenolics, than non-polar, such as carotenoids [35]. Moreover, the high extractability of phenolic compounds from DESs and their water mixtures has been attributed to the H-bonding interactions that can be formed between the phenolic molecules and those of the DESs [42].

Mahmood et al., 2019 [38] used different polyol-based DESs for the extraction of polyphenols from *C. vulgaris*, among which the ChCl:glycerol (1:2) and ChCl:ethylene glycol (1:2). The TPC value reported by the same authors for the ChCl:glycerol (1:2), extract (5.27 mg$_{GAE}$ g^{-1}$_{DW}$) is comparable to the TPC measured at the ChCl:glycerol (1:2)/H$_2$O 70/30 (DES2/w) extract of the present study. Moreover, the ChCl:ethylene glycol (1:2)/H$_2$O 70/30 (DES3/w) extract of the present work exhibited approximately 5 times higher total phenolics content in comparison to the ChCl:ethylene glycol (1:2) extract reported by Mahmood et al., 2019 [40]. However, it should be mentioned that a direct comparison of results given by different studies of microalgae biomass extractions is rather difficult. The biochemical composition and other characteristics of microalgae biomass can differ significantly due to the type of the cultivated strain, the growth conditions, the growth phase, and the composition of the cultivating medium. Additionally, other parameters, such as the biomass drying method used, its treatment prior to the extraction, and of course, the extraction conditions and the solvent used can have a significant impact on the obtained extracts' composition and antioxidant activity [50–52]. However, it can be claimed that the addition of water in ChCl:ethylene glycol (1:2) positively affected the extracts' total phenolic content. This can be attributed to the selective extraction of water-soluble phenolic compounds present in *C. vulgaris* biomass [53].

The extract obtained from the conventional solvent EtOH/w exhibited the smallest IC50 value (Table 2), hence the highest antioxidant activity. The DES-based solvents follow the order DES1/w > DES2/w > DES3/w as per their antioxidant activity. Carotenoids and phenolics are potent antioxidants, and the contribution of both these bioactive compounds in the measured antioxidant activity of microalgae extracts is significant [50,52]. Considering that the DES1/w delivered an extract with considerably higher TCC and TPC values in comparison to the other two DES-based solvents, its superior antioxidant activity can be justified. Despite the higher phenolics yield exhibited by the DES1/w extract in comparison to EtOH/w one, its lower carotenoid content seems to have an impact on the measured IC50 value.

According to the solvent screening results analyzed above, the DES1/w was proven to be the most convenient for the purpose of our study between the three DES-based solvents that were tested. As mentioned, it led to the *C. vulgaris* extract with the highest content of carotenoid and phenolic compounds and the highest antioxidant activity. Consequently, the ChCl:1,2 butanediol (1:4)/H$_2$O 70/30 w/w solvent was further exploited for the optimization of the *C. vulgaris* microalgae extraction process.

2.3. Experimental Design Results

The experimental results of *C. vulgaris* biomass extractions performed using as a solvent the mixture ChCl:1,2 butanediol (1:4)/H$_2$O 70/30 w/w (DES1/w) are listed in Table 3. It is observed that the variation of the three independent variables (X_1, X_2, X_3) affected the dependent (Y_1, Y_2, Y_3) ones. The TCC (Y_1) fluctuated between 1.868 and 3.709 mg g$^{-1}_{DW}$, and the TPC (Y_2) between 7.468 and 12.768 mg$_{GAE}$ g$^{-1}_{DW}$. The IC50 (Y_3) exhibited a minimum value of 0.118 and a maximum of 0.332 g$_{DW}$ mL$^{-1}_{sol}$. The greatest TCC value was exhibited by the extract of Run 16 (T = 60 °C, t = 13.5 h, r = 30:1 g$_{SW}$g$^{-1}_{DW}$), while for the extract of Run 18 (T = 60 °C, t = 3 h, r = 40:1 g$_{SW}$g$^{-1}_{DW}$), the highest TPC and the lowest IC50 values were found. On the contrary, for the extract of Run 4 (T = 30 °C, t = 24 h, r = 40:1 g$_{SW}$g$^{-1}_{DW}$), the highest IC50 was measured, rendering it the least potent one among the eighteen, as far as the antioxidant activity is concerned. The lowest carotenoids content was reported for the *C. vulgaris* extract of Run 3 (T = 30 °C, t = 3 h, r = 20:1 g$_{SW}$g$^{-1}_{DW}$), while that of Run 13 (T = 45 °C, t = 3 h, r = 30:1 g$_{SW}$g$^{-1}_{DW}$) showed the lowest phenolic content.

Table 3. Experimental results of *C. vulgaris* extraction with DES1/w regarding TCC, TPC, and IC50.

Run	Experimental Design Conditions			Experimental Results		
	X_1: T [°C]	X_2: t [h]	X_3: r [g$_{SW}$ g$^{-1}_{DW}$]	Y_1: TCC [mg g$^{-1}_{DW}$]	Y_2: TPC [mg$_{GAE}$ g$^{-1}_{DW}$]	Y_3: IC50 [g$_{DW}$ mL$^{-1}_{sol}$]
1	45	13.5	20:1	3.102	9.257	0.215
2	45	13.5	40:1	3.517	10.787	0.207
3	30	3	20:1	1.868	8.696	0.255
4	30	24	40:1	3.257	8.897	0.332
5	60	24	20:1	2.872	9.438	0.181
6	45	13.5	30:1	3.268	8.904	0.201
7	30	24	20:1	2.332	9.383	0.237
8	30	3	40:1	2.721	7.667	0.241
9	45	24	30:1	3.256	8.058	0.220
10	60	3	20:1	3.462	8.553	0.180
11	45	13.5	30:1	3.152	8.643	0.166

Table 3. Cont.

Run	Experimental Design Conditions			Experimental Results		
	X_1: T [°C]	X_2: t [h]	X_3: r [$g_{SW}\ g^{-1}_{DW}$]	Y_1: TCC [$mg\ g^{-1}_{DW}$]	Y_2: TPC [$mg_{GAE}\ g^{-1}_{DW}$]	Y_3: IC50 [$g_{DW}\ mL^{-1}_{sol}$]
12	45	13.5	30:1	3.415	9.074	0.164
13	45	3	30:1	2.693	7.468	0.189
14	60	24	40:1	3.137	12.586	0.170
15	30	13.5	30:1	2.818	8.062	0.241
16	60	13.5	30:1	3.709	10.131	0.147
17	45	13.5	30:1	3.277	8.894	0.164
18	60	3	40:1	3.571	12.768	0.118

2.4. Statistical Analysis of Experimental Design Results

In order to draw conclusions about the responses' dependence on the factors, a regression analysis of the experimental design results was carried out. Moreover, the ANOVA test was used to evaluate the resulting regression models. The statistical analysis of experimental data led to the development of reduced quadratic multiple regression models for each of the three responses investigated.

The second-order polynomial quadratic functions of the TCC (Y_1), TPC (Y_2), and IC50 (Y_3) and the factors T (X_1), t (X_2), and r (X_3) are shown below:

$$Y_1 = -1.84295 + 0.081819 \cdot X_1 + 0.166893 \cdot X_2 + 0.078320 \cdot X_3 - 0.001606 \cdot X_1 \cdot X_2 - 0.001170 \cdot X_1 \cdot X_3 - 0.003314 \cdot X_2^2 \quad (1)$$

$$Y_2 = 25.65774 - 0.150048 \cdot X_1 + 0.206034 \cdot X_2 - 1.18459 \cdot X_3 + 0.007376 \cdot X_1 \cdot X_3 - 0.006498 \cdot X_2^2 + 0.015426 \cdot X_3^2 \quad (2)$$

$$Y_3 = 0.0467107 + 0.000450 \cdot X_1 - 0.004219 \cdot X_2 - 0.013056 \cdot X_3 - 0.000128 \cdot X_1 \cdot X_3 + 0.000190 \cdot X_2 \cdot X_3 + 0.000271 \cdot X_3^2 \quad (3)$$

Details regarding the ANOVA results are included in Table 4. ANOVA investigation indicated that all models were significant and accurate since their F-values were high and their p-values were lower than 0.0001 (Table 4). A factor is considered impactful to a given response when $p < 0.005$, hence it is concluded that TCC and TPC were most influenced by temperature (X_1) and solvent-to-biomass ratio (X_3), while the antioxidant activity was most affected by temperature (X_1). Moreover, for TCC, two interaction terms ($X_1 X_2$ and $X_1 X_3$) and one quadratic X_2^2 were also significant, and for TPC one interaction term ($X_1 X_3$) and two quadratics X_2^2, X_3^2. Non-significant lack of fit was found for all the developed predictive models.

Table 4. Analysis of variance (ANOVA) and measures of the model's prediction accuracy.

RESPONSE Y_1-TCC					
Source	Sum of Squares	df	Mean Square	F-Value	p-Value Prob > F
Model	3.45	6	0.5750	31.55	<0.0001
X_1-T	1.41	1	1.41	77.36	<0.0001
X_2-t	0.0291	1	0.0291	1.59	0.2329
X_3-r	0.6589	1	0.6589	36.15	<0.0001
$X_1 X_2$	0.5121	1	0.5121	28.10	0.0003
$X_1 X_3$	0.2464	1	0.2464	13.52	0.0036
X_2^2	0.5932	1	0.5932	32.55	0.0001
Residual	0.2005	11	0.0182		
Lack of fit	0.1657	8	0.0207	1.79	0.3432
Std. Dev.	0.1350		R^2	0.9451	
Mean	3.08		Adj R^2	0.9151	
C.V. %	4.38		Pred R^2	0.8464	
			Adeq Precision	22.0389	

Table 4. Cont.

	RESPONSE Y_1-TCC				
Source	Sum of Squares	df	Mean Square	F-Value	p-Value Prob > F
	RESPONSE Y_2-TPC				
Source	Sum of Squares	df	Mean Square	F-Value	p-Value Prob > F
Model	35.39	6	5.90	47.85	<0.0001
X_1-T	11.61	1	11.61	94.16	<0.0001
X_2-t	1.03	1	1.03	8.37	0.0146
X_3-r	5.45	1	5.45	44.19	<0.0001
$X_1 X_3$	9.85	1	9.85	79.87	<0.0001
X_2^2	1.59	1	1.59	12.90	0.0042
X_3^2	7.38	1	7.38	59.85	<0.0001
Residual	1.36	11	0.1233		
Lack of fit	1.26	8	0.1577	5.00	0.1064
Std. Dev.	0.3511		R^2	0.9631	
Mean	9.29		Adj R^2	0.9430	
C.V. %	3.78		Pred R^2	0.8731	
			Adeq Precision	23.5207	
	RESPONSE Y_3-IC50				
Source	Sum of Squares	df	Mean Square	F-Value	p-Value Prob > F
Model	0.0379	6	0.0063	23.39	<0.0001
X_1-T	0.0260	1	0.0260	96.30	<0.0001
X_2-t	0.0025	1	0.0025	9.13	0.0116
X_3-r	0.0000	1	0.0000	0.0000	1.0000
$X_1 X_3$	0.0030	1	0.0030	10.98	0.0069
$X_2 X_3$	0.0032	1	0.0032	11.85	0.0055
X_3^2	0.0033	1	0.0033	12.08	0.0052
Residual	0.0030	11	0.0003		
Lack of fit	0.0020	8	0.0002	0.7473	0.6714
Std. Dev.	0.0164		R^2	0.9273	
Mean	0.2016		Adj R^2	0.8877	
C.V. %	8.15		Pred R^2	0.8156	
			Adeq Precision	20.676	

The models' precision accuracy measures, which are also included in Table 4, indicated that the predictive models were reliable since their R^2 values were greater than 0.9, and their R^2 Adjusted, and R^2 Predicted values were in reasonable agreement, i.e., the difference between them was less than 0.2. Moreover, values of adequate precision greater than 4 indicate that a model can be used to navigate the design space, something that was confirmed for all the obtained models of the present work.

2.5. Study of the Factors' Combined Effects

The 3D surface plots obtained by the models can contribute to the investigation of the interactions between the different independent variables regarding their effect on the dependent ones. In Figure 1, the combined effects of temperature and time, as well as temperature and solvent-to-biomass ratio on the carotenoid content, are presented. In Figure 2, the dependence of the phenolic content on the extraction time and solvent-to-biomass ratio is shown. The dependence of the extracts' IC50 values on extraction temperature combined with a solvent-to-biomass ratio, as well as extraction time combined with solvent-to-biomass ratio, is depicted in Figure 3.

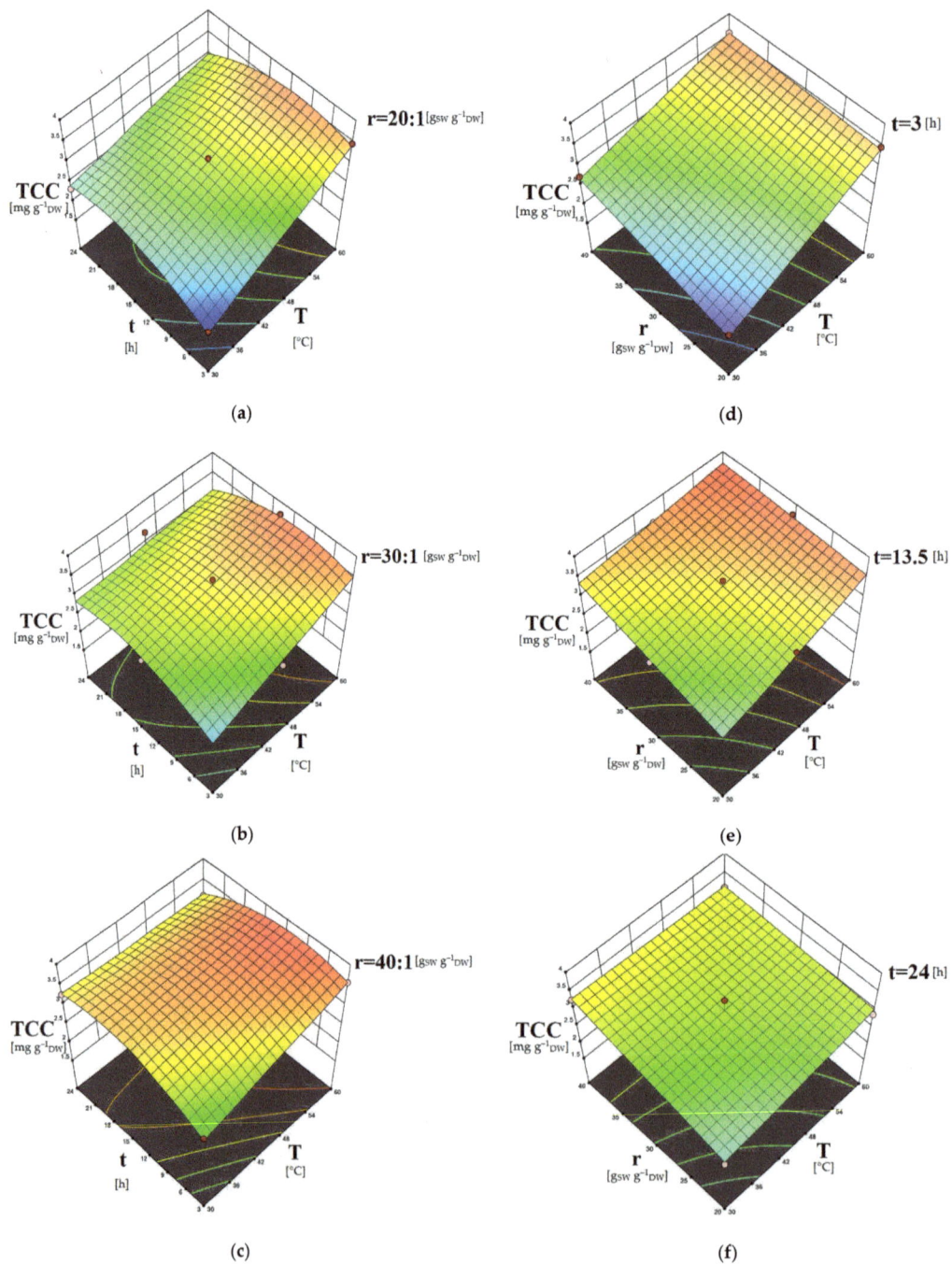

Figure 1. Three-dimensional surface plots showing the combined effects of temperature and time (**a**–**c**) and temperature and solvent-to-biomass ratio (**d**–**f**) on TCC [mg g^{-1}$_{DW}$] of the *C. vulgaris* extracts obtained using DES1/w solvent.

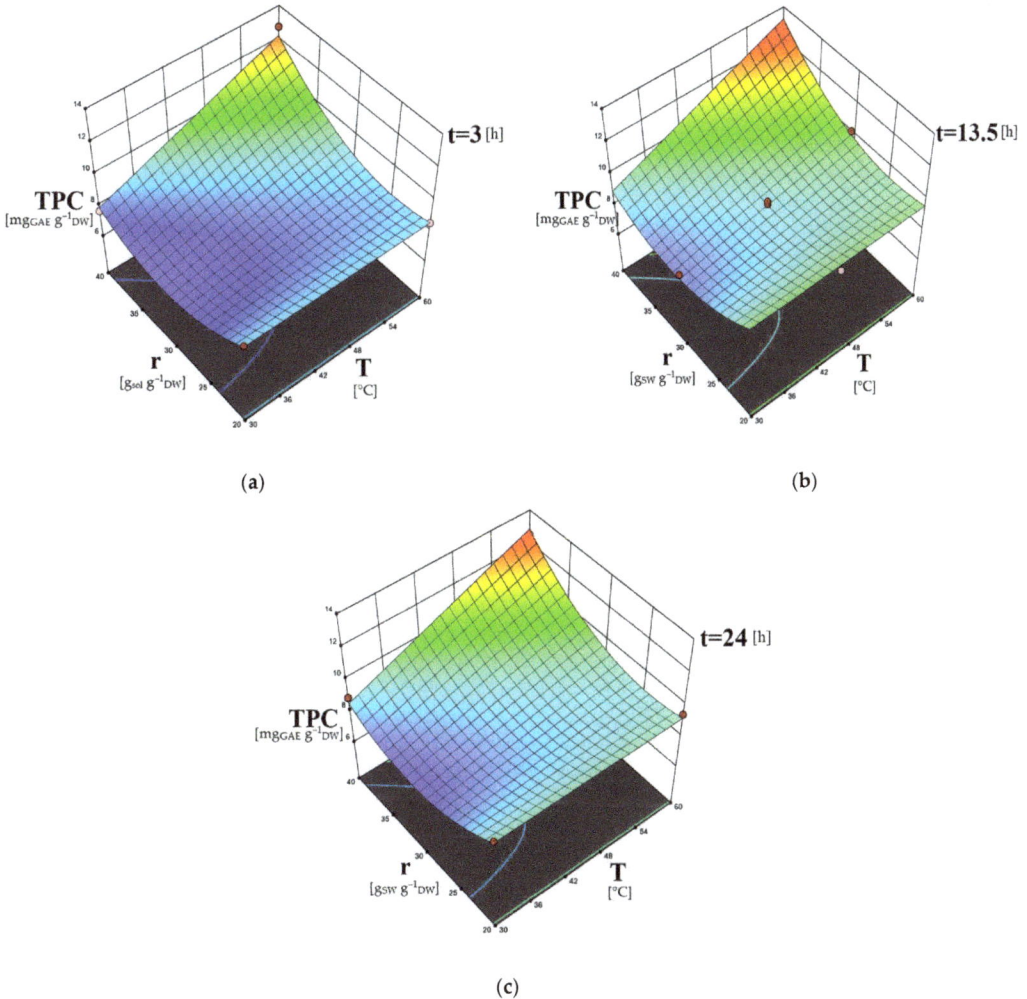

Figure 2. Three-dimensional surface plots showing the combined effects of temperature and solvent–biomass ratio (**a**–**c**) on TPC [$mg_{GAE} g^{-1}_{DW}$] of the *C. vulgaris* extracts obtained using DES1/w solvent.

The extraction should last long enough to achieve effective contact of the solvent with the biomass, saturation of the biomass, as well as diffusion of the target biomolecules from the biomass to the solvent. Consequently, when performing solid–liquid extractions, the application of limited extraction times cannot assist remarkably in the recovery of biomolecules from their natural sources [54,55]. Figures 1a–c and 2a–c indicate that the application of low extraction time (3 h) facilitated the least recovery of carotenoids and phenolics. Consequently, increasing the extraction duration is expected to give rise to the extraction of the solutes of interest [56]. By Figure 1a–c, it is evident that increasing extraction time assisted the carotenoids' extraction; however, from a certain point, it did not contribute to a further increase in the TCC values. Particularly, it is observed in Figure 1d–f that for all T-r combinations, a rise from t = 3 h to t = 13.5 h augmented the TCC values, while a further increase to t = 24 h resulted in their reduction. Similar are the results for the phenolics recovery as shown in Figure 2a–c. These findings imply that the extracted carotenoids and phenolics sustained degradation under prolonged exposure to oxygen and light [49,55]. An initially increasing trend of the carotenoids content, followed by a decreasing one versus extraction time, was also reported for the *C. vulgaris* extractions using a conventional solvent [37]. A maximum extraction time above which the TPC of the *C. vulgaris* extracts decreases due to oxidation of the phenolics compounds was found by Mahmood et al., 2019 [40]; and Zakaria et al., 2017 [55], as well.

Higher extraction temperature facilitates the mass transport phenomena of the biomolecules of interest from the biomass cells to the solvent, as well as their solubilization [54,57]. By Figure 1d–f, it is observed that rising T and r positively impacted the carotenoids recovery. It can be supported that the concentration gradient of the specific biomolecules between the *C. vulgaris* cells and the solvent is increased with increasing solvent-to-biomass ratio [40]. This phenomenon, combined with the helpful influence of increasing temperature contributed to the acceleration of the carotenoids' diffusion from the cells to the liquid phase and enabled their solubilization, resulting in greater TCC values for these extracts. Regarding the phenolics recovery (Figure 2a–c), according to the aforementioned mechanism of the combined effect of T-r, the beneficial impact of the rising extraction temperature is pronounced only under higher solvent-to-biomass ratio values.

Figure 3a–c indicate that the extracts obtained at low extraction temperature (30 °C) combined with a high solvent-to-biomass ratio (40:1 g_{SW} $g^{-1}{}_{DW}$) had the worst antioxidant activity. According to the analysis that preceded, at low extraction temperatures, extracts with a relatively lower content of carotenoids are obtained. Moreover, the application of a high solvent-to-biomass ratio, i.e., the use of a larger amount of solvent at a low extraction temperature, might have led to the extraction of other molecules present in the *C. vulgaris* cells that did not contribute to the antioxidant activity of the extracts [37]. Due to the presence of these compounds, along with the lower carotenoid concentrations, the extracts' antioxidant potency was reduced, and higher IC50 values were measured. Furthermore, Figure 3d–f show that increasing extraction temperature enhances the antioxidant capacity of the obtained extracts for all r-t combinations. This finding could be attributed to the increased carotenoids and phenolics contents obtained under higher extraction temperatures. Figure 3a–c show also that the increase in extraction time for all r-T combinations negatively impacted the extracts' antioxidant activity. As explained, prolonged extraction durations had a negative effect on the recovered carotenoids and phenolics integrity due to oxidative reactions favored by long exposure to air and light.

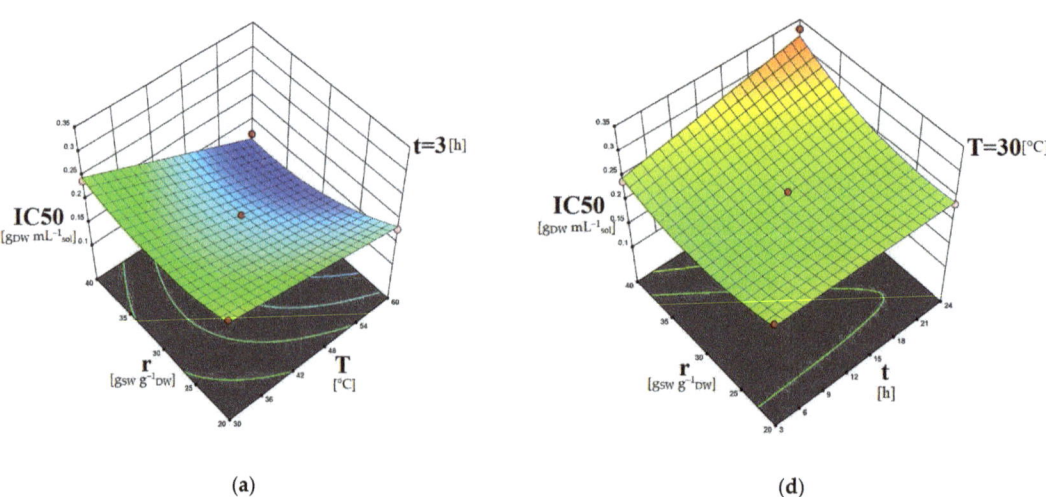

Figure 3. *Cont.*

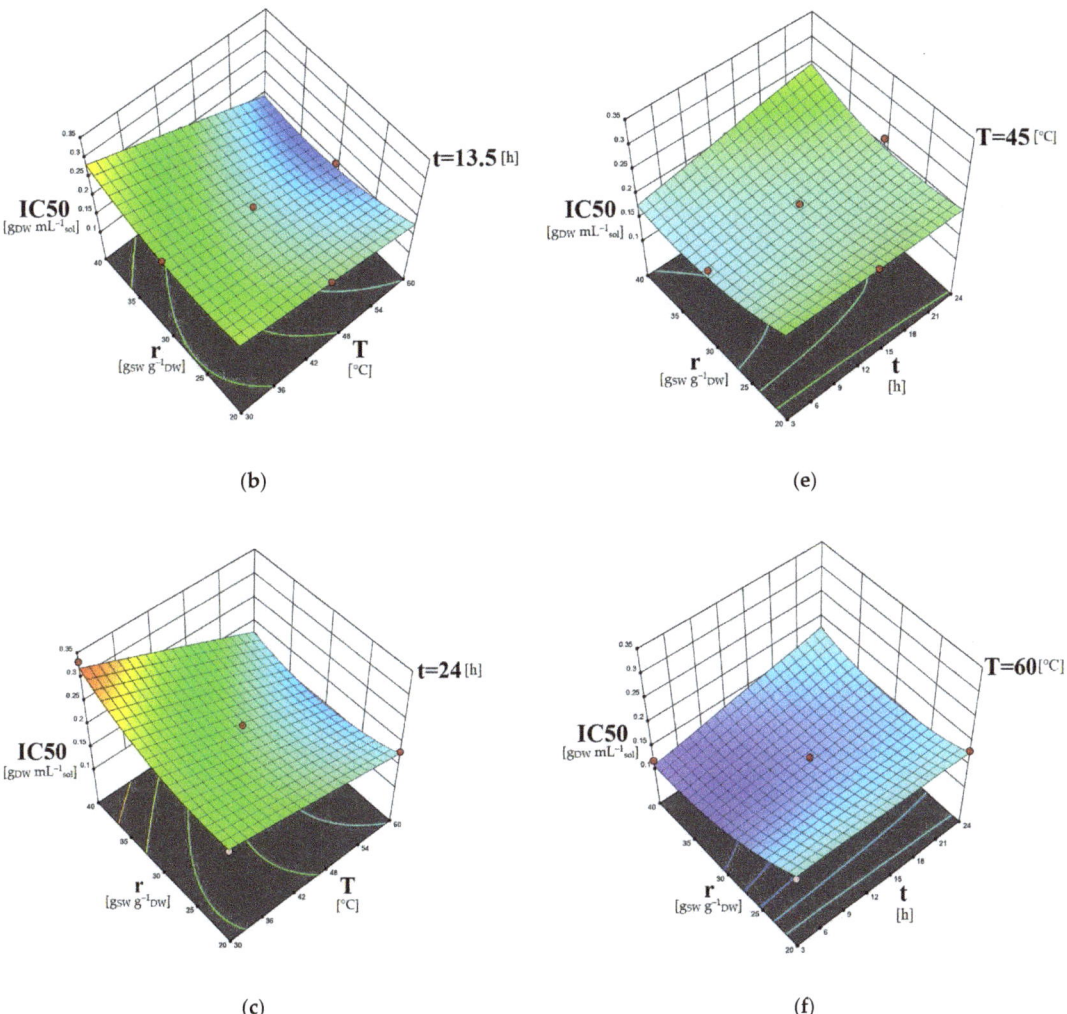

Figure 3. Three-dimensional surface plots showing the combined effects of (**a**–**c**) temperature and solvent-to-biomass ratio and (**d**–**f**) time and solvent-to-biomass ratio on the IC50 [g_{DW} mL$^{-1}_{sol}$] of the *C. vulgaris* extracts obtained using DES1/w solvent.

2.6. Experimental Validation of the Models

The predictive capacity of the models was validated experimentally. Specifically, two different extractions of *C. vulgaris* with DES1/w were carried out under conditions that were chosen randomly to avoid any bias (Table 5). The experimental procedure and the analysis of the obtained extracts were performed in the exact same way that was employed for the implementation of the experimental design. The experimental TCC, TPC, and IC50 values of the extracts and their predicted values, which were calculated by using the aforementioned models (Equations (1)–(3)), are presented in Table 5.

It is observed that the calculated TCC, TPC, and IC50 values agree very well with the experimentally measured ones. Hence, it is concluded that the models reproduce the experimental results satisfactorily, and thus, they can be safely used for prediction purposes within the range of the examined conditions.

Table 5. Experimental conditions of the extractions conducted for models' validation and the corresponding experimental and calculated TCC, TPC, and IC50 values.

Extraction	Experimental Values			Calculated Values		
	TCC [mg g^{-1}$_{DW}$]	TPC [mg$_{GAE}$ g^{-1}$_{DW}$]	IC50 [g$_{DW}$ mL^{-1}$_{sol}$]	TCC [mg g^{-1}$_{DW}$]	TPC [mg$_{GAE}$ g^{-1}$_{DW}$]	IC50 [g$_{DW}$ mL^{-1}$_{sol}$]
1 (30 °C, 6 h, 20:1 g$_{SW}$ g^{-1}$_{DW}$)	1.927	9.103	0.238	2.069	9.075	0.249
2 (45 °C, 24 h, 20:1 g$_{SW}$ g^{-1}$_{DW}$)	2.706	8.256	0.194	2.714	9.243	0.209

2.7. Optimization of Extraction Process

The optimum extraction conditions of *C. vulgaris* biomass using DES1/w were defined by employing Design–Expert Ver. 13.0.5.0 software using the models of Equations (1)–(3). The independent variables, i.e., extraction temperature (X_1), time (X_2), and solvent-to-biomass ratio (X_3), were set to vary within the ranges that were initially chosen for each one of them for the implementation of the experimental design. The responses TCC (Y_1) and TPC (Y_2) were set to maximize, while IC50 (Y_3) was to minimize. Moreover, a weight factor of 1 was applied for TCC and TPC and 2.5 for IC50. The weight factors resulted from the optimization study, according to which the extracts' antioxidant activity had to obtain a greater weight factor in comparison to the other two responses in order to obtain an overall optimized solution. The described optimization process and its results are depicted graphically in Figure 4.

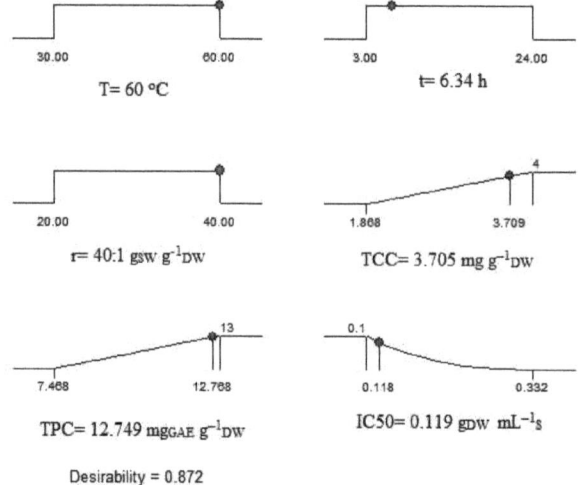

Desirability = 0.872

Figure 4. Optimization of the *C. vulgaris* biomass extraction using the solvent DES1/w.

As seen, the predicted optimum conditions for the simultaneous maximization of TCC and TPC and minimization of IC50 were T = 60 °C, t = 6.34 h, and r = 40:1 g$_{SW}$ g^{-1}$_{DW}$, under which the dependent variables TCC, TPC, and IC50 were determined 3.705 mg g^{-1}$_{DW}$, 12.749 mg$_{GAE}$ g^{-1}$_{DW}$ and 0.119 g$_{DW}$ mL^{-1}$_S$, respectively. Consequently, the highest temperature (60 °C) and solvent-to-biomass ratio (40:1 g$_{SW}$ g^{-1}$_{DW}$), and an extraction time in-between the lowest (3 h) and intermediate (13.5 h) values that were studied were concluded to be the most convenient extraction conditions for the purpose of the present work. By comparing the values of the three studied responses obtained under the optimum conditions to the experimental ones in Table 3, it is seen that the recovered carotenoids yield is a bit lower than the experimental maximum. This small compromise was inevitable in order to obtain an overall optimized process.

3. Materials and Methods

3.1. Chemicals

The chemical reagents used for the preparation of the DESs and the extracts' analyses were all of the analytical grade. Details regarding their provider and purity degree are included in Table 6.

Table 6. Chemical reagents.

Chemical Reagents	Provider	Purity
2,2-Diphenyl-1-picrylhydrazyl	Alfa Aesar	95%
Folin Ciocalteu's reagent	Carlo Erba reagents	Special grade
Methanol	Fisher Scientific	≥99.8%
Ethanol	Fisher Scientific	≥99.8%
Water	Fisher Scientific	HPLC grade
Choline chloride	Sigma Aldrich	≥98%
1,2 Butanediol	Sigma Aldrich	98%
Glycerol	Sigma Aldrich	≥99.0%
Ethylene glycol	Sigma Aldrich	99.8%
β–carotene	Alfa Aesar	99%
Gallic acid	Acros Organics	98%

3.2. Microalgae Culture

The microalga *Chlorella Vulgaris* UTEX 1809 was obtained from the Algal Culture Collection at the University of Texas, Austin, USA. Autotrophic cultivation of *C. vulgaris* was performed in a 12 L stirred tank photobioreactor with a 10 L working volume (Bioengineering, Switzerland) continuously illuminated with cool white light at 65 µmol m^{-2}sec^{-1} using light emitting diodes (LED). The temperature was maintained at 25 °C, and the stirring speed at 150 rpm. The bioreactor was filled with 9 L of Bold's modified basal medium (BBM) containing NaNO$_3$ (250 mg L^{-1}), KH$_2$PO$_4$ (175 mg L^{-1}), K$_2$HPO$_4$ (75 mg L^{-1}), NaCl (25 mg L^{-1}), MgSO$_4$ 7H$_2$O (75 mg L^{-1}), anhydrous EDTA (50 mg L^{-1}), CaCl$_2$ 2H$_2$O (25 mg L^{-1}), FeSO$_4$ 7H$_2$O (4.98 mg L^{-1}), MnCl$_2$ 4H$_2$O (1.44 mgL^{-1}), ZnSO$_4$ 7H$_2$O (8.82 mg L^{-1}), CuSO$_4$ 5H$_2$O (1.57 mg L^{-1}), and KOH (31 mg L^{-1}). It was then autoclaved at 120 °C for 20 min. Then, 1 L preculture of microalgal cells grown in 250 mL Erlenmeyer flasks under the same conditions on an orbital shaker at 180 rpm was used as inoculum. The pH at the beginning of the culture was adjusted to 6.8, and the sterile air supply to 0.25 vvm. After 15 days of cultivation, the microalgal biomass reached over 1.5 g L^{-1} dry weight. The liquid algal culture was concentrated by centrifugation (5000 rpm, for 10 min), and the concentrated algal mass was stored at 4 °C until used in further experiments.

3.3. Preparation of DESs

The synthesis of the DESs was conducted by mixing their two components under a determined molar ratio using constant magnetic agitation at 80 °C until a clear and homogeneous liquid was formed (3–4 h). Three different DES systems were obtained by using choline chloride (ChCl) as HBA and the polyols 1,2 butanediol, glycerol, and ethylene glycol as HBD, at molar ratios 1:4, 1:2, and 1:2, respectively (Table 7). The DES constituents were selected due to their biodegradability, biocompatibility, availability, and relatively low cost. After synthesis, the DESs were dried in an oven at 50 °C under vacuum for 24 h. Subsequently, they were stored in dark in a desiccator.

Table 7. Synthesized DESs and their abbreviation.

DES	HBA	HBD	HBA:HBD Ratio
DES1		1,2 butanediol	1:4
DES2	choline chloride	glycerol	1:2
DES3		ethylene glycol	1:2

3.4. Measurement of DES Physical Properties

Dynamic viscosity and density measurements for all pure DES and their mixtures with water (70/30 w/w) were conducted at 60 °C. A Brookfield digital viscometer (LV–DVI–E) connected to a thermostatic bath was employed for the viscosity measurements. An SC4–13R chamber was used to carry the solvent sample, along with an SC4–18 spindle attached to the moving shaft of the viscometer. Density measurements were performed with a KEM KYOTO density/specific gravity meter (DA–640). Prior to measurements, all DESs were dried in an oven at 50 °C under vacuum for 1 h to eliminate any adsorbed water. All measurements were performed in triplicate.

3.5. Extraction Process

Mixtures of ChCl:1,2 butanediol (1:4), ChCl:ethylene glycol (1:2), and ChCl:glycerol (1:2) with water (70/30 w/w) were employed as extraction solvents. According to the literature [16,58–60], the addition of water at this concentration reduces the viscosity enough without negatively affecting the hydrogen bond interactions between the constituents of the DES. S. Rozas et al., 2021 [60], in particular, concluded that in the case of ChCl:glycerol (1:2), a 10 to 30 wt% water content does not affect the main properties of the DES. The ethanol/water 70/30 w/w mixture (EtOH/w), a non-toxic, efficient, conventional solvent, was used for comparison [49]. The extraction conditions for the solvent screening were selected based on the relevant published work of our scientific group [37,49]. Specifically, the extractions were carried out at a 20:1 solvent-to-biomass ratio (r, [g_{SW} g^{-1}_{DW}]) for a time period of three hours (t, [h]). Biomass and solvents were weighed and added in a stoppled, double-walled glass vial that was connected to a thermostatic bath to keep the extraction temperature at 60 °C. Constant magnetic agitation was employed for the stirring of the biomass–solvent mixture. Upon the completion of the extractions, the resulting mixture was centrifuged at 4430 rcf for 10 min to separate the extract from the biomass. Subsequently, the extracts' total carotenoid (TCC, [mg g^{-1}_{DW}]) and phenolic (TPC, [mg_{GAE} g^{-1}_{DW}]) contents, as well as their antioxidant activity (IC50, [g_{DW} mL^{-1}_{sol}]) were determined.

Whereupon the solvent screening process, the most convenient one for the purpose of our study, was employed for further investigation using the same experimental configuration and procedure. Particularly, three of the most influential extraction parameters, temperature (T = 30–60 °C), solvent-to-biomass ratio (r = 20:1–40:1 g_{sol} g^{-1}_{DW}), and time (t = 3–24 h) were studied regarding their effect on the extracts' total carotenoids content, total phenolics content and on their antioxidant capacity.

3.6. Determination of the Extracts' Total Phenolic and Carotenoid Contents

Total phenolic content was estimated using the Folin–Ciocalteu reagent as described by Singleton et al., 1965 [61]. In particular, 7.9 mL of distilled water and 0.1 mL of extract were homogenized before the addition of 0.5 mL of Folin-Ciocalteu reagent. After vortexing, the resulting mixture, 1.5 mL Na$_2$CO$_3$ solution (20% w/v), was added, and the final mixture was incubated for 30 min in a water bath at 40 °C. Its absorbance was subsequently measured at 765 nm and compared to a gallic acid calibration curve.

Total carotenoid content was determined according to the Association of Official Analytical Collaboration (AOAC) [62] methods. Following the extraction, the absorbance of 3 mL of extract was measured at 450 nm, and total carotenoids content was calculated from Equation (4) which was acquired by the β–carotene calibration curve:

$$TCC = 6.9691 \cdot Abs_{450\ nm} - 0.1286 \qquad (4)$$

3.7. Determination of the Extracts' Antioxidant Activity

The antioxidant activity of the extracts was assessed using the 2,2–Diphenyl–1–Picrylhydrazyl (DPPH) assay. In total, 100 µL of the extract was added to 3 mL of a DPPH ethanolic solution (0.03% w/v). The absorbance of the mixture was measured at 515 nm after its incubation for 20 min at room temperature. The calculated

IC50 values refer to the sample concentration that is required to scavenge 50% of DPPH free radicals [63].

The spectrophotometric measurements were conducted In a SHIMADZU UV–1900, UV–VIS spectrophotometer. All measurements described above were conducted in triplicate, and the reported values of total carotenoids, total phenolics, and IC50 are the calculated average values.

3.8. Experimental Design and Statistical Analysis

Design–Expert Ver. 13.0.5.0 (Statease Inc., Minneapolis, MN, USA, test version) was employed for performing Response Surface Methodology (RSM) in order to optimize the extraction process of *C. vulgaris* biomass using DES-based solvents. A three-factor, three-level Central Composite Design (CCD) was followed. The influence of extraction temperature (X_1), extraction time (X_2), and solvent-to-biomass ratio (X_3) on TCC (Y_1), TPC (Y_2), and the IC50 (Y_3) value of the extracts was examined. Eighteen experiments in total, composed of six axial, eight factorial, and four central points, were realized randomly (Table 8). The obtained experimental data were subjected to regression analysis and Analysis of Variance (ANOVA). Thus, the determination of the significance of the influence of the independent variables (Factors) on every dependent variable (Response) was allowed, and fitting mathematical models were developed.

Table 8. Experimental design of three-factor, three-level CCD.

Run	Space Type	Factor 1/Level X_1: T [°C]	Factor 2/Level X_2: t [h]	Factor 3/Level X_3: r [$g_{sol} g^{-1}_{DW}$]
1	Axial	45/0	13.5/0	20/−1
2	Axial	45/0	13.5/0	40/+1
3	Factorial	30/−1	3/−1	20/−1
4	Factorial	30/−1	24/+1	40/+1
5	Factorial	60/+1	24/+1	20/−1
6	Center	45/0	13.5/0	30/0
7	Factorial	30/−1	24/+1	20/−1
8	Factorial	30/−1	3/−1	40/+1
9	Axial	45/0	24/+1	30/0
10	Factorial	60/+1	3/−1	20/−1
11	Center	45/0	13.5/0	30/0
12	Center	45/0	13.5/0	30/0
13	Axial	45/0	3/−1	30/0
14	Factorial	60/+1	24/+1	40/+1
15	Axial	30/−1	13.5/0	30/0
16	Axial	60/+1	13.5/0	30/0
17	Center	45/0	13.5/0	30/0
18	Factorial	60/+1	3/−1	40/+1

4. Conclusions

In this study, the extraction of *C. vulgaris* using DES-based solvents was examined. Among the three DESs that were synthesized in the present work, ChCl: glycerol (1:2) (DES2) was found to be the most viscus one, followed by ChCl:1,2 butanediol (1:4) (DES1) and ChCl: Ethylene glycol (1:2) (DES3). In order to overcome the drawback of high viscosity, which is essential for industrial applications, DES/water 70/30 w/w mixtures were also tested. Their viscosities followed the order that the pure DESs exhibited as well, i.e., DES2/w > DES1/w > DES3/w, but they were significantly lower than those of pure DESs. Regarding the measured densities of the pure DESs, their values diminish following the same order as the viscosity values. DES2/w and DES3/w demonstrated lower densities in comparison to pure DES2 and DES3, respectively, though this reduction with water addition was not significant. The density of DES1/w was found to be marginally higher than DES1.

From the solvent screening procedure, it was found that DES1/w solvent outperformed the two others, i.e., DES2/w and DES3/w, in extracting both carotenoid and phenolic compounds from *C. vulgaris* biomass. The TCC value measured at the DES1/w extract was 93.7% and 91.5% greater than the corresponding values measured for the DES2/w and DES3/w extracts, respectively. The TPC value that was determined for the DES1/w extract was almost double the TPC values found for the DES2/w and DES3/w extracts. It should be noted that all the DES-based solvents used in the present study performed significantly better in extracting phenolic compounds than carotenoids. DES1/w exhibited a higher phenolics extraction capacity than ethanol/water (control solvent) by approximately 10%. DES-based solvents follow the order DES1/w> DES2/w > DES3/w as per their antioxidant activity. According to the solvent screening results described above, it was concluded that the most convenient for the purpose of our study was DES1/w, and it was exploited for the *C. vulgaris* extraction process optimization.

To this purpose, a three-factor, three-level Central Composite Design (CCD) was performed, and the influence of the extraction parameters temperature (X_1), time (X_2), and solvent-to-biomass ratio (X_3) on the responses TCC (Y_1), TPC (Y_2) and IC50 (Y_3) was examined by performing 18 extractions. RSM and ANOVA assessment of the experimental data led to the development of reduced quadratic multiple regression models for each of the three dependent parameters that were studied. Non-significant lack of fit was found for all the developed models, and their predictive capacity was confirmed experimentally via two different experiments within the range of the examined experimental conditions. TCC and TPC were most influenced by temperature and solvent-to-biomass ratio, while the antioxidant activity was most affected by temperature. Carotenoid and phenolic extraction were enhanced under higher T and r values and increasing extraction temperature boosted the extracts' antioxidant capacity. The increase in extraction time had a negative impact on the extracts' antioxidant activity due to the oxidation of the recovered carotenoids and phenolics under long exposure to air and light. The highest TCC and TPC and the lowest IC50 values were found at the temperature of 60 °C, a solvent-to-biomass mass ratio of 40:1, and extraction time of about six and a half hours.

Author Contributions: Conceptualization, E.V., K.M. and H.S.; methodology, S.P., M.M.D. and M.G.S.; validation, V.L. and M.M.D.; investigation, M.M.D., S.P., M.G.S. and P.K.; writing—original draft preparation, M.M.D. and S.P; writing—review and editing, E.V., V.L. and M.G.S.; visualization, M.M.D.; supervision, E.V., K.M. and H.S.; project administration, E.V. and V.L. All authors have read and agreed to the published version of the manuscript.

Funding: This research has been co-financed by the European Union and Greek national funds through the Operational Program Competitiveness, Entrepreneurship, and Innovation, under the call "Special Actions, Aquaculture-Industrial Materials-Open Innovation in Culture" (project code: T6YBP-00033).

Institutional Review Board Statement: Not applicable.

Informed Consent Statement: Not applicable.

Data Availability Statement: Not applicable.

Conflicts of Interest: The authors declare no conflict of interest.

Sample Availability: Samples of the compounds are available from the authors.

References

1. Yang, T.P.; Lee, H.J.; Ou, T.T.; Chang, Y.J.; Wang, C.J. Mulberry leaf polyphenol extract induced apoptosis involving regulation of adenosine monophosphate–activated protein kinase/fatty acid synthase in a p53–negative hepatocellular carcinoma cell. *J. Agric. Food Chem.* **2012**, *60*, 6891–6898. [CrossRef] [PubMed]
2. Zhang, H.; Tsao, R. Dietary polyphenols, oxidative stress and antioxidant and anti–inflammatory effects. *Curr. Opin. Food Sci.* **2016**, *8*, 33–42. [CrossRef]

3. Caban, M.; Owczarek, K.; Chojnacka, K.; Lewandowska, U. Overview of polyphenols and polyphenol–rich extracts as modulators of IGF–1, IGF–1R, and IGFBP expression in cancer diseases. *J. Funct. Foods* **2019**, *52*, 389–407. [CrossRef]
4. Hidalgo, G.I.; Almajano, M.P. Red fruits: Extraction of antioxidants, phenolic content, and radical scavenging determination: A review. *Antioxidants* **2017**, *6*, 7. [CrossRef]
5. Gao, M.Z.; Cui, Q.; Wang, L.T.; Meng, Y.; Yu, L.; Li, Y.Y.; Fu, Y.J. A green and integrated strategy for enhanced phenolic compounds extraction from mulberry (Morus alba L.) leaves by deep eutectic solvent. *Microchem. J.* **2020**, *154*, 104598. [CrossRef]
6. Serna–Vázquez, J.; Ahmad, M.Z.; Boczkaj, G.; Castro–Muñoz, R. Latest Insights on Novel Deep Eutectic Solvents (DES) for Sustainable Extraction of Phenolic Compounds from Natural Sources. *Molecules* **2021**, *26*, 5037. [CrossRef]
7. Lasarte–Aragonés, G.; Lucena, R.; Cárdenas, S.; Valcárcel, M. Use of switchable solvents in the microextraction context. *Talanta* **2015**, *131*, 645–649. [CrossRef]
8. Tang, W.; Row, K.H. Evaluation of CO_2–induced azole–based switchable ionic liquid with hydrophobic/hydrophilic reversible transition as single solvent system for coupling lipid extraction and separation from wet microalgae. *Bioresour. Technol.* **2020**, *296*, 122309. [CrossRef]
9. Tang, W.; Row, K.H. Design and evaluation of polarity controlled and recyclable deep eutectic solvent based biphasic system for the polarity driven extraction and separation of compounds. *J. Clean. Prod.* **2020**, *268*, 122306. [CrossRef]
10. Abbott, A.P.; Capper, G.; Davies, D.L.; Rasheed, R.K.; Tambyrajah, V. Novel solvent properties of choline chloride/urea mixtures. *Chem. Comm.* **2003**, *1*, 70–71. [CrossRef]
11. Shishov, A.; Bulatov, A.; Locatelli, M.; Carradori, S.; Andruch, V. Application of deep eutectic solvents in analytical chemistry. A review. *Microchem. J.* **2017**, *135*, 33–38. [CrossRef]
12. Ling, J.K.U.; Chan, Y.S.; Nandong, J. Extraction of antioxidant compounds from the wastes of Mangifera pajang fruit: A comparative study using aqueous ethanol and deep eutectic solvent. *SN Appl. Sci.* **2020**, *2*, 1–12. [CrossRef]
13. Gullón, P.; Gullón, B.; Romaní, A.; Rocchetti, G.; Lorenzo, J.M. Smart advanced solvents for bioactive compounds recovery from agri-food by-products: A review. *Trends Food Sci. Technol.* **2020**, *101*, 182–197. [CrossRef]
14. Fernández, M.; Espino, M.; Gomez, F.J.V.; Silva, M.F. Novel approaches mediated by tailor–made green solvents for the extraction of phenolic compounds from agro-food industrial by-products. *Food Chem.* **2018**, *239*, 671–678. [CrossRef]
15. Zhao, B.Y.; Xu, P.; Yang, F.X.; Wu, H.; Zong, M.H.; Lou, W.Y. Biocompatible deep eutectic solvents based on choline chloride: Characterization and application to the extraction of rutin from Sophora japonica. *ACS Sustain. Chem. Eng.* **2015**, *3*, 2746–2755. [CrossRef]
16. Cui, Q.; Peng, X.; Yao, X.H.; Wei, Z.F.; Luo, M.; Wang, W.; Zu, Y.G. Deep eutectic solvent–based microwave–assisted extraction of genistein, genistein and apigenin from pigeon pea roots. *Sep. Purif. Technol.* **2015**, *150*, 63–72. [CrossRef]
17. Wang, Y.; Ma, C.; Liu, C.; Lu, X.; Feng, X.; Ji, X. Thermodynamic Study of Choline Chloride–Based Deep Eutectic Solvents with Water and Methanol. *Chem. Eng. Data* **2020**, *65*, 2446–2457. [CrossRef]
18. Coelho de Andrade, D.; Aquino Monteiro, S.; Merib, J. A review on recent applications of deep eutectic solvents in microextraction techniques for the analysis of biological matrices. *Adv. Sample Prep.* **2022**, *1*, 100007. [CrossRef]
19. Rodríguez-Álvarez, M.J.; García-Garrido, S.E.; Perrone, S.; García-Álvarez, J.; Capriati, V. Deep Eutectic Solvents and Heterogeneous Catalysis with Metallic Nanoparticles: A Powerful Partnership in Sustainable Synthesis. *Curr. Opin. Green Sustain. Chem.* **2022**, *39*, 100723. [CrossRef]
20. Chang, X.X.; Mubarak, N.M.; Mazari, S.A.; Jatoi, A.S.; Ahmad, A.; Khalid, M.; Nizamuddin, S. A review on the properties and applications of chitosan, cellulose and deep eutectic solvent in green chemistry. *J. Ind. Eng. Chem.* **2021**, *104*, 362–380. [CrossRef]
21. Perna, F.M.; Vitale, P.; Capriati, V. Deep eutectic solvents and their applications as green solvents. *Curr. Opin. Green Sustain. Chem.* **2020**, *21*, 27–33. [CrossRef]
22. Pätzold, M.; Siebenhaller, S.; Kara, S.; Liese, A.; Syldatk, C.; Holtmann, D. Deep eutectic solvents as efficient solvents in biocatalysis. *Trends Biotechnol.* **2019**, *37*, 943–959. [CrossRef] [PubMed]
23. Tomé, L.I.; Baião, V.; Da Silva, W.; Brett, C.M. Deep eutectic solvents for the production and application of new materials. *Appl. Mater. Today* **2018**, *10*, 30–50. [CrossRef]
24. Smith, E.L.; Abbott, A.P.; Ryder, K.S. Deep eutectic solvents (DESs) and their applications. *Chem. Rev.* **2014**, *114*, 11060–11082. [CrossRef]
25. Zainal-Abidin, M.H.; Hayyan, M.; Hayyan, A.; Jayakumar, N.S. New horizons in the extraction of bioactive compounds using deep eutectic solvents: A review. *Anal. Chim. Acta* **2017**, *979*, 1–23. [CrossRef]
26. Kalyniukova, A.; Holuša, J.; Musiolek, D.; Sedlakova–Kadukova, J.; Płotka–Wasylka, J.; Andruch, V. Application of deep eutectic solvents for separation and determination of bioactive compounds in medicinal plants. *Ind. Crops. Prod.* **2012**, *172*, 114047. [CrossRef]
27. Orejuela-Escobar, L.M.; Landázuri, A.C.; Goodell, B. Second generation biorefining in Ecuador: Circular bioeconomy, zero waste technology, environment and sustainable development: The nexus. *J. Bioresour. Bioprod.* **2021**, *6*, 83–107. [CrossRef]
28. Shao, J.; Ni, Y.; Yan, L. Oxidation of furfural to maleic acid and fumaric acid in deep eutectic solvent (DES) under vanadium pentoxide catalysis. *J. Bioresour. Bioprod.* **2021**, *6*, 39–44. [CrossRef]
29. Barbieri, J.B.; Goltz, C.; Cavalheiro, F.B.; Toci, A.T.; Igarashi-Mafra, L.; Mafra, M.R. Deep eutectic solvents applied in the extraction and stabilization of rosemary (Rosmarinus officinalis L.) phenolic compounds. *Ind. Crops Prod.* **2020**, *144*, 112049. [CrossRef]

30. Nam, M.W.; Zhao, J.; Lee, M.S.; Jeong, J.H.; Lee, J. Enhanced extraction of bioactive natural products using tailor–made deep eutectic solvents: Application to flavonoid extraction from Flos sophorae. *Green Chem.* **2015**, *17*, 1718–1727. [CrossRef]
31. Das, A.K.; Sharma, M.; Mondal, D.; Prasad, K. Deep eutectic solvents as efficient solvent system for the extraction of κ–carrageenan from Kappaphycus alvarezii. *Carbohydr. Polym.* **2016**, *136*, 930–935. [CrossRef] [PubMed]
32. Vilková, M.; Płotka-Wasylka, J.; Andruch, V. The role of water in deep eutectic solvent–base extraction. *J. Mol. Liq.* **2020**, *304*, 112747. [CrossRef]
33. Skarpalezos, D.; Detsi, A. Deep eutectic solvents as extraction media for valuable flavonoids from natural sources. *Appl. Sci.* **2019**, *9*, 4169. [CrossRef]
34. Kafyra, M.S.G.; Papadaki, S.; Chronis, M.; Krokida, M. Microalgae based innovative animal fat and proteins replacers for application in functional baked products. *Open Agric.* **2018**, *3*, 427–436. [CrossRef]
35. Peng, X.; Duan, M.-H.; Yao, X.-H.; Zhang, Y.-H.; Zhao, C.-J.; Zu, Y.-G.; Fu, Y.-J. Green extraction of five target phenolic acids from Lonicerae japonicae Flos with deep eutectic solvent. *Sep. Purif. Technol.* **2016**, *157*, 249–257. [CrossRef]
36. Pradhan, B.; Patra, S.; Dash, S.R.; Nayak, R.; Behera, C.; Jena, M. Evaluation of the anti–bacterial activity of methanolic extract of *Chlorella vulgaris* Beyerinck [Beijerinck] with special reference to antioxidant modulation. *Future J. Pharm. Sci.* **2021**, *7*, 1–11. [CrossRef]
37. Georgiopoulou, I.; Tzima, S.; Pappa, G.D.; Louli, V.; Voutsas, E.; Magoulas, K. Experimental Design and Optimization of Recovering Bioactive Compounds from *Chlorella vulgaris* through Conventional Extraction. *Molecules* **2021**, *27*, 29. [CrossRef]
38. Kulkarni, S.; Nikolov, Z. Process for selective extraction of pigments and functional proteins from Chlorella vulgaris. *Algal Res.* **2018**, *35*, 185–193. [CrossRef]
39. Dimova, D.; Dobreva, D.; Panayotova, V.; Makedonski, L. DPPH antiradical activity and total phenolic content of methanol and ethanol extracts from macroalgae (Ulva rigida) and microalgae (Chlorella). *Scr. Sci. Pharm.* **2019**, *6*, 37–41. [CrossRef]
40. Mahmood, W.; Lorwirachsutee, W.M.A.; Theodoropoulos, A.C.; Gonzalez-Miquel, M. Polyol–based deep eutectic solvents for extraction of natural polyphenolic antioxidants from Chlorella vulgaris. *ACS Sustain. Chem. Eng.* **2019**, *7*, 5018–5026. [CrossRef]
41. Guo, W.; Hou, Y.; Ren, S.; Tian, S.; Wu, W. Formation of Deep Eutectic Solvents by Phenols and Choline Chloride and Their Physical Properties. *J. Chem. Eng. Data* **2013**, *58*, 866–872. [CrossRef]
42. Dai, Y.; Witkamp, G.J.; Verpoorte, R.; Choi, Y.H. Tailoring properties of natural deep eutectic solvents with water to facilitate their applications. *Food Chem.* **2015**, *187*, 14–19. [CrossRef] [PubMed]
43. Van Osch, D.J.; Dietz, C.H.; Van Spronsen, J.; Kroon, M.C.; Gallucci, F.; van Sint Annaland, M.; Tuinier, R. A Search for Natural Hydrophobic Deep Eutectic Solvents Based on Natural Components. *ACS Sustain. Chem. Eng.* **2019**, *7*, 2933–2942. [CrossRef]
44. Basaiahgari, A.; Panda, S.; Gardas, R.L. Effect of Ethylene, Diethylene, and Triethylene Glycols and Glycerol on the Physicochemical Properties and Phase Behavior of Benzyltrimethyl and Benzyltributylammonium Chloride Based Deep Eutectic Solvents at 283.15–343.15. *J. Chem. Eng. Data* **2018**, *63*, 2613–2627. [CrossRef]
45. Abbott, A.P. Application of Hole Theory to the Viscosity of Ionic and Molecular Liquids. *ChemPhysChem* **2004**, *5*, 1242–1246. [CrossRef] [PubMed]
46. Florindo, C.; Oliveira, F.S.; Rebelo, L.P.N.; Fernandes, A.M.; Marrucho, I.M. Insights into the synthesis and properties of deep eutectic solvents based on cholinium chloride and carboxylic acids. *ACS Sustain. Chem. Eng.* **2014**, *2*, 2416–2425. [CrossRef]
47. Ijardar, S.P. Deep eutectic solvents composed of tetrabutylammonium bromide and PEG: Density, speed of sound and viscosity as a function of temperature. *J. Chem. Thermodyn.* **2020**, *140*, 105897. [CrossRef]
48. Yadav, A.; Trivedi, S.; Rai, R.; Pandey, S. Densities and dynamic viscosities of (choline chloride + glycerol) deep eutectic solvent and its aqueous mixtures in the temperature range (283.15–363.15) K. *Fluid. Phase Equilib.* **2014**, *367*, 135–142. [CrossRef]
49. Pappou, S.; Dardavila, M.M.; Savvidou, M.G.; Louli, V.; Magoulas, K.; Voutsas, E. Extraction of Bioactive Compounds from Ulva lactuca. *Appl. Sci.* **2022**, *12*, 2117. [CrossRef]
50. Bulut, O.; Akın, D.; Sönmez, Ç.; Öktem, A.; Yücel, M.; Öktem, H.A. Phenolic compounds, carotenoids, and antioxidant capacities of a thermo–tolerant Scenedesmus sp. (Chlorophyta) extracted with different solvents. *J. Appl. Phycol.* **2019**, *31*, 1675–1683. [CrossRef]
51. Goiris, K.; Muylaert, K.; Fraeye, I.; Foubert, I.; De Brabanter, J.; De Cooman, L. Antioxidant potential of microalgae in relation to their phenolic and carotenoid content. *J. Appl. Psychol.* **2012**, *24*, 1477–1486. [CrossRef]
52. Corrêa, P.S.; Morais Júnior, W.G.; Martins, A.A.; Caetano, N.S.; Mata, T.M. Microalgae Biomolecules: Extraction, Separation and Purification Methods. *Processes* **2021**, *9*, 10. [CrossRef]
53. Ali, M.C.; Chen, J.; Zhang, H.; Li, Z.; Zhao, L.; Qiu, H. Effective extraction of flavonoids from Lycium barbarum L. fruits by deep eutectic solvents–based ultrasound–assisted extraction. *Talanta* **2019**, *203*, 16–22. [CrossRef] [PubMed]
54. Strati, I.F.; Oreopoulou, V. Effect of extraction parameters on the carotenoid recovery from tomato waste. *Int. J. Food Sci.* **2011**, *46*, 23–29. [CrossRef]
55. Zakaria, S.M.; Kamal, S.M.M.; Harun, M.R.; Omar, R.; Siajam, S.I. Subcritical Water Technology for Extraction of Phenolic Compounds from Chlorella sp. Microalgae and Assessment on Its Antioxidant Activity. *Molecules* **2017**, *22*, 1105. [CrossRef]
56. Rodrigues, L.A.; Pereira, C.V.; Leonardo, I.C.; Fernández, N.; Gaspar, F.B.; Silva, J.M.; Reis, R.L.; Duarte, A.R.C.; Paiva, A.; Matias, A.A. Terpene-Based Natural Deep Eutectic Systems as Efficient Solvents To Recover Astaxanthin from Brown Crab Shell Residues. *ACS Sustain. Chem. Eng.* **2020**, *8*, 2246–2259. [CrossRef]

57. Ozturk, B.; Parkinson, C.; Gonzalez–Miquel, M. Extraction of polyphenolic antioxidants from orange peel waste using deep eutectic solvents. *Sep. Purif. Technol.* **2018**, *206*, 1–13. [CrossRef]
58. Bajkacz, S.; Adamek, J. Evaluation of new natural deep eutectic solvents for the extraction of isoflavones from soy products. *Talanta* **2017**, *168*, 329–335. [CrossRef]
59. Yao, X.H.; Zhang, D.Y.; Duan, M.H.; Cui, Q.; Xu, W.J.; Luo, M.; Li, C.Y.; Zu, Y.G.; Fu, Y.J. Preparation and determination of phenolic compounds from Pyrola incarnata Fisch. with a green polyols based–deep eutectic solvent. *Sep. Purif. Technol.* **2015**, *149*, 116–123. [CrossRef]
60. Rozas, S.; Benito, C.; Alcalde, R.; Atilhan, M.; Aparicio, S. Insights on the water effect on deep eutectic solvents properties and structuring: The archetypical case of choline chloride + ethylene glycol. *J. Mol. Liq.* **2021**, *344*, 117717. [CrossRef]
61. Singleton, V.; Rossi, J. Colorimetry of total phenolics with phosphomolybdic–phosphotungstic acid reagents. *Am. J. Enol. Vitic.* **1965**, *16*, 144–158.
62. Association of Official Analytical Chemists (AOAC). *Official Methods of Analysis of the Association of Official Analytical Chemists International*, 16th ed.; The Association of Official Analytical Chemists: Washington, DC, USA, 1995.
63. Brand–Williams, W.; Cuvelier, M.; Berset, C. Use of a free radical method to evaluate antioxidant activity. *LWT* **1995**, *28*, 25–30. [CrossRef]

Disclaimer/Publisher's Note: The statements, opinions and data contained in all publications are solely those of the individual author(s) and contributor(s) and not of MDPI and/or the editor(s). MDPI and/or the editor(s) disclaim responsibility for any injury to people or property resulting from any ideas, methods, instructions or products referred to in the content.

Article

Valorization of Hemp-Based Packaging Waste with One-Pot Ionic Liquid Technology

Julius Choi [1,2], Alberto Rodriguez [1,2], Blake A. Simmons [1,3] and John M. Gladden [1,2,*]

[1] Deconstruction Division, Joint BioEnergy Institute, 5885 Hollis Street, Emeryville, CA 94608, USA
[2] Department of Biomaterials and Biomanufacturing, Sandia National Laboratories, 7011 East Avenue, Livermore, CA 94550, USA
[3] Biological Systems and Engineering Division, Lawrence Berkeley National Laboratory, 1 Cyclotron Road, Berkeley, CA 94720, USA
* Correspondence: jmgladden@lbl.gov

Abstract: The range of applications for industrial hemp has consistently increased in various sectors over the years. For example, hemp hurd can be used as a resource to produce biodegradable packaging materials when incorporated into a fungal mycelium composite, a process that has been commercialized. Although these packaging materials can be composted after usage, they may present an opportunity for valorization in a biorefinery setting. Here, we demonstrate the potential of using this type of discarded packaging composite as a feedstock for biofuel production. A one-pot ionic liquid-based biomass deconstruction and conversion process was implemented, and the results from the packaging material were compared with those obtained from untreated hemp hurd. At a 120 °C reaction temperature, 7.5% ionic liquid loading, and 2 h reaction time, the packaging materials showed a higher lignocellulosic sugar yield and sugar concentrations than hemp hurd. Hydrolysates prepared from packaging materials also promoted production of higher titers (1400 mg/L) of the jet-fuel precursor bisabolene when used to cultivate an engineered strain of the yeast *Rhodosporidium toruloides*. Box–Behnken experiments revealed that pretreatment parameters affected the hemp hurd and packaging materials differently, evidencing different degrees of recalcitrance. This study demonstrated that a hemp hurd-based packaging material can be valorized a second time once it reaches the end of its primary use by supplying it as a feedstock to produce biofuels.

Keywords: biomass pretreatment; cholinium lysinate; *Rhodosporidium toruloides*; bisabolene

Citation: Choi, J.; Rodriguez, A.; Simmons, B.A.; Gladden, J.M. Valorization of Hemp-Based Packaging Waste with One-Pot Ionic Liquid Technology. *Molecules* 2023, 28, 1427. https://doi.org/10.3390/molecules28031427

Academic Editors: Slavica Ražić, Aleksandra Cvetanović Kljakić and Enrico Bodo

Received: 23 December 2022
Revised: 30 January 2023
Accepted: 30 January 2023
Published: 2 February 2023

Copyright: © 2023 by the authors. Licensee MDPI, Basel, Switzerland. This article is an open access article distributed under the terms and conditions of the Creative Commons Attribution (CC BY) license (https:// creativecommons.org/licenses/by/ 4.0/).

1. Introduction

Lignocellulosic biomass can be a renewable carbon resource for the production of fuels and chemicals to replace non-renewable fossil carbon sources such as natural gas, oil, and coal [1]. Lignocellulosic biomass can be converted into a wide range of chemicals and energy carriers through a closed-loop biorefinery [2,3]. Despite decades of scientific and technological advances in this field, many technical challenges associated with the recalcitrance of lignocellulose to deconstruction into fermentable carbon still hamper the establishment of economically feasible biorefineries. These include the lack of efficient and inexpensive processes to depolymerize all components of biomass and convert them to fuels and chemicals at high yields by microbial fermentation [4,5]. To improve the accessibility of the biomass carbohydrate polymers to hydrolytic enzymes, a pretreatment step is frequently required. Numerous pretreatment technologies have been developed, including physical methods (e.g., milling, grinding, irradiation, and sonication) [6,7], chemical methods (e.g., alkali, acid, oxidizing agents, organic solvents, ionic liquids, and deep-eutectic solvents) [8,9], combined physico-chemical methods (e.g., hydrothermolysis) [10,11], or biological methods (e.g., bacteria and fungi) [12].

The discovery and implementation of biocompatible ionic liquids such as cholinium lysinate ([Ch][Lys]) provide a compelling alternative to the aforementioned approaches because of their high pretreatment efficiency and lower inhibitory effect to enzymes and microbes [13–15]. These features have allowed the development of a one-pot reaction system where all the process steps, such as pretreatment, enzymatic hydrolysis and fermentation can be consolidated into a single vessel without any separation, potentially reducing operating costs. Our group has demonstrated the applicability of this process by generating hydrolysates with high concentrations of monomeric sugars and organic acids from several feedstocks like grasses, hardwoods, and softwoods, and converting them to terpene-based jet-fuel molecules using engineered strains of the yeast *Rhodosporidium toruloides* [16–18]. Nevertheless, it is important to expand the range of lignocellulosic feedstocks used in this process to evaluate its versatility to advance towards the goal of developing a truly lignocellulosic feedstock-agnostic biorefinery.

Hemp is an attractive crop due to its fast growth, bioremediation potential, and diverse agricultural applications, including the production of natural fibers, grains, essential oils, and other commodities [19]. This biomass is composed of an outer fiber that represents approximately 30% of the weight and an inner core known as hurd that accounts for the remaining 70% [20]. The hemp fiber is utilized in the textile industry as insulation material and for the production of bioplastics in the automotive industry, while hemp hurd is used for low value applications such as animal bedding, concrete additives, or disposed of by combustion and landfill accumulation [21–23]. This indicates that approximately 70 wt% of hemp biomass has the potential to be valorized into higher-value products and applications, which would improve the economics of the hemp industry and increase its sustainability footprint to promote a green economy. Mycelium-based composites are emerging as cheap and environmentally sustainable materials generated by fungal growth on a scaffold made of agricultural waste materials [24]. The mycelium composite can replace foams, timber, and plastics for applications like insulation, packaging, flooring, and other furnishings [24]. For example, the company Ecovative Design LLC (Green Island, NY, USA) produces a foam-like packaging material made of hemp hurd and fungal mycelia, which is fully compostable. Anticipating the possibility of an increased demand of eco-friendly packaging materials in the near future, we are interested in evaluating the feasibility of diverting this used packing material away from landfills or composting facilities towards higher value applications, such as feedstock for biofuels. It is known that fungal enzymes can reduce the recalcitrance of the biomass to deconstruction [25,26], likely through modification of polysaccharides and lignin in plant biomass. Therefore, we hypothesized that the mycelium composite material could be more easily deconstructed and converted into higher value fuels and chemicals than the raw hemp hurd.

In this study, hemp hurd and the mycelium-based packaging material were tested as biomass feedstocks for the production of the jet-fuel precursor bisabolene, using a one-pot ionic liquid technology and microbial conversion. First, we examined the deconstruction efficiency of the packaging material compared to hemp hurd, when subjugated to a one-pot ionic liquid pretreatment process. Second, the influence of the pretreatment process parameters on the sugar yields was investigated by using a Box–Behnken statistical design. Finally, the generated hydrolysates were fermented to evaluate the bioconversion of the depolymerized components by a bisabolene-producing *R. toruloides* strain.

2. Results and Discussion

2.1. Biomass Composition

The composition of the hemp hurd and packaging material was determined as shown in Table 1. The total extractives of the hemp hurd and packaging material comprised 8.3 and 14.7% of the biomass, respectively. The higher extractive content of the packaging material may be a result of the fungal (mycelium) growth stage in the packaging construction process. For the polysaccharide content, hemp hurd had higher glucan (30.3%) and xylan (13.5%) contents than the glucan (28.6%) and xylan (11.9%) content of the packaging material. Combining glucan and xylan content, the total fermentable sugars of the hemp hurd and packaging material was 43.7% and 40.4% of the hemp hurd biomass, respectively.

This indicates that a small fraction of the polysaccharides may have been consumed and converted into extractives during mycelial growth. However, both types of biomass contain a substantial amount of polymeric carbohydrates that can be depolymerized into simple sugars for fermentation. The lignin content for both materials was the same (22.4%); however, it is possible that the mycelial growth in the packaging material could have altered the structure of lignin and made the polysaccharides more accessible to hydrolysis [27–29]. We used the one-pot ionic liquid process on hemp hurd and package materials to test this hypothesis.

Table 1. Chemical composition of hemp hurd and packaging materials.

	Hemp Hurd (wt%)	Packaging Material (wt%)
Extractives	8.3 ± 3.2	14.7 ± 1.1
Glucan	30.3 ± 0.9	28.6 ± 0.1
Xylan	13.5 ± 0.5	11.9 ± 0.1
Klason lignin	22.4 ± 0.7	22.4 ± 3.0
Ash	0.6 ± 0.4	0.6 ± 0.3

2.2. Hydrolysate Generation Using a One-Pot Ionic Liquid Process

The raw hemp hurd and packaging material were deconstructed into fermentable sugars under the same reaction conditions using 10 wt% [Ch][Lys] and 20 wt% biomass loading. Fermentable sugar concentrations and yields are shown in Figure 1. We observed glucose and xylose concentrations from packaging materials of 43.0 ± 2.9 g/L and 19.3 ± 2.3 g/L, respectively, while hemp hurd released 35.2 ± 5.3 g/L and 16.2 ± 2.9 g/L. These represent glucose and xylose yields of 80.4 ± 5.4 and 87.1 ± 9.6% for the packaging material and 66.4 ± 6.8 and 68.3 ± 9.3% for hemp hurd, as shown in Figure 1B. These results indicate that the process used to generate the packaging material renders both cellulose and hemicellulose 10–20% more digestible than in raw hemp hurd, probably due to the decreased biomass recalcitrance caused by mycelium growth.

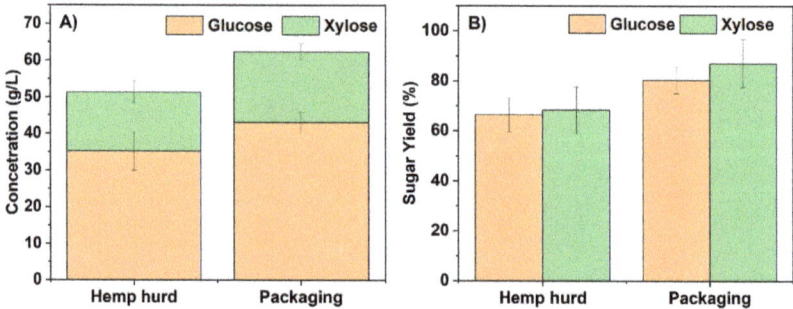

Figure 1. (**A**) glucose and xylose concentration and (**B**) glucose and xylose yield from enzymatic hydrolysis of [Ch][Lys]-pretreated hemp hurd and packaging materials.

2.3. Biocompatibility of Hydrolysates

One of the bottlenecks for the efficient conversion of lignocellulosic hydrolysates is the presence of compounds generated during the pretreatment and enzymatic hydrolysis stages that are toxic to biofuel-producing microbes [30,31]. The degree of toxicity mainly depends on the type of biomass, pretreatment conditions, and the identity of the microorganism that will be used for fermenting the depolymerized substrates. Therefore, we performed a biocompatibility test with the hydrolysates prepared from hemp hurd and packaging materials, using an engineered strain of the yeast *R. toruloides* known to be tolerant to ILs and biomass-derived compounds, and convert glucose and xylose to the jet fuel precursor bisabolene [32,33].

When the strain was inoculated directly in concentrated hydrolysates, negligible sugar consumption and very little growth was observed, as shown in Figure 2. Therefore, we prepared 50% diluted hydrolysates for further testing. Under these conditions, more than 90% of glucose and xylose conversion was observed in both hydrolysates, and the cells were able to grow and produce bisabolene (Figure 2). This result suggests that there is some degree of toxicity present in these hydrolysates. The utilization of hydrolysate with higher concentrations is beneficial for the economically feasible biorefinery development [34,35]. Therefore, other strategies such as hydrolysate culture adaptation or detoxification may be required to improve biocompatibility [36].

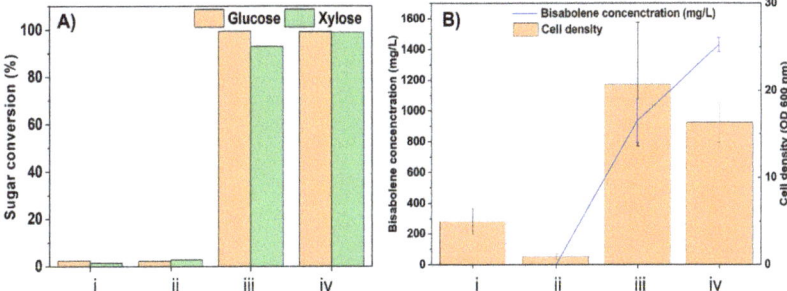

Figure 2. (**A**) sugar conversion and (**B**) cell density and bisabolene titer after fermentation of as-prepared hydrolysates from (i) hemp hurd and (ii) packaging material, and 50% diluted hydrolysates from (iii) hemp hurd and (iv) packaging material.

As shown in Figure 2B, the cell growth in 50% diluted hemp hurd hydrolysates (OD600 = 20) was higher than the growth observed in 50% diluted hydrolysates from packaging material (OD600 = 16). Nevertheless, the bisabolene titer produced in 50% diluted hydrolysates from the packaging was higher than 50% diluted hydrolysates from hemp hurd hydrolysates (1400 mg/L in 50% diluted hydrolysate from packaging vs. 600 mg/L in 50% diluted hydrolysate from hemp hurd). These results show that the hydrolysate with higher initial sugars resulted in higher bisabolene concentration, as expected, and the lower cell biomass in hydrolysates from the packaging material may be caused by growth inhibitors produced by mycelial growth during the packaging process.

2.4. Effect of Process Parameters on Sugar Yield and Optimization of Pretreatment Conditions

To optimize the pretreatment condition and investigate the effect of the process factors such as reaction time, ionic liquid loading, and reaction time, we employed a response surface methodology (RSM) (Table 2).

Table 2. Experimental variables and their code levels in the Box–Behnken design.

Variables	Factor Code	Level of Factor		
		−1	0	1
Temperature (°C)	X1	100	120	140
Time (h)	X2	1	2	3
Ionic liquid loading (%)	X3	5	7.5	10

Tables S1 and S2 show the glucose and xylose yield from hemp hurd and packaging materials obtained from Box–Behnken-designed experiments. The glucose yield from hemp hurd ranged from 28.2% to 81.6%, and the packaging material ranged from 51.2% to 75%. Xylose yields from hemp hurd and packaging materials varied in the range of 22.7–90.5% and 51.8–80%, respectively. The quadratic regression models for glucose yield and xylose yields from hemp hurd and packaging materials are shown as Equations (1)–(4) below:

$$Y_{glucose}^{Hemp\ hurd} = 0.68 + 0.14X_1 + 0.053X_2 - 0.073X_3 - 0.125X_1X_2 + 0.05X_1X_3 \\ -0.01X_2X_3 + 0.11X_1^2 - 0.14X_2^2 - 0.18X_3^2 \quad (1)$$

$$Y_{xylose}^{Hemp\ hurd} = 0.69 + 0.17X_1 + 0.036X_2 - 0.055X_3 - 0.099X_1X_2 + 0.043X_1X_3 \\ 0.025X_2X_3 + 0.14X_1^2 - 0.14X_2^2 - 0.20X_3^2 \quad (2)$$

$$Y_{glucose}^{packaging} = 0.74 + 0.051X_1 + 0.024X_2 - 0.01X_3 - 0.038X_1X_2 + 0.015X_1X_3 \\ -0.01X_2X_3 - 0.072X_1^2 - 0.052X_2^2 - 0.04X_3^2 \quad (3)$$

$$Y_{xylose}^{packaging} = 0.80 + 0.05X_1 + 0.028X_2 + 0.027X_3 - 0.035X_1X_2 - 0.021X_1X_3 \\ -0.03X_3 - 0.055X_1^2 - 0.086X_2^2 - 0.055X_3^2 \quad (4)$$

As shown in Figure 3 and summary of fit in Tables S3–S6, the models fitted well with experimental data with R2.

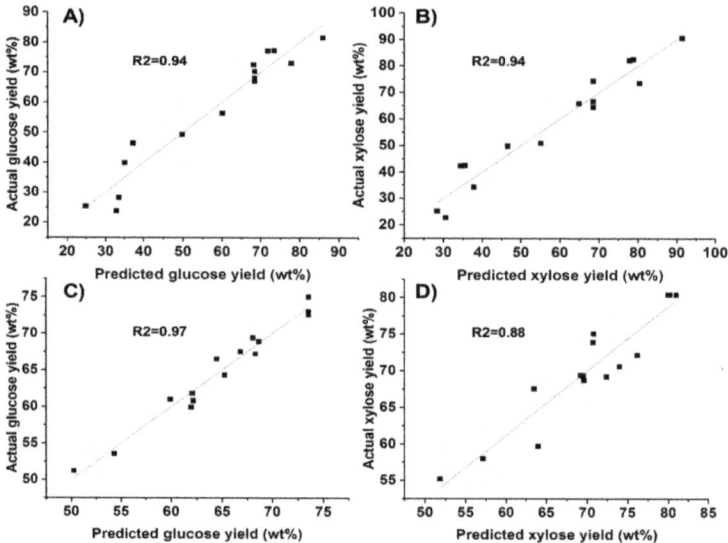

Figure 3. Actual yield versus predicted yield of glucose and xylose from (A,B) hemp hurd and (C,D) packaging material.

The optimum levels of parameters for glucose and xylose yields from packaging materials recommended by the model were: reaction temperature of 126 and 128 °C, reaction time of 2.1 and 2.0 h, and ionic liquid loading of 7.3% and 7.9%, corresponding to a predicted glucose and xylose yield of 74.6% and 81.7%. However, this optimal condition did not significantly improve the yields compared to the center point, even though the reaction conditions required a 4% higher temperature than the center point, a rather small difference in temperature. This result suggests that other process parameters such as agitation and biomass solid loading percentage should be tested for further improvement in the yield.

The model for hemp hurd found a saddle point instead of optimum levels, which means that the optimum process condition was not aligned within the current experimental conditions [37]. Further investigation into the different range of reaction conditions such as higher reaction temperature is required to optimize the reaction condition for hemp hurd. If operating with a limited budget and time, the reaction condition having the highest glucose and xylose yield can be chosen [37]. The highest glucose yield (81%) in the current reaction condition was obtained from hemp hurd at 140 °C, 1 h reaction time and 7.5% ionic liquid loading, which has higher severity in reaction condition than the optimized reaction condition of packaging materials. This result indicates that the reaction parameter affects the sugar yield differently according to the biomass type, implying that the biomass properties change by mycelium growth.

Regarding the packaging materials, the combined effects of reaction temperature, reaction time and ionic liquid loading on glucose yields are illustrated in Figure 4(A-I–A-III) and xylose yields in Figure 4(B-I–B-III). Response surface plots show that the glucose yield increased with the reaction temperature up to 133 °C with subsequent decrease in yield at a higher temperature. The xylose yields showed a similar trend. Additionally, the glucose and xylose yield increased with the reaction time up to 2 h and 7.5% ionic liquid loading. After those points, the glucose and xylose yield decreased, probably due to the loss of enzyme activity caused by the higher ionic liquid concentration [38]. Additionally, the longer reaction time and the higher ionic liquid concentration might facilitate the production of other compounds such as furan derivatives or organic acids, which inhibits the enzyme activity during the pretreatment [2,39]. Moreover, the production of other components probably led to a decrease in accessible carbohydrates to the enzyme [40]. Further tests may be necessary to improve the sugar yield. ANOVA results shown in Table S5 indicate that reaction temperature and reaction time has statistically significant effects on glucose yield ($p < 0.0008$ and 0.025, respectively), while ionic liquid loading was not significant ($p > 0.2134$). Additionally, the statistically significant interaction effects of reaction temperature with reaction time and ionic liquid loadings ($p < 0.0012$ and $p < 0.0153$) were confirmed. ANOVA results associated with xylose yield (Table S6) show that reaction temperature had a significant effect on the yield ($p < 0.0318$), while reaction time and ionic liquids had no effect ($p > 0.1702$ and 0.1768). Additionally, the interaction effect of reaction temperature with reaction time and ionic liquid loading was not significant, while the interaction effects of reaction time with ionic liquid were significant ($p < 0.0185$).

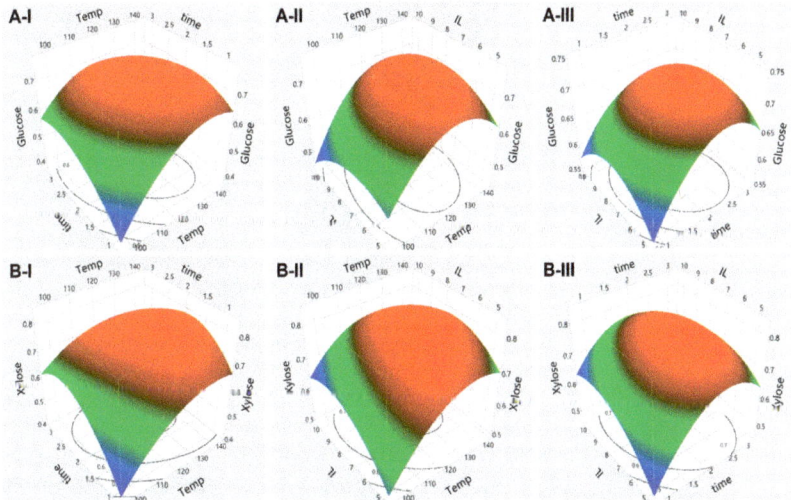

Figure 4. Surface response plot for the combined effect of (I) reaction time and temperature, (II) ionic liquid loading and temperature and (III) ionic liquid loading and reaction time on (**A**) glucose and (**B**) xylose yield from packaging material.

Process parameter effects on the glucose yield and xylose yield of hemp hurd examined the combined effects of reaction time with ionic liquid loading and the combined effect of ionic liquid with reaction time on glucose (Figure 5(A-I–A-III)) and xylose yield (Figure 5(B-I–B-III)). We can observe that the glucose and xylose yields increased with the increased temperature. However, the glucose and xylose yield increased with higher reaction time and ionic liquid loading up to 2 h and 7.5 wt%. After those points, the glucose and xylose yield decreased with the increased reaction time and ionic liquid loading, as observed in the hemp hurd test. ANOVA results shown in Tables S3 and S4 confirmed that reaction temperature had a significant effect on the glucose yield ($p < 0.0049$) and xylose yield ($p < 0.0024$). The interaction effect of reaction temperature with ionic liquid loading and reaction time with ionic liquid loading had a significant effect on glucose yield ($p < 0.0318$ and $p < 0.0199$), while the interaction effect of reaction temperature with reaction time had no statistically significant effect on glucose yield ($p > 0.0520$). Additionally, xylose yields were significantly affected by the combined effect of reaction temperature with reaction time and reaction time with ionic liquid loading ($p < 0.0278$ and $p < 0.0261$). Quadratic effects of ionic liquids on glucose yield and xylose yield were confirmed ($p < 0.0091$ and 0.0065).

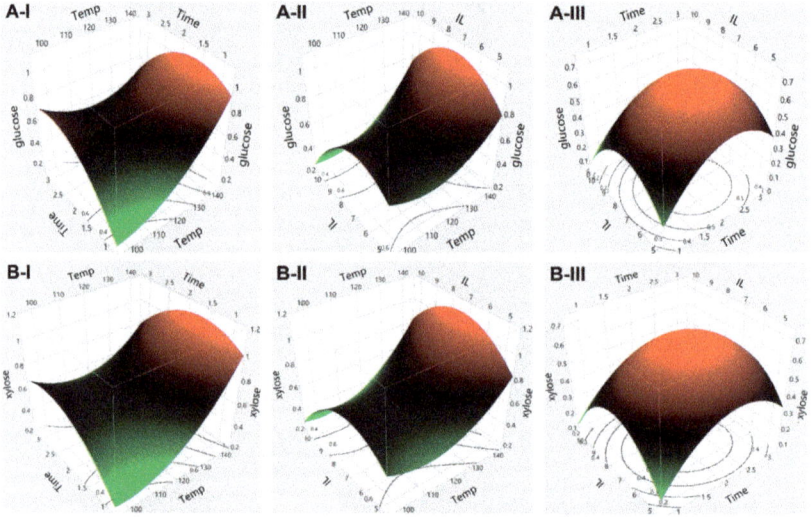

Figure 5. Surface response plot for the combined effect of (I) reaction time and temperature, (II) ionic liquid loading and temperature and (III) ionic liquid loading and reaction time on (**A**) glucose and (**B**) xylose yield from hemp hurd.

In summary, these results demonstrate that the process parameters have different effects on the fermentable sugar yield of hemp hurd compared to the packaging material and implies that mycelium growth affects the hemp hurd material properties. However, no significant improvement in the sugar yield was observed from packaging materials compared to hemp hurd. Even though packaging materials had higher sugar yield in less severe reaction conditions, the highest sugar yield was confirmed from hemp hurd in the harsher reaction condition. Therefore, economic evaluation combined with the evaluation of the reaction condition severity needs to be performed to determine which process parameter condition will be more beneficial. One important consideration is that the packaging materials were less dense than hemp hurd, possibly affecting the degree of mixing of the materials with ionic liquids during pretreatment. Therefore, further investigation of other process parameters such as biomass loading and agitation method is required [34,41,42].

3. Materials and Methods

3.1. Materials and Chemicals

Hemp hurd and packaging materials were donated by Ecovative Design LLC (Green Island, NY, USA) for evaluation. Both materials were sundried for 24 h and knife-milled with a 2 mm screen (Thomas-Wiley model 4, Swedesboro, NJ, USA). For biomass pretreatment, cholinium lysinate [Ch][Lys] was obtained from Proionic (Grambach, Styria). Commercial cellulase (Cellic CTec3) and hemicellulase (Cellic HTec3) were provided by Novozymes (Franklinton, NC, USA). Sulfuric acid (72% and ACS reagent, >99%), glucose (>99.5%), xylose (>99%), and arabinose (>98%) were purchased from Sigma-Aldrich (St. Louis, MO, USA).

3.2. Compositional Analysis

The biomass composition analysis of the hemp hurd and packaging materials was performed to determine glucan, xylan, lignin, and ash contents following the procedure described by NREL [43]. In summary, 0.3 g of biomass was soaked with 3 mL of 72% w/w H_2SO_4 at 30 °C for 1 h, followed by secondary hydrolysis at 121 °C for 1 h after adding 84 mL of DI water. After the two-step acidic hydrolysis, the mixture was filtered to separate glucan, xylan, and acid-soluble lignin from acid-insoluble lignin through filter crucibles. Acid-insoluble lignin was determined by subtracting the weight of residual solids dried in the oven at 105 °C and the weight of ash formed after burning at 575 °C. Acid-soluble lignin was determined by UV-VIS at 240 nm using a Nanodrop spectrophotometer (Thermo Scientific, Waltham, MA, USA). Monomeric sugars (glucose and xylose) were determined by HPLC using an Agilent 1200 series instrument (Agilent Technologies, Santa Clara, CA, USA) equipped with a refractive index detector and an HPX-87H column (Bio-Rad, Hercules, CA, USA). The instrument was operated at 0.6 mL/min flow rate using 4 mM H_2SO_4 as mobile phase and 60 °C column oven temperature. Extractives in the hemp hurd and packaging materials were removed through a water and ethanol extraction process using Dionex ASE 350 (Thermo Scientific, Waltham, MA, USA) following a procedure described by NREL [44]. The extractive content was determined by the dry weight differences before and after the extraction.

3.3. Biomass Pretreatment and Enzymatic Hydrolysis

For the comparison between hemp hurd and packaging materials, biomass pretreatment was carried out at a 20 wt% biomass loading in a one-pot (separation-free) configuration. An amount of 1 g of the biomass was mixed thoroughly with 5 g of solvent (10 wt% [Ch][Lys], and 90 wt% water) in a pressure tube (15 mL, Ace Glass Inc., Vineland, NJ, USA). The mixture was heated in an oil bath at 120 °C for 3 h. After pretreatment, the pH of biomass slurry was adjusted to 5 by adding 10 M HCl and 2 M NaOH in preparation for enzymatic hydrolysis. Subsequently, 20 mg of a commercial enzyme mixture (Cellic CTec3 and HTec3, 9:1 v/v) per g of biomass was added to the biomass slurry. Enzymatic hydrolysis was conducted at 50 °C for 72 h at 50 rpm in a rotary incubator. After hydrolysis, samples were collected and centrifuged at 4500 rpm for 10 min followed by the filtration of the supernatant with 0.45 μm centrifuge filters before sugar analysis by HPLC. All the pretreatment experiments were performed in triplicate and the standard deviation was estimated to represent errors.

3.4. Process Optimization

A Box–Behnken design (JMP Pro 14, SAS Institute, Inc., Cary, NC, USA) was used to explore the effect of each pretreatment process parameter on the saccharification yield and to optimize the pretreatment conditions. A preliminary test was conducted at a temperature of 120 °C, 10% ionic liquid loading and 3 h reaction time, which is the harshest pretreatment condition in our setup. To determine the severity of each process parameter on the yield and identify milder reaction conditions for potentially improving the economic viability of the process, three levels for the factors such as temperature (X_1), ionic liquid loading (X_2) and pretreatment time (X_3) were selected to 100–140 °C, 5–10%, and 1–3 h, respectively (Table 2). Low and high levels of each factor were coded as −1 and +1. A total of 14 experiments were designed and performed in duplicate. Three center points were used to determine the experimental error and reproducibility of the responses. Glucose yield and xylose yield were selected as the response. A secondary-degree polynomial Equation (5) was fitted to the experimental results as a function of the factors for optimizing pretreatment conditions and investigating the effect of parameters, as follows:

$$Y = \beta_0 + \beta_1 X_1 + \beta_2 X_2 + \beta_3 X_3 + \beta_{12} X_1 X_2 + \beta_{13} X_1 X_3 + \beta_{23} X_2 X_3 + \beta_{11} X_1^2 + \beta_{22} X_2^2 + \beta_{33} X_3^2 \quad (5)$$

where Y is the response, β_0 is the intercept, β_1, β_2, β_3 and β_{12}, β_{13}, β_{23} and β_{11}, β_{22}, β_{33} are linear, interaction and quadratic effect regression coefficients, respectively. The JMP statistical package was used to formulate the design and analyze the obtained data.

3.5. Yeast Strain and Cultivation Conditions

An engineered *Rhodosporidium toruloides* strain called GB2 that produces the sesquiterpene bisabolene was used in the bioconversion experiments. Details on strain construction and characterization have been previously reported [33], and the strain is deposited in the Agile BioFoundry public registry https://public-registry.agilebiofoundry.org (accessed on 29 January 2023) under the ID number ABFPUB_000319.

For microbial growth experiments, the pretreated and saccharified hydrolysates were pH-adjusted to 7.5 using 10 N NaOH, supplemented with ammonium sulfate to reach a final concentration of 5 g/L, and filtered through 0.45 μm surfactant-free cellulose-acetate membranes. A fraction of the hydrolysates was diluted 50% by adding water, before pH adjusting, nitrogen supplementation and filtering. For fermentations, cultures were started by adding 1 mL of a frozen glycerol stock to 49 mL of yeast peptone dextrose broth (YPD) in a 500 mL baffled flask and grown at 30 °C and 200 rpm for 24 h. 20 μL of cells in the grown cultures were combined with 780 μL of hydrolysate per reaction in 48-well FlowerPlates (m2p labs, Aachen, Germany). A dodecane overlay (200 μL per well) was added to capture bisabolene from the aqueous phase throughout the fermentation. The plates were covered with sterile AeraSeal films (Excel Scientific, Victorville, CA, USA) and incubated for 7 days in a humidity-controlled incubator with orbital shaking at 999 rpm. At the end of the fermentation, the entire contents of each well were collected in 1.5 mL tubes and centrifuged to separate the overlay, supernatant, and cell fractions. The cell pellets were resuspended in 800 μL of water, diluted with water, and 100 μL per sample were transferred to a Costar black 96-well plate with a flat, clear bottom (Corning, Glendale, AZ, USA) to measure optical density at 600 nm with a SpectraMax Plus 384 reader (Molecular Devices, San Jose, CA, USA). Substrates in the supernatant were analyzed by HPLC with the same method described in Section 3.3. Bisabolene was quantified by GC-MS using previously published methods [39]. The bisabolene concentrations reported here represent the concentrations that would be present in the aqueous phase of the cultivations.

4. Conclusions

This work demonstrates the feasibility of hemp hurd and packaging materials made of mycelium grown on hemp hurd to be used as feedstocks for bioconversion to a jet-fuel precursor using a one-pot ionic liquid technology. During the initial test (120 °C, 7.5 wt% ionic liquid loading and 2 h reaction time), the packaging materials produced higher sugar concentrations (43 g/L of glucose and 19.3 g/L of xylose) and yield (80.4% for glucose and 87.1% for xylose) than the hemp hurd (35.2 g/L of glucose and 16.2 g/L of xylose and 66.4% and 68.3% glucose and xylose yield, respectively). However, the Box–Behnken experimental design showed that the reaction conditions for the maximum sugar yields from each material was different and that the significance of the process parameter effect on the fermentable sugar yield was dependent on the biomass properties, suggesting that the mycelial growth affected the deconstructability of the hemp hurd. Furthermore, the fermentation test to convert fermentable sugar into bisabolene showed that hydrolysates from the packaging material resulted in a higher bisabolene titer (1400 mg/L) than hydrolysates from the hemp hurd, probably due to the higher sugar concentrations generated form the packaging material.

To fully take advantage of these packaging materials to produce biofuels after they are used and discarded, a more detailed correlation study between the fermentable sugar yield and physicochemical properties of biomass and packaging materials or packaging process parameters is required by testing different hemp material sources. In addition, methods to overcome hydrolysate toxicity will need to be employed to enable utilization of concentrated hydrolysate for increased product titers and a reduction in water consumption. Finally, further investigation into other process parameters such as agitation and biomass loadings are merited to fully optimize the pretreatment conditions, as well as performing pilot scale tests to generate data that can help assess the economic feasibility of this new conceptual process. Overall, this study indicates that it is possible to produce lignocellulosic supply chains for production of biofuels and biochemicals that include both raw biomass and biomass that has been first processed and valorized as commercial products, such as packaging materials, enabling the carbon in these lignocellulosic products to generate value multiple times in their life cycle.

Supplementary Materials: The following supporting information can be downloaded at: https://www.mdpi.com/article/10.3390/molecules28031427/s1. Table S1: Glucose yields from hydrolysis of one-pot pretreated hemp hurd and packaging material under different experimental conditions. Table S2: Xylose yields from hydrolysis of one-pot pretreated hemp hurd and packaging material under different experimental conditions. Table S3: ANOVA, summary of fit and significance of regression coefficients for glucose yield model of hemp hurd. Table S4: ANOVA, summary of fit and significance of regression coefficients for xylose yield model of hemp hurd. Table S5: ANOVA, summary of fit and significance of regression coefficients for glucose yield model of packaging material. Table S6: ANOVA, summary of fit and significance of regression coefficients for xylose yield model of packaging material.

Author Contributions: Conceptualization, J.C. and J.M.G.; methodology, J.C., A.R. and J.M.G.; formal analysis, J.C.; writing—original draft preparation, J.C. and A.R.; writing—review and editing, J.C., A.R., B.A.S. and J.M.G.; supervision, B.A.S. and J.M.G. All authors have read and agreed to the published version of the manuscript.

Funding: This research was funded by the U.S. Department of Energy, Office of Science, Office of Biological and Environmental Research, through contract DE-AC02-05CH11231 between Lawrence Berkeley National Laboratory and the U.S. Department of Energy.

Institutional Review Board Statement: Not applicable.

Informed Consent Statement: Not applicable.

Data Availability Statement: All data are contained within the article and Supplementary Materials.

Acknowledgments: The authors thank Ecovative Design LLC (Green Island, NY, USA, https://www.ecovative.com/ (accessed on 29 January 2023)) for providing samples for this study. This work was part of the DOE Joint BioEnergy Institute (http://www.jbei.org (accessed on 29 January 2023)) supported by the U.S. Department of Energy, Office of Science, Office of Biological and Environmental Research, through contract DE-AC02-05CH11231 between Lawrence Berkeley National Laboratory and the U.S. Department of Energy. The United States Government retains and the publisher, by accepting the article for publication, acknowledges that the United States Government retains a non-exclusive, paid-up, irrevocable, worldwide license to publish or reproduce the published form of this manuscript, or allow others to do so, for United States Government purposes. The Department of Energy will provide public access to these results of federally sponsored research in accordance with the DOE Public Access Plan. Sandia National Laboratories is a multimission laboratory managed and operated by National Technology & Engineering Solutions of Sandia, LLC, a wholly owned subsidiary of Honeywell International Inc., for the U.S. Department of Energy's National Nuclear Security Administration under contract DE-NA0003525. This paper describes objective technical results and analysis. Any subjective views or opinions that might be expressed in the paper do not necessarily represent the views of the U.S. Department of Energy or the United States Government.

Conflicts of Interest: The authors declare no conflict of interest.

References

1. Choi, J.; Won, W.; Capareda, S.C. The economical production of functionalized Ashe juniper derived-biochar with high hazardous dye removal efficiency. *Ind. Crops Prod.* **2019**, *137*, 672–680. [CrossRef]
2. Galbe, M.; Wallberg, O. Pretreatment for biorefineries: A review of common methods for efficient utilisation of lignocellulosic materials. *Biotechnol. Biofuels* **2019**, *12*, 294. [CrossRef]
3. Abu-Omar, M.M.; Barta, K.; Beckham, G.T.; Luterbacher, J.S.; Ralph, J.; Rinaldi, R.; Román-Leshkov, Y.; Samec, J.S.M.; Sels, B.F.; Wang, F. Guidelines for performing lignin-first biorefining. *Energy Environ. Sci.* **2021**, *14*, 262–292. [CrossRef]
4. Himmel, M.E.; Ding, S.-Y.; Johnson, D.K.; Adney, W.S.; Nimlos, M.R.; Brady, J.W.; Foust, T.D. Biomass recalcitrance: Engineering plants and enzymes for biofuels production. *Science* **2007**, *315*, 804–807. [CrossRef]
5. Wyman, C.E.; Dale, B.E.; Elander, R.T.; Holtzapple, M.; Ladisch, M.R.; Lee, Y.Y. Comparative sugar recovery data from laboratory scale application of leading pretreatment technologies to corn stover. *Bioresour. Technol.* **2005**, *96*, 2026–2032. [CrossRef]
6. Nitsos, C.; Lazaridis, P.; Mach-Aigner, A.; Matis, K.; Triantafyllidis, K. Increasing the efficiency of lignocellulosic biomass enzymatic hydrolysis: Hydrothermal pretreatment, extraction of surface lignin, wet milling and production of cellulolytic enzymes. *ChemSusChem* **2019**, *12*, 1179–1195. [CrossRef]
7. Kumari, D.; Singh, R. Pretreatment of lignocellulosic wastes for biofuel production: A critical review. *Renew. Sustain. Energy Rev.* **2018**, *90*, 877–891. [CrossRef]
8. Li, C.; Knierim, B.; Manisseri, C.; Arora, R.; Scheller, H.V.; Auer, M.; Vogel, K.P.; Simmons, B.A.; Singh, S. Comparison of dilute acid and ionic liquid pretreatment of switchgrass: Biomass recalcitrance, delignification and enzymatic saccharification. *Bioresour. Technol.* **2010**, *101*, 4900–4906. [CrossRef]
9. Mosier, N.; Wyman, C.; Dale, B.; Elander, R.; Lee, Y.Y.; Holtzapple, M.; Ladisch, M. Features of promising technologies for pretreatment of lignocellulosic biomass. *Bioresour. Technol.* **2005**, *96*, 673–686. [CrossRef] [PubMed]
10. Liggenstoffer, A.S.; Youssef, N.H.; Wilkins, M.R.; Elshahed, M.S. Evaluating the utility of hydrothermolysis pretreatment approaches in enhancing lignocellulosic biomass degradation by the anaerobic fungus Orpinomyces sp. strain C1A. *J. Microbiol. Methods* **2014**, *104*, 43–48. [CrossRef]
11. Liu, K.; Atiyeh, H.K.; Pardo-Planas, O.; Ezeji, T.C.; Ujor, V.; Overton, J.C.; Berning, K.; Wilkins, M.R.; Tanner, R.S. Butanol production from hydrothermolysis-pretreated switchgrass: Quantification of inhibitors and detoxification of hydrolyzate. *Bioresour. Technol.* **2015**, *189*, 292–301. [CrossRef]
12. Hastrup, A.C.S.; Howell, C.; Larsen, F.H.; Sathitsuksanoh, N.; Goodell, B.; Jellison, J. Differences in crystalline cellulose modification due to degradation by brown and white rot fungi. *Fungal Biol.* **2012**, *116*, 1052–1063. [CrossRef] [PubMed]
13. Sun, N.; Parthasarathi, R.; Socha, A.M.; Shi, J.; Zhang, S.; Stavila, V.; Sale, K.L.; Simmons, B.A.; Singh, S. Understanding pretreatment efficacy of four cholinium and imidazolium ionic liquids by chemistry and computation. *Green Chem.* **2014**, *16*, 2546–2557. [CrossRef]
14. Das, L.; Achinivu, E.C.; Barcelos, C.A.; Sundstrom, E.; Amer, B.; Baidoo, E.E.K.; Simmons, B.A.; Sun, N.; Gladden, J.M. Deconstruction of woody biomass via protic and aprotic ionic liquid pretreatment for ethanol production. *ACS Sustain. Chem. Eng.* **2021**, *9*, 4422–4432. [CrossRef]
15. Xu, F.; Sun, J.; Konda, N.V.S.N.M.; Shi, J.; Dutta, T.; Scown, C.D.; Simmons, B.A.; Singh, S. Transforming biomass conversion with ionic liquids: Process intensification and the development of a high-gravity, one-pot process for the production of cellulosic ethanol. *Energy Environ. Sci.* **2016**, *9*, 1042–1049. [CrossRef]
16. Shi, J.; Gladden, J.M.; Sathitsuksanoh, N.; Kambam, P.; Sandoval, L.; Mitra, D.; Zhang, S.; George, A.; Singer, S.W.; Simmons, B.A.; et al. One-pot ionic liquid pretreatment and saccharification of switchgrass. *Green Chem.* **2013**, *15*, 2579. [CrossRef]
17. Das, L.; Geiselman, G.M.; Rodriguez, A.; Magurudeniya, H.D.; Kirby, J.; Simmons, B.A.; Gladden, J.M. Seawater-based one-pot ionic liquid pretreatment of sorghum for jet fuel production. *Bioresour. Technol. Rep.* **2020**, *13*, 100622. [CrossRef]
18. Rigual, V.; Papa, G.; Rodriguez, A.; Wehrs, M.; Kim, K.H.; Oliet, M.; Alonso, M.V.; Gladden, J.M.; Mukhopadhyay, A.; Simmons, B.A.; et al. Evaluating protic ionic liquid for woody biomass one-pot pretreatment + saccharification, followed by *Rhodosporidium toruloides* cultivation. *ACS Sustain. Chem. Eng.* **2019**, *8*, 782–791. [CrossRef]
19. Das, L.; Li, W.; Dodge, L.A.; Stevens, J.C.; Williams, D.W.; Hu, H.; Li, C.; Ray, A.E.; Shi, J. Comparative evaluation of industrial hemp cultivars: Agronomical practices, feedstock characterization, and potential for biofuels and bioproducts. *ACS Sustain. Chem. Eng.* **2020**, *8*, 6200–6210. [CrossRef]
20. Cranshaw, W.; Schreiner, M.; Britt, K.; Kuhar, T.P.; McPartland, J.; Grant, J. Developing insect pest management systems for hemp in the United States: A work in progress. *J. Integr. Pest Manag.* **2019**, *10*, 26. [CrossRef]
21. Shahzad, A. Impact and fatigue properties of hemp–glass fiber hybrid biocomposites. *J. Reinf. Plast. Compos.* **2011**, *30*, 1389–1398. [CrossRef]
22. González-García, S.; Hospido, A.; Feijoo, G.; Moreira, M.T. Life cycle assessment of raw materials for non-wood pulp mills: Hemp and flax. *Resour. Conserv. Recycl.* **2010**, *54*, 923–930. [CrossRef]
23. Murugu Nachippan, N.; Alphonse, M.; Bupesh Raja, V.K.; Shasidhar, S.; Varun Teja, G.; Harinath Reddy, R. Experimental investigation of hemp fiber hybrid composite material for automotive application. *Mater. Today Proc.* **2021**, *44*, 3666–3672. [CrossRef]

24. Jones, M.; Mautner, A.; Luenco, S.; Bismarck, A.; John, S. Engineered mycelium composite construction materials from fungal biorefineries: A critical review. *Mater. Des.* **2020**, *187*, 108397. [CrossRef]
25. Vasco-Correa, J.; Luo, X.; Li, Y.; Shah, A. Comparative study of changes in composition and structure during sequential fungal pretreatment of non-sterile lignocellulosic feedstocks. *Ind. Crops Prod.* **2019**, *133*, 383–394. [CrossRef]
26. Salvachúa, D.; Katahira, R.; Cleveland, N.S.; Khanna, P.; Resch, M.G.; Black, B.A.; Purvine, S.O.; Zink, E.M.; Prieto, A.; Martínez, M.J.; et al. Lignin depolymerization by fungal secretomes and a microbial sink. *Green Chem.* **2016**, *18*, 6046–6062. [CrossRef]
27. Sun, Z.; Fridrich, B.; de Santi, A.; Elangovan, S.; Barta, K. Bright side of lignin depolymerization: Toward new platform chemicals. *Chem. Rev.* **2018**, *118*, 614–678. [CrossRef] [PubMed]
28. Eudes, A.; George, A.; Mukerjee, P.; Kim, J.S.; Pollet, B.; Benke, P.I.; Yang, F.; Mitra, P.; Sun, L.; Cetinkol, O.P.; et al. Biosynthesis and incorporation of side-chain-truncated lignin monomers to reduce lignin polymerization and enhance saccharification. *Plant Biotechnol. J.* **2012**, *10*, 609–620. [CrossRef]
29. Wu, W.; Dutta, T.; Varman, A.M.; Eudes, A.; Manalansan, B.; Loqué, D.; Singh, S. Lignin valorization: Two hybrid biochemical routes for the conversion of polymeric lignin into value-added chemicals. *Sci. Rep.* **2017**, *7*, 8420. [CrossRef]
30. Kim, D. Physico-Chemical Conversion of Lignocellulose: Inhibitor Effects and Detoxification Strategies: A Mini Review. *Molecules* **2018**, *23*, 309. [CrossRef]
31. Frederix, M.; Mingardon, F.; Hu, M.; Sun, N.; Pray, T.; Singh, S.; Simmons, B.A.; Keasling, J.D.; Mukhopadhyay, A. Development of an E. coli strain for one-pot biofuel production from ionic liquid pretreated cellulose and switchgrass. *Green Chem.* **2016**, *18*, 4189–4197. [CrossRef]
32. Yaegashi, J.; Kirby, J.; Ito, M.; Sun, J.; Dutta, T.; Mirsiaghi, M.; Sundstrom, E.R.; Rodriguez, A.; Baidoo, E.; Tanjore, D.; et al. *Rhodosporidium toruloides*: A new platform organism for conversion of lignocellulose into terpene biofuels and bioproducts. *Biotechnol. Biofuels* **2017**, *10*, 241. [CrossRef] [PubMed]
33. Kirby, J.; Geiselman, G.M.; Yaegashi, J.; Kim, J.; Zhuang, X.; Tran-Gyamfi, M.B.; Prahl, J.-P.; Sundstrom, E.R.; Gao, Y.; Munoz, N.; et al. Further engineering of R. toruloides for the production of terpenes from lignocellulosic biomass. *Biotechnol. Biofuels* **2021**, *14*, 101. [CrossRef] [PubMed]
34. Olivieri, G.; Wijffels, R.H.; Marzocchella, A.; Russo, M.E. Bioreactor and bioprocess design issues in enzymatic hydrolysis of lignocellulosic biomass. *Catalysts* **2021**, *11*, 680. [CrossRef]
35. Li, C.; Tanjore, D.; He, W.; Wong, J.; Gardner, J.L.; Sale, K.L.; Simmons, B.A.; Singh, S. Scale-up and evaluation of high solid ionic liquid pretreatment and enzymatic hydrolysis of switchgrass. *Biotechnol. Biofuels* **2013**, *6*, 154. [CrossRef]
36. Mohamed, E.T.; Wang, S.; Lennen, R.M.; Herrgård, M.J.; Simmons, B.A.; Singer, S.W.; Feist, A.M. Generation of a platform strain for ionic liquid tolerance using adaptive laboratory evolution. *Microb. Cell Fact.* **2017**, *16*, 204. [CrossRef]
37. Chan, K.S.; Greaves, S.J.; Rahardja, S. Techniques for addressing saddle points in the response surface methodology (RSM). *IEEE Access* **2019**, *7*, 85613–85621. [CrossRef]
38. He, Y.-C.; Liu, F.; Gong, L.; Di, J.-H.; Ding, Y.; Ma, C.-L.; Zhang, D.-P.; Tao, Z.-C.; Wang, C.; Yang, B. Enzymatic in situ saccharification of chestnut shell with high ionic liquid-tolerant cellulases from *Galactomyces* sp. CCZU11-1 in a biocompatible ionic liquid-cellulase media. *Bioresour. Technol.* **2016**, *201*, 133–139. [CrossRef]
39. Magurudeniya, H.D.; Baral, N.R.; Rodriguez, A.; Scown, C.D.; Dahlberg, J.; Putnam, D.; George, A.; Simmons, B.A.; Gladden, J.M. Use of ensiled biomass sorghum increases ionic liquid pretreatment efficiency and reduces biofuel production cost and carbon footprint. *Green Chem.* **2021**, *23*, 3127–3140. [CrossRef]
40. Fu, D.; Mazza, G. Optimization of processing conditions for the pretreatment of wheat straw using aqueous ionic liquid. *Bioresour. Technol.* **2011**, *102*, 8003–8010. [CrossRef]
41. Rodriguez, A.; Salvachúa, D.; Katahira, R.; Black, B.A.; Cleveland, N.S.; Reed, M.; Smith, H.; Baidoo, E.E.K.; Keasling, J.D.; Simmons, B.A.; et al. Base-Catalyzed Depolymerization of Solid Lignin-Rich Streams Enables Microbial Conversion. *ACS Sustain. Chem. Eng.* **2017**, *5*, 8171–8180. [CrossRef]
42. Benz, G.T. *Agitator Design Technology for Biofuels and Renewable Chemicals*; Wiley: Hoboken, NJ, USA, 2022; ISBN 9781119815495.
43. NREL. *NREL: Preparation of Samples for Compositional Analysis. Laboratory Analytical Procedure*; NREL: Golden, CO, USA, 2008.
44. NREL. *NREL: Determination of Extractives in Biomass: Laboratory Analytical Procedure*; NREL: Golden, CO, USA, 2005.

Disclaimer/Publisher's Note: The statements, opinions and data contained in all publications are solely those of the individual author(s) and contributor(s) and not of MDPI and/or the editor(s). MDPI and/or the editor(s) disclaim responsibility for any injury to people or property resulting from any ideas, methods, instructions or products referred to in the content.

Article

Extraction of Gold Based on Ionic Liquid Immobilized in UiO-66: An Efficient and Reusable Way to Avoid IL Loss Caused by Ion Exchange in Solvent Extraction

Xinyu Cui, Yani Wang, Yanfeng Wang, Pingping Zhang and Wenjuan Lu *

Institute of Materia Medica, Shandong First Medical University & Shandong Academy of Medical Sciences, Jinan 250062, China
* Correspondence: luwenjuan@sdfmu.edu.cn

Abstract: Ionic liquids (ILs) have received considerable attention as a promising green solvent for extracting metal ions from aqueous solutions. However, the recycling of ILs remains difficult and challenging because of the leaching of ILs, which is caused by the ion exchange extraction mechanism and hydrolysis of ILs in acidic aqueous conditions. In this study, a series of imidazolium-based ILs were confined in a metal–organic framework (MOF) material (UiO-66) to overcome the limitations when used in solvent extraction. The effect of the various anions and cations of the ILs on the adsorption ability of $AuCl_4^-$ was studied, and 1-hexyl-3-methylimidazole tetrafluoroborate ($[HMIm]^+[BF_4]^-$@UiO-66) was used for the construction of a stable composite. The adsorption properties and mechanism of $[HMIm]^+[BF_4]^-$@UiO-66 for Au(III) adsorption were also studied. The concentrations of tetrafluoroborate ($[BF_4]^-$) in the aqueous phase after Au(III) adsorption by $[HMIm]^+[BF_4]^-$@UiO-66 and liquid–liquid extraction by $[HMIm]^+[BF_4]^-$ IL were 0.122 mg/L and 18040 mg/L, respectively. The results reveal that Au(III) coordinated with the N-containing functional groups, while $[BF_4]^-$ was effectively confined in UiO-66, instead of undergoing anion exchange in liquid–liquid extraction. Electrostatic interactions and the reduction of Au(III) to Au(0) were also important factors determining the adsorption ability of Au(III). $[HMIm]^+[BF_4]^-$@UiO-66 could be easily regenerated and reused for three cycles without any significant drop in the adsorption capacity.

Keywords: ionic liquids; UiO-66; gold recovery; regenerability; selective adsorption

Citation: Cui, X.; Wang, Y.; Wang, Y.; Zhang, P.; Lu, W. Extraction of Gold Based on Ionic Liquid Immobilized in UiO-66: An Efficient and Reusable Way to Avoid IL Loss Caused by Ion Exchange in Solvent Extraction. *Molecules* **2023**, *28*, 2165. https://doi.org/10.3390/molecules28052165

Academic Editors: Slavica Ražić, Aleksandra Cvetanović Kljakić and Enrico Bodo

Received: 31 January 2023
Revised: 20 February 2023
Accepted: 22 February 2023
Published: 25 February 2023

Copyright: © 2023 by the authors. Licensee MDPI, Basel, Switzerland. This article is an open access article distributed under the terms and conditions of the Creative Commons Attribution (CC BY) license (https:// creativecommons.org/licenses/by/ 4.0/).

1. Introduction

Ionic liquids (ILs) represent a class of environmentally friendly "green" solvents with unusual physical and chemical properties, such as a low vapor pressure and the absence of volatilization, and they can be used for the extraction of metal ions from aqueous solutions [1–9]. The most commonly proposed mechanism for the extraction of metal ions into the hydrophobic ionic liquid phase is ion exchange [2–9]. The simultaneous release of IL ions to the aqueous phase (resulting in IL losses and aqueous phase pollution) when metal ions are extracted into the IL phase has been reported in many cases. Moreover, the chemical stability of ILs has already aroused attention. The hydrolysis of fluorine-based anions of ILs has been reported. Fernandes and co-workers found that the hydrolysis of $[BF_4]^-$ even occurred at room temperature [10]. The hydrolysis of fluorine-based anions generates an abundance of toxic hydrofluoric acid (HF), which results in environmental pollution.

Supported ionic liquids (SILs) are a new type of solid material prepared through physical adsorption, chemical bonding, or loading of ILs onto porous supports, and they have the characteristics of both ILs and carriers. The process can significantly improve the utilization of ILs, solve the problems of the high viscosity, separation, and mass transfer of ILs, and expand the application field of ILs. Various porous materials have been used as supports, such as covalent organic frameworks [11], porous celluloses [12,13], molecular sieves [14], magnetic materials [15], resins [16], porous silica [17], and biopolymers [18]. SILs have been widely used in various applications, such as catalysis [19], adsorption separation [12,13,15–18], gas storage [14,20], and electrode materials [11]. However, the ion exchange mechanism is responsible for the adsorption of metal ions in SILs in many cases [13,16–18].

Immobilizing ILs in metal–organic frameworks (symbolized as IL@MOFs) is favorable for the adsorption of metal ions from the aqueous phase because of the strong bond between ILs and MOFs, as reported in [19,21,22]. UiO-66 is a porous crystalline material with Zr^{2+} as the excess metal ion and terephthalic acid as the organic ligand. The Zr_6 cluster ([$Zr_6O_4(OH)_4$]) has 12 terephthalic acids (H_2BDC), the maximum number of organic ligands and metal coordination clusters in MOFs. Compared with other MOFs, the dense structural unit and Zr-O property make UiO-66 the most stable porous material in aqueous solutions and other common solvents [23–27]. It possesses high thermal and mechanical stability, as well as good resistance to strong acids [28,29]. Moreover, UiO-66 contains tetrahedral and octahedral pore cages, and this 3D pore structure facilitates the dispersion of ILs [30]. The [$Zr_6O_4(OH)_4$] metal ion clusters of UiO-66 are easily approached by the spherical anions of ILs, such as [BF_4]$^-$ and hexafluorophosphate ([PF_6]$^-$) [31]. Therefore, UiO-66, with a stable and well-defined structure, becomes a promising host material for immobilizing IL anions.

Gold is valuable, and the recovery of Au(III) from waste CPUs has become attractive due to the limited availability of gold. Gold can be extracted from electronic waste technology via pyrometallurgy and hydrometallurgy [32]. The selectivity of pyrometallurgy is poor, the process is complex, the cost is high, and the generation of volatile metals and dust causes environmental pollution. Therefore, hydrometallurgy is the main method used to leach metals from e-waste [33]. Currently, commonly used methods for gold recovery from e-waste leaching solutions include solvent extraction [4,34], chemical precipitation [35], membrane separation [36], and adsorption [37–39]. In contrast, adsorption has the advantages of mild conditions, simple operation, low cost, and recyclability. Adsorption is the most economical and practical method for gold recovery, with great potential. Additionally, in solvent extraction, it is worth noting that ILs themselves extract [$AuCl_4$]$^-$ through anion exchange. Based on the above, a study of gold recovery from waste CPUs using composite materials composed of ILs encapsulated in UiO-66 (ILs@UiO-66) was conducted. The composites combine the excellent physicochemical properties of ILs and the advantages of the high surface area and high porosity of UiO-66. However, the investigations on ILs@UiO-66 composites available thus far are mainly focused on examining the effect of the loading of ILs on their extraction performance, while the effects of stronger interactions of IL anions with UiO-66 are rarely explored. Therefore, in the current study, an effort was made to address this issue. We expect that the knowledge gained will not only contribute to a better understanding of the properties of ILs@UiO-66 composites for gold recovery, but will also be beneficial in providing guidance on how to overcome the limitations of ILs used in acidic conditions.

2. Results
2.1. IL Dependence of Adsorption Ratio

In this study, UiO-66, originally synthesized by Lillerud and co-workers [40], was used as a support for ILs. Previous studies have shown that UiO-66 is one of the most stable MOFs. It possesses high thermal and mechanical stability, as well as good resistance to strong acids [28]. The unique robustness of UiO-66 derives from the 12-coordinated Zr_6 clusters that constitute the framework. It is built up from [$Zr_6O_4(OH)_4$] units interlinked via terephthalate linkers, leading to a 3D porous framework with two types of cages, namely, tetrahedral and octahedral cages, with diameters of 8 and 11 Å, respectively [22].

In order to select ionic liquids with an appropriate structure, the influence of the anions and cations of ILs on the Au(III) adsorption ability of ILs@UiO-66 was examined at first. Several ILs were transported into UiO-66, and their Au(III) adsorption ability was investigated, including [HMIm]$^+$-based ILs with different anions (tetrafluoroborate [BF_4]$^-$, hexafluorophosphate [PF_6]$^-$, acetate [OAc]$^-$, and bistrifluoromethane sulfonimide [NTf_2]$^-$) and

[BF$_4$]$^-$-based ILs with different cations (1-ethyl-3-methylimidazolium [EMIm]$^+$, 1-butyl-3-methylimidazole [BMIm]$^+$, [HMIm]$^+$, and 1-octyl-3-methylimidazolium [OMIm]$^+$). The results are shown in Figure 1a, b.

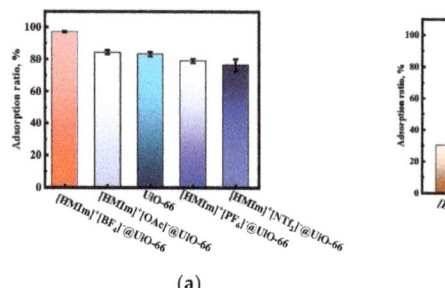
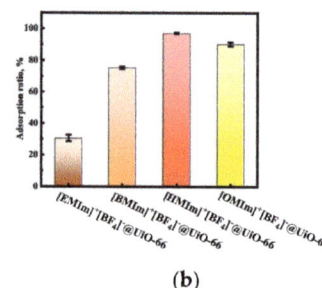
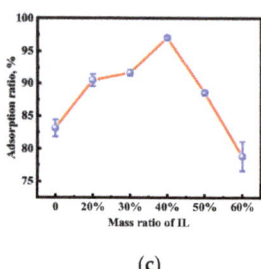

(a) (b) (c)

Figure 1. (a) IL cation dependence of adsorption ratio; (b) IL anion dependence of adsorption ratio (c $_{Au(III)}$ = 60 mg/L, V $_{Au(III)}$ = 10 mL, m $_{[HMIm]+[BF4]-@UiO-66}$ = 10 mg, t = 6 h, T = 35 °C, pH = 2); (c) mass ratios of [HMIm]$^+$[BF$_4$]$^-$ (c $_{Au(III)}$ = 60 mg/L, V $_{Au(III)}$ = 10 mL, m $_{[HMIm]+[BF4]-@UiO-66}$ = 10 mg, t = 6 h, T = 35 °C, pH = 2).

According to the reported studies, the composition of ILs is an important factor that affects the extractability of AuCl$_4$$^-$ when ILs are used in solvent extraction or immobilized on solid supports. The extractability of AuCl$_4$$^-$ was found to be greater for ILs composed of more hydrophilic anions and more hydrophobic cations [13,41]. In this work, with the [HMIm]$^+$ cation, the Au(III) adsorption ability of ILs/UiO-66 was in the order of [BF$_4$]$^-$ > [OAc]$^-$ > [PF$_6$]$^-$ > [NTf$_2$]$^-$, while with the [BF$_4$]$^-$ anion, it was in the order of [HMIm]$^+$ > [OMIm]$^+$ > [BMIm]$^+$ > [EMIm]$^+$. These results are roughly related to the hydrophilicity of the anions and hydrophobicity of the cations. The hydrophilicity of the used anions in our work decreased in the order of [OAc]$^-$ > [BF$_4$]$^-$ > [PF$_6$]$^-$ > [NTf$_2$]$^-$, and the hydrophobicity of the cations decreased in the order of [OMIm]$^+$ > [HMIm]$^+$ > [BMIm]$^+$ > [EMIm]$^+$. Therefore, ILs composed of more hydrophilic anions and hydrophobic cations are more conducive to adsorbing Au(III), except for [OAc]$^-$, [BF$_4$]$^-$, [OMIm]$^+$, and [HMIm]$^+$.

For [OAc]$^-$ and [BF$_4$]$^-$, the quasi–spherical structure of the [BF$_4$]$^-$ anion allows it to fit more closely in the Zr sites of UiO-66 than [OAc]$^-$, which has a chain-like structure [31]. As a result of this, the Au(III) adsorption ability showed the order of [HMIm]$^+$[BF$_4$]$^-$@UiO-66 > [HMIm]$^+$[OAc]$^-$@UiO-66. Hence, the effect of anions on the adsorption ratio of Au(III) resulted from the hydrophilic effect in combination with the anionic structure.

For [OMIm]$^+$ and [HMIm]$^+$, considering the very large [OMIm]$^+$ cation transported into UiO-66 cavities, it is possible to block the pores of UiO-66, resulting in difficult Au(III) adsorption [42]. Thus, the effect of cations on the adsorption ratio of Au(III) was caused by the hydrophobic effect in combination with the cationic size. Since [HMIm]$^+$[BF$_4$]$^-$@UiO-66 showed the best adsorption result, the rest of the study was carried out with this adsorbent.

Subsequently, the influence of [HMIm]$^+$[BF$_4$]$^-$ IL loading on the adsorption ratio of Au(III) was investigated (by conducting experiments on the synthesis of [HMIm]$^+$[BF$_4$]$^-$@UiO-66 with different mass ratios of [HMIm]$^+$[BF$_4$]$^-$ to UiO-66). UiO-66 was loaded with [HMIm]$^+$[BF$_4$]$^-$ from [HMIm]$^+$[BF$_4$]$^-$-ethanol solutions. As reported, the IL mass in the solution is essentially proportional to the IL loading in the MOF pores [43]. Thus, the investigations of the effect of [HMIm]$^+$[BF$_4$]$^-$ loading were conducted by varying the mass ratio of [HMIm]$^+$[BF$_4$]$^-$ to UiO-66 in ethanol solutions. As shown in Figure 1c, with an increase in the mass ratio of [HMIm]$^+$[BF$_4$]$^-$, the adsorption ratio of Au(III) increased at first and then decreased when the mass ratio was higher than 40%. The maximum adsorption ratio maintained above 95% was achieved at medium loadings, when the mass ratio of [HMIm]⁺[BF₄]⁻ was 40%. At low [HMIm]⁺[BF₄]⁻ loadings, incorporating more ILs into UiO-66 provided more sites that can interact with Au(III); thus, the adsorption ability increased. At high [HMIm]⁺[BF₄]⁻ loadings, where the percentage of the free volume of the adsorbent decreased when more ILs were transported into UiO-66, the pore blockage hindered the passage of gold ions [16,43].

2.2. Characterization of [HMIm]⁺[BF₄]⁻@UiO-66

The resulting immobilized [HMIm]⁺[BF₄]⁻ in UiO-66 was characterized by various microscopic and spectroscopic techniques, such as scanning electron microscopy (SEM), Fourier transform infrared spectra (FT-IR), and X-ray diffraction spectrometry (XRD). The specific surface area and porous structures of UiO-66 and [HMIm]⁺[BF₄]⁻@UiO-66 were investigated via nitrogen adsorption–desorption isotherms and pore size distribution curves.

The surface morphology of UiO-66 before and after IL loading was observed via SEM. Figure S1a,b show the SEM images of UiO-66 and [HMIm]⁺[BF₄]⁻@UiO-66, indicating that no detectable IL phase formed on the surface of UiO-66 [42]. A large number of irregularly rounded particles can be found closely attached to the characterized surfaces of UiO-66 and [HMIm]⁺[BF₄]⁻@UiO-66, and this rough surface can enhance the adsorption capacity. This is the same form as previously reported for UiO-66 [25]. At the same time, many pores and voids can be observed in Figure S1, and this porous tertiary structure is beneficial for the following adsorption performance. The dimensions of both samples are the same, indicating that the morphological characteristics of the samples are the same before and after synthesis.

To confirm the successful immobilization of [HMIm]⁺[BF₄]⁻ in UiO-66, FT-IR was performed. The FT-IR spectra of UiO-66 and [HMIm]⁺[BF₄]⁻@UiO-66 are shown in Figure 2a. The band at 1581 cm⁻¹ in the spectra of UiO-66 and [HMIm]⁺[BF₄]⁻@UiO-66 is due to the stretching vibration of the C=O bond on the BDC, the appearance of the characteristic band at 1398 cm⁻¹ indicates the typical framework vibration of the benzene ring, and the stretching vibrations of the Zr-O and Zr-O₂ bonds are indicated by the bands at 747 cm⁻¹ and 663 cm⁻¹ [26,27], which indicate the successful synthesis of UiO-66. The strong band at 1105 cm⁻¹ can be attributed to C-N stretching, and the band at 1296 cm⁻¹ can be attributed to the imidazole ring stretching [42], which indicates the successful sequestration of IL onto UiO-66.

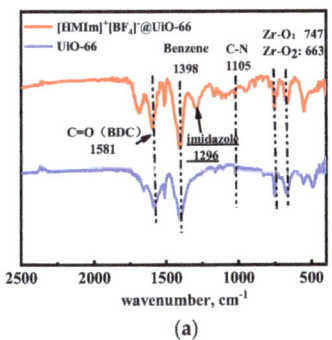
(a)

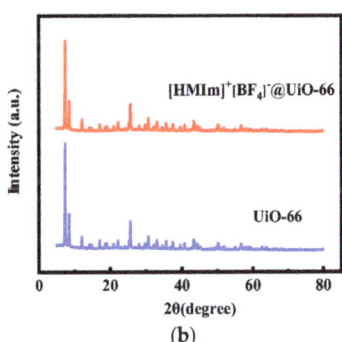
(b)

Figure 2. (a) FT-IR spectra of UiO-66 and [HMIm]⁺[BF₄]⁻@UiO-66; (b) PXRD patterns of UiO-66 and [HMIm]⁺[BF₄]⁻@UiO-66.

The powder XRD patterns of UiO-66 and [HMIm]⁺[BF₄]⁻@UiO-66 are shown in Figure 2b. The prominent characteristic peaks of UiO-66 are located at 2θ = 7.5°, 8.5°, and 25.8°, which are consistent with the UiO-66 patterns in the literature [20,26,29,42]. In addition, [HMIm]⁺[BF₄]⁻@UiO-66 has the same typical peaks and intensities, indicating that the IL-supported UiO-66 crystal pore cage structure remains intact. However, there are no characteristic peaks of [HMIm]$^+$[BF$_4$]$^-$ in the XRD pattern of [HMIm]$^+$[BF$_4$]$^-$@UiO-66, probably because [HMIm]$^+$[BF$_4$]$^-$ is entrapped in the framework of UiO-66 and there is no crystalline phase of [HMIm]$^+$[BF$_4$]$^-$ to be detected.

The N$_2$ adsorption–desorption isotherms of the adsorbent materials are shown in Figure 3a. The characteristics of UiO-66 and [HMIm]$^+$[BF$_4$]$^-$@UiO-66 both follow an I-shaped curve [27], characteristic of microporous materials, proving the existence of microporosity in the materials. IL curing can still maintain the microporous properties of UiO-66, and the stable microporous structure benefits the diffusion of Au(III). The pore size distribution parameters were obtained based on the Brunauer–Emmett–Teller (BET) pore size distribution curves. The BET surface area and pore volume of UiO-66 and [HMIm]$^+$[BF$_4$]$^-$@UiO-66 were 1211.178 m^2/g and 0.455 cm^3/g, and 1207.26 m^2/g and 0.442 cm^3/g, respectively. Before and after IL curing, the high specific surface area possessed by UiO-66 can provide more adsorption sites for Au(III). In addition, the pore diameters of [HMIm]$^+$[BF$_4$]$^-$@UiO-66 are mainly distributed around 0.94 nm and 1.51 nm. These diameters are much larger than those of gold species. Thus, this facilitates the diffusion of gold ions from the surface of the material into the pores, so that there are enough adsorption sites on the surface to adsorb gold ions. The reduction in the specific surface area and pore volume of UiO-66 before and after loading also confirms that the IL was successfully transported into UiO-66.

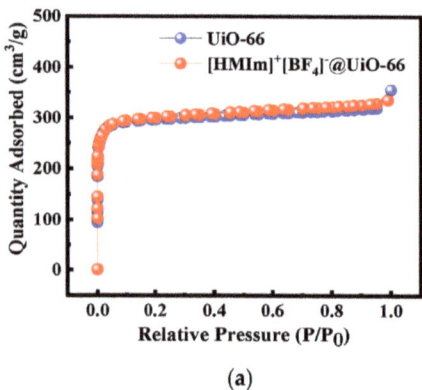 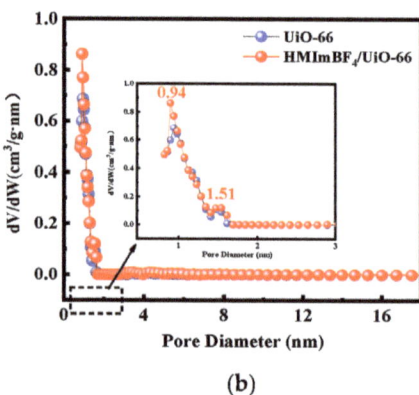

(a) (b)

Figure 3. (a) Nitrogen adsorption–desorption isotherms of UiO-66 before and after [HMIm]$^+$[BF$_4$]$^-$ modification; (b) aperture distribution curve of UiO-66 before and after [HMIm]$^+$[BF$_4$]$^-$ modification.

2.3. Adsorption Mechanism

2.3.1. The Anion Influence on Adsorption

In liquid–liquid extraction, ILs extract [AuCl$_4$]$^-$ through the anion exchange mechanism [3–8,44]. A series of SILs as adsorbents have been developed; however, the anion exchange mechanism is responsible for the adsorption of [AuCl$_4$]$^-$ [13,16,18]. The loss of IL anions not only leads to water pollution, but also challenges the regeneration of ILs.

By introducing an IL into the porous framework of an MOF, it has been reported that the interionic interactions become stronger due to the interaction of the anions of the IL with the metal sites of the MOF, and the direct interaction between the imidazolium ring of the IL with either the MOF or the anion of the IL [21]. Another computational study concluded that anions of ILs ([NTf$_2$]$^-$, [PF$_6$]$^-$, [BF$_4$]$^-$, and [SCN]$^-$) have a stronger interaction than [BMIm]$^+$ cations with MOFs. Thus, UiO-66 possesses application potential in avoiding the anion exchange mechanism because of the interactions between IL anions and the Zr atoms of UiO-66.

ILs with [PF$_6$]$^-$ or [NTf$_2$]$^-$ anions are more widely used in the liquid–liquid extraction of metal ions, despite their higher cost compared to ILs with [BF$_4$]$^-$ anions [1,2,4–8]. The applications of $[BF_4]^-$ ILs in the extraction of metal ions are limited due to the higher water miscibility compared to ILs with $[PF_6]^-$ or $[NTf_2]^-$. To figure out whether UiO-66 can effectively confine $[BF_4]^-$ to restrain the exchange process of $[BF_4]^-$ with $[AuCl_4]^-$ and the dissolution of $[HMIm]^+[BF_4]^-$ in water, the concentrations of $[BF_4]^-$ in the aqueous phase after $[AuCl_4]^-$ adsorption or liquid–liquid extraction were quantified. An amount of 150ppm of $[AuCl_4]^-$ at pH = 2 was used for the study at first. The adsorption of $[AuCl_4]^-$ by $[HMIm]^+[BF_4]^-$@UiO-66 and the liquid–liquid extraction of $[AuCl_4]^-$ by $[HMIm]^+[BF_4]^-$ IL itself were performed under the same conditions ($c_{Au(III)}$ = 60 mg/L, $V_{Au(III)}$ = 10 mL, T = 35 °C, t = 6 h). The ion chromatograms for $[BF_4]^-$ in the aqueous phase are shown in Figure S3. When the adsorption and extraction ratios of Au(III) were 78.85% and 96.84%, the $[BF_4]^-$ concentrations in the aqueous phase were 0.122 mg/L and 18,040 mg/L, respectively. Although $[HMIm]^+[BF_4]^-$ IL has a higher enrichment efficiency for Au(III), there was almost a 45% loss of the IL composed of $[BF_4]^-$ from the IL phase into the aqueous phase, leading to serious environmental pollution and difficult reuse of $[HMIm]^+[BF_4]^-$. Meanwhile, there were only ultra-trace amounts of $[BF_4]^-$ found in the aqueous phase after the adsorption of Au(III) by immobilizing $[HMIm]^+[BF_4]^-$ on UiO-66. Obviously, the loss of $[BF_4]^-$ was restrained because $[BF_4]^-$ anions were effectively confined in UiO-66 cages as we initially expected.

To further explore the stronger interaction between UiO-66 and the anions of $[BF_4]^-$ IL, the effect of $[BF_4]^-$ concentrations on Au(III) adsorption was investigated. As shown in Figure 4, the adsorption ratio of Au(III) decreased significantly from 83.72% to 17.10% with increasing $[BF_4]^-$ concentrations from 0 to 0.01 mol/L. The effect of $[BF_4]^-$ concentrations can be understood by considering the stronger interaction between the $[BF_4]^-$ anions and Zr^{4+} metal sites in UiO-66. The adsorbent preferred to adsorb $[BF_4]^-$ rather than $[AuCl_4]^-$, which was mainly affected by the anion radius. $[BF_4]^-$ with a smaller anionic radius (0.232 nm) is more conducive to transport into the pores of UiO-66. The increased amount of $[BF_4]^-$ in the adsorbent occupied adsorption sites and pore channels in UiO-66; thus, the adsorption of Au(III) on $[HMIm]^+[BF_4]^-$@UiO-66 was significantly reduced.

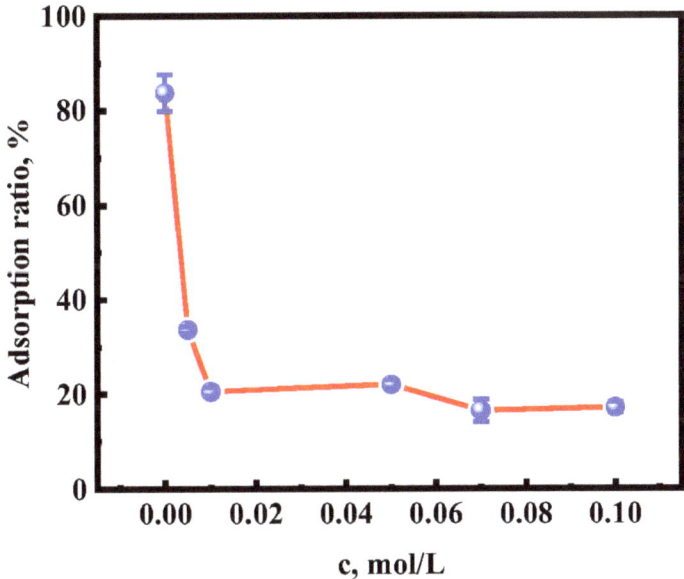

Figure 4. Effects of different concentrations of BF_4^- on Au(III) adsorption ($c_{Au(III)}$ = 60 mg/L, $V_{Au(III)}$ = 20 mL, $m_{[HMIm]+[BF4]-@UiO-66}$ = 10 mg, t = 6 h, T = 35 °C, pH = 2).

Since the results confirm that [HMIm]$^+$[BF$_4$]$^-$@UiO-66 adsorbed Au(III) certainly without ion exchange, to further explore the adsorption mechanism of Au(III), the effect of pH on adsorption was investigated, and the X-ray photoelectron spectroscopy (XPS) analyses of [HMIm]$^+$[BF$_4$]$^-$@UiO-66 before and after adsorption were examined.

2.3.2. Effect of pH

The pH effect on the surface charge of the adsorbent and metal ions is one of the fundamental factors affecting the adsorption rate. Here, the effect of pH varying from 1 to 9 on Au(III) adsorption was evaluated. The adsorption ratio of Au(III) was in the range of 88.28% to 94.37% when the pH was 1 to 2. Meanwhile, it decreased from 94.37% to 73.48% with increasing pH from 2 to 9. As far as we know, the species of Au(III) in solutions at pH 1–9 have negative charges, such as [AuCl$_4$]$^-$, [AuCl$_3$(OH)]$^-$, [AuCl$_2$(OH)$_2$]$^-$, [AuCl(OH)$_3$]$^-$, and [Au(OH)$_4$]$^-$ [38–40]. As shown in Figure 5, the zero point charge (pH$_{zpc}$) of [HMIm]$^+$[BF$_4$]$^-$@UiO-66 is 5.3, indicating that the surface charge of [HMIm]$^+$[BF$_4$]$^-$@UiO-66 is positively charged when the pH is lower than 5.3. The positively charged [HMIm]$^+$[BF$_4$]$^-$@UiO-66 is conducive to the adsorption of the negatively charged Au(III) species. The decrease in the zeta potential of [HMIm]$^+$[BF$_4$]$^-$@UiO-66 with the increase in the pH value is consistent with the lower pH value, facilitating Au(III) adsorption [45–47]. Thus, electrostatic interaction plays an important role in the adsorption process, and a pH of 2.0 was chosen in the following adsorption experiments. However, the adsorption ratio of Au(III) is 73.48% when the surface charge of [HMIm]$^+$[BF$_4$]$^-$@UiO-66 is negatively charged at pH = 9, which indicates that [HMIm]$^+$[BF$_4$]$^-$@UiO-66 has other attractions to Au(III).

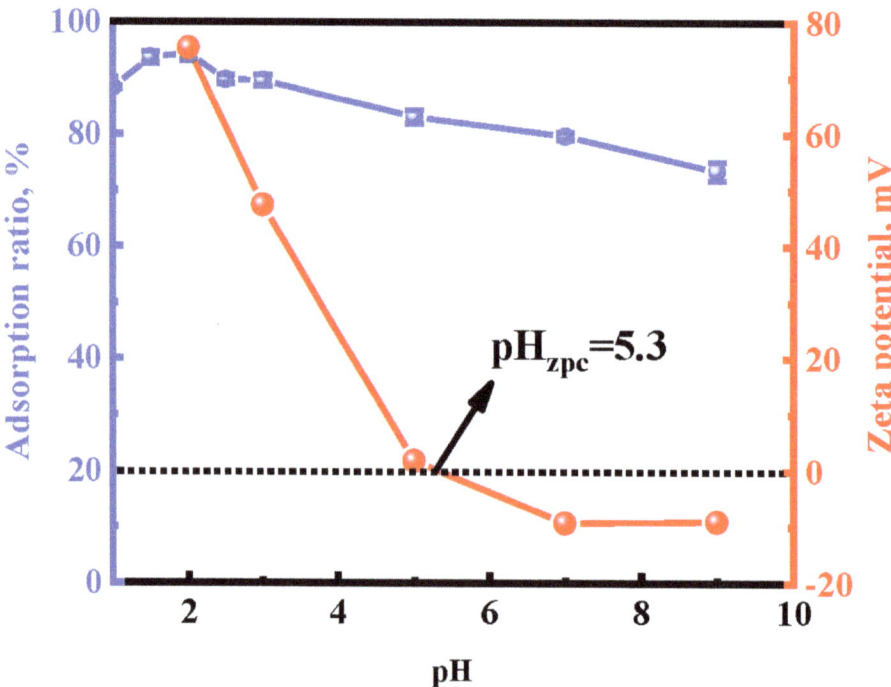

Figure 5. Effects of pH and zeta potential of [HMIm]$^+$[BF$_4$]$^-$@UiO-66.

2.3.3. XPS analysis of [HMIm]$^+$[BF$_4$]$^-$@UiO-66

To investigate the adsorption mechanism of Au(III), [HMIm]$^+$[BF$_4$]$^-$@UiO-66 before and after the adsorption of Au(III) was analyzed via XPS. The complete XPS spectra of [HMIm]$^+$[BF$_4$]$^-$@UiO-66 and [HMIm]$^+$[BF$_4$]$^-$@UiO-66/Au are shown in Figure 6. Compared to the spectrum of [HMIm]$^+$[BF$_4$]$^-$@UiO-66, a new Au4f peak was found for [HMIm]$^+$[BF$_4$]$^-$@UiO-66/Au. This indicates that Au(III) was successfully adsorbed by this material. The peaks in the high-resolution Au4f spectrum of the gold adsorbent can be divided into Au4f 7/2 and Au4f 5/2, as shown in Figure 6b. The two peaks at 87.54 eV (Au4f 5/2) and 83.87 eV (Au4f 7/2) correspond to Au(0), while the two peaks at 88.12 eV (Au4f 5/2) and 84.47 eV (Au4f 7/2) correspond to Au(I) [25,45,46,48]. The results suggest that Au(III) was reduced to Au(0) and Au(I) by [HMIm]$^+$[BF$_4$]$^-$@UiO-66, and that a redox mechanism exists during the adsorption process. The Au(0) area ratio is 64.68%, and the size of the peak area ratio indicates that gold mainly exists on the adsorbent in the form of Au(0) [46].

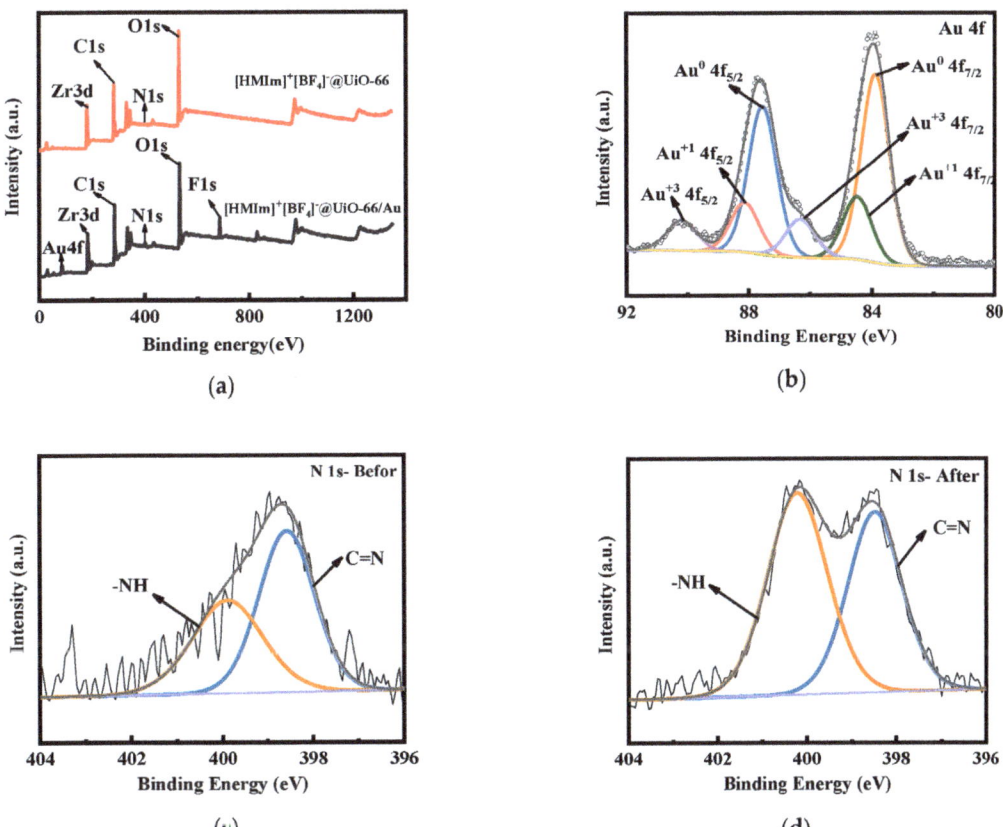

Figure 6. XPS spectra: (**a**) full spectra of [HMIm]$^+$[BF$_4$]$^-$@UiO-66 and [HMIm]$^+$[BF$_4$]$^-$@UiO-66/Au; (**b**) Au 4f spectra of [HMIm]$^+$[BF$_4$]$^-$@UiO-66/Au; (**c**,**d**) N1s spectra of [HMIm]$^+$[BF$_4$]$^-$@UiO-66 and [HMIm]$^+$[BF$_4$]$^-$@UiO-66/Au.

To understand the interactions between the gold and N atoms, the XPS N1s spectra of [HMIm]$^+$[BF$_4$]$^-$@UiO-66 and [HMIm]$^+$[BF$_4$]$^-$@UiO-66/Au were studied. In Figure 6c, the N1s spectra can be divided into C=N and -NH. The peak of the C=N binding energy changes from 398.57 eV to 398.47 eV, while the peak of the -NH binding energy changes from 399.88 eV to 400.2 eV, after adsorption. The peak area of C=N decreases while the peak area of NH increases relatively after adsorption. The results indicate that the electrons were transferred from N to Au(III). The N-containing functional groups were bound to Au(III) through complexation, and Au(III) was reduced to Au(I) and Au(0) [47,48]. The results prove the mechanism of Au(III) adsorption after IL immobilization, which not only avoids ion exchange in solvent extraction, but also provides a new adsorption site for Au(III).

In summary, the mechanism of the adsorption of Au(III) by $[HMIm]^+[BF_4]^-$@UiO-66 has three parts: electrostatic interaction, coordination between Au(III) and N-containing functional groups, and a reduction of Au(III) to Au(I) and Au(0). The loss of $[HMIm]^+[BF_4]^-$ IL in the aqueous phase caused by the miscibility of the water and anion exchange was effectively restrained because of the strong interaction between the IL and UiO-66.

2.4. Adsorption Kinetics

The adsorption equilibrium time is one of the critical indicators for the evaluation of adsorbents. The effect of time on the adsorption of Au(III) by UiO-66 and $[HMIm]^+[BF_4]^-$@UiO-66 was investigated. The results are shown in Figure 7a. It can be seen that the adsorption curve maintains the same trend. The adsorption of Au(III) by UiO-66 and $[HMIm]^+[BF_4]^-$@UiO-66 increased rapidly from 0 to 10 min, reaching 61.09% and 92.60%, respectively, and as time increased, the adsorption ratio reached equilibrium at 50 and 25 min, respectively. Au(III) was rapidly adsorbed in the first 10 min. The adsorption ratio of Au(III) by $[HMIm]^+[BF_4]^-$@UiO-66 was much higher than that of UiO-66. This indicates that the sequestration of the IL not only inhibited the exchange and decomposition of $[BF_4]^-$, but also provided a large number of adsorption sites for Au(III).

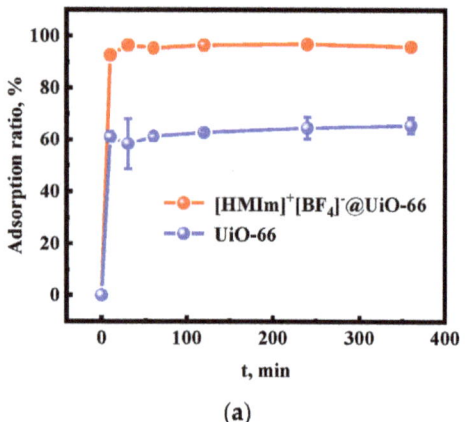 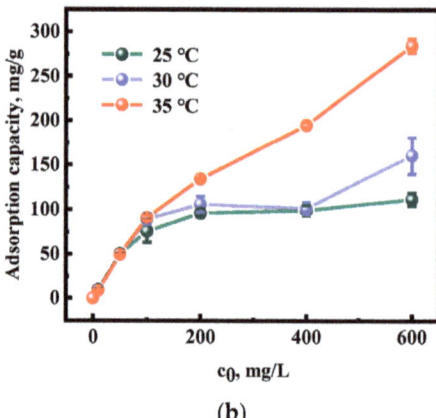

Figure 7. (a) Effect of adsorption time; (b) effect of the initial Au(III) concentration on the adsorption capacity of $[HMIm]^+[BF_4]^-$@UiO-66.

The data obtained were further fitted using pseudo-first-order (PFO, Equation (1)), pseudo-second-order (PSO, Equation (2)), and intraparticle diffusion (Id, Equation (3)) models to describe the adsorption behavior. The pseudo-first-order kinetic model assumes [49] that the adsorption process is physical adsorption, and its rate-limiting step is related to pore diffusion. The pseudo-second-order kinetic model assumes [39] that the adsorption process is chemical adsorption, and its rate-limiting step is a chemical reaction. The intraparticle diffusion model assumes [50] that the external mass transfer process leads to either rapid intraparticle diffusion or rate control steps.

$$\ln(q_e - q_t) = \ln q_e - K_1 t, \tag{1}$$

$$\frac{t}{q_t} = \frac{1}{K_2 q_e^2} + \frac{t}{q_e},\qquad(2)$$

$$q_t = K_3 t^{1/2} + C,\qquad(3)$$

where q_e and q_t (mg/g) are the metal amounts when adsorption equilibrium is reached and at time t, respectively, K_1 (1/min) is the pseudo-first-order rate constant, K_2 (mg/g·min) is the pseudo-second-order rate constant, K_3 (mg/g·min$^{0.5}$) is the particle diffusion rate constant, and C is the intercept, representing the thickness of the boundary layer.

The experimental data were analyzed using pseudo-first-order, pseudo-second-order, and intraparticle diffusion models, and the parameters obtained are shown in Figure 8 and Table 1. For UiO-66 and [HMIm]$^+$[BF$_4$]$^-$@UiO-66, the pseudo-second-order model has the highest correlation coefficient ($R^2_{\text{UiO-66}}$ = 0.9998 and $R^2_{\text{[HMIm]+[BF4]-@UiO-66}}$ = 0.9934), meaning it has a good linear correlation. In addition, the experimental values of Q in the PSO kinetics are closer to the theoretical values, which further demonstrates that the adsorption process of Au(III) can be better reflected by the PSO kinetic model. This suggests that the rate-limiting step of the Au(III) adsorption behavior of UiO-66 and [HMIm]$^+$[BF$_4$]$^-$@UiO-66 is a chemical reaction. In contrast, the adsorption capacity of [HMIm]$^+$[BF$_4$]$^-$@UiO-66 is positively correlated with the number of adsorption sites [51–53].

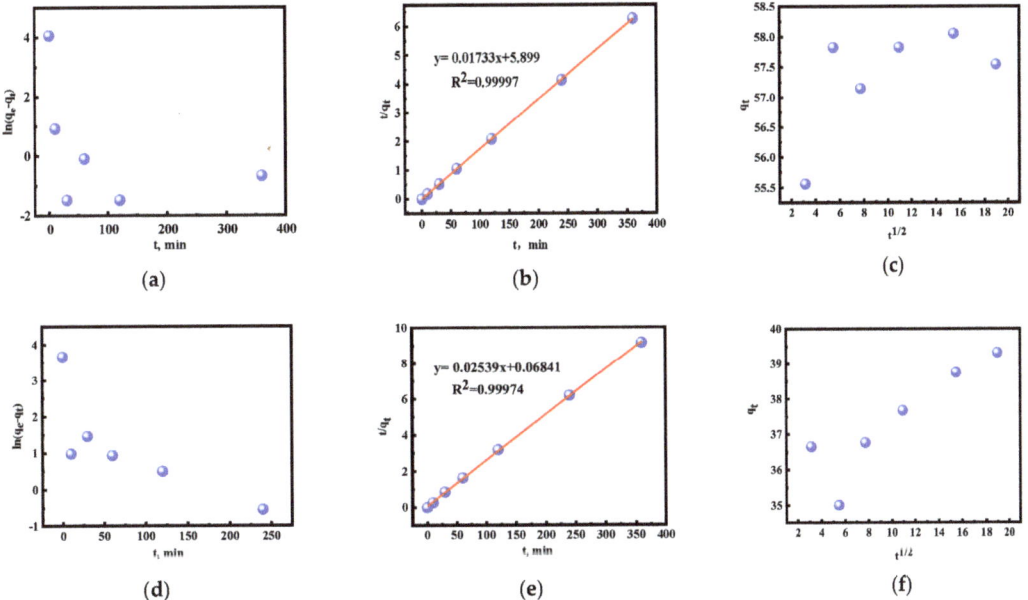

Figure 8. The fitting curves of UiO-66 and [HMIm]$^+$[BF$_4$]$^-$@UiO-66: (a–c) UiO-66; (d–f) [HMIm]$^+$[BF$_4$]$^-$@UiO-66; (a,d) pseudo-first-order; (b,e) pseudo-second-order; and (c,f) particle diffusion.

Table 1. The kinetic parameters of pseudo-second-order model.

	Pseudo-Second-Order		
	R^2	qe(fit)	Qe(exp)
UiO-66	0.9998	39.326	39.541
[HMIm]$^+$[BF$_4$]$^-$@UiO-66	0.9934	64.185	59.844

2.5. Isotherm Study

The adsorption isotherm is an important indicator to evaluate the maximum adsorption capacity of the adsorbent. The variation in the adsorption capacity with the initial Au(III) concentration at three temperatures (25, 30, and 35 °C) is shown in Figure 7b. Furthermore, the adsorption capacity increases with increasing initial Au(III) concentration. It can be adsorbed entirely at low concentrations because a sufficient number of adsorption sites are provided by the adsorbent. Due to the limited number of adsorption sites, the adsorption capacity tends to increase gradually and slowly with increasing Au(III) concentration, which leads to adsorption saturation. The obtained data were fitted and analyzed with the Langmuir model (Equation (4)), Freundlich model (Equation (5)), and Dubinin–Radushkevich (D–R) model (Equation (6)). The Langmuir model assumes [54] that adsorption is monolayer adsorption on a uniform surface, and that the adsorption and desorption are in dynamic equilibrium. The Freundlich isotherm model assumes [55] that adsorption is multilayer adsorption and occurs on heterogeneous surfaces. The D–R isotherm model assumes [56] that the adsorption process is not layer-by-layer adsorption on the adsorbent surface but related to the micropore volume.

$$\frac{c_e}{q_e} = \frac{1}{K_L q_m} + \frac{c_e}{q_m}, \quad (4)$$

$$\ln q_e = \ln K_F + \frac{1}{n} c_e, \quad (5)$$

$$\ln q_e = \ln q_m - \beta \varepsilon^2, \quad \varepsilon = RT \ln\left(1 + \frac{1}{c_e}\right), \quad (6)$$

where K_L (L/mg) is the Langmuir constant related to the affinity of the binding site, K_F ((mg/g)/(L/mg)$^{1/n}$) shows the Freundlich constant associated with the adsorption strength, q_m (mg/g) and n express the highest adsorption capacity and the coefficient of the Freundlich model, respectively, β represents the D–R isotherm constant, R represents the universal gas constant (8.314 J/mol·K), and T represents the temperature (K).

The separation factor (R_L) describes the basic characteristics and feasibility of the Langmuir isotherm:

$$R_L = \frac{1}{1 + K_L c_0} \quad (7)$$

The experimental results were analyzed using the Langmuir, Freundlich, and D–R models, and the parameters obtained are shown in Figure 9a–i and Table 2. The results show that the adsorption isotherms of Au(III) were more consistent with the Langmuir model (R^2 = 0.995, 0.91634, 0.92302), and the theoretical maximum adsorption amounts of Au(III) at the three temperatures were 109.89, 142.05, and 279.33 mg/g, which are very close to the actual maximum adsorption amounts of 111.57, 160.77, and 284.64 mg/g, indicating that the adsorption of Au(III) is monolayer adsorption at a specific homogeneous location on the adsorbent surface [47,48,57]. Moreover, the R_L values of Au(III) in the Langmuir model were all below 1, which indicates that the adsorption is appropriate [53]. In addition, the maximum adsorption of [HMIm]$^+$[BF$_4$]$^-$@UiO-66 (284.64 mg/g) was higher than that of UiO-66 (203.42 mg/g) (see SI for detailed results).

Table 2. The parameters of the Langmuir model.

	Langmuir		
	K_L	q_m	R^2
25 °C	0.11685	109.89	0.995
30 °C	0.05213	142.05	0.91634
35 °C	0.02790	279.33	0.92302

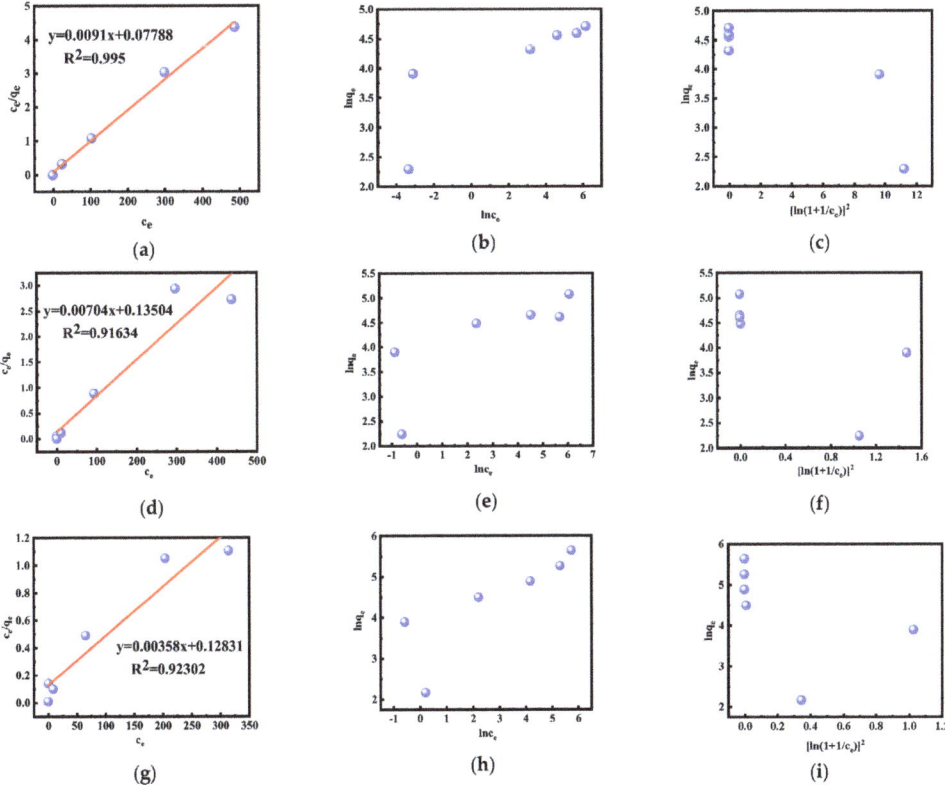

Figure 9. Isotherm fitting: (**a**) Langmuir model at 25 °C; (**b**) Freundlich model at 25 °C; (**c**) D–R model at 25 °C; (**d**) Langmuir model at 30 °C; (**e**) Freundlich model at 30 °C; (**f**) D–R model at 30 °C; (**g**) Langmuir model at 35 °C; (**h**) Freundlich model at 35 °C; and (**i**) D–R model at 35 °C.

2.6. Thermodynamic Study

The influence of the adsorption temperature on the adsorption process is significant and explains the adsorption thermodynamics with relevant thermodynamic parameters. The data obtained were evaluated with the following equations (Equations (8)–(10)) containing classical thermodynamic parameters [25]:

$$K_c = \frac{q_e}{c_e}, \tag{8}$$

$$\Delta G = -RT \ln K_c, \tag{9}$$

$$\ln K_c = \frac{\Delta S}{R} - \frac{\Delta H}{RT}, \tag{10}$$

where K_c, T (K), and R (8.314 J/mol·K) are the thermodynamic equilibrium constant, the adsorption temperature, and the gas constant, respectively, ΔS (J/mol/K), ΔH (KJ/mol), and ΔG (KJ/mol), respectively, are the changes in the entropy, enthalpy, and Gibb's free energy.

As shown in Figure 10, the Au(III) adsorption of [HMIm]$^+$[BF$_4$]$^-$@UiO-66 was enhanced with increasing temperature. The thermodynamic parameters at different temperatures are summarized in Table 3. The increase in temperature favors the increase in the number of active molecules. Thus, the adsorption of Au(III) by the adsorbent is promoted, indicating that the adsorption process is heat absorption. It was found that ΔG was negative at different temperature conditions, indicating that the reaction process is spontaneous and feasible [20,58]. Additionally, the negative value of ΔG was increased with increasing temperature, indicating that the higher the temperature, the more spontaneous and favorable the adsorption of Au(III). The positive value of ΔH indicates that the adsorption is a heat-absorbing reaction. On the contrary, a positive value of ΔS indicates that the system's degrees of freedom and disorder are increased, which favors an increase in the frequency of collisions between Au(III) and the adsorbent [42,45,48,59].

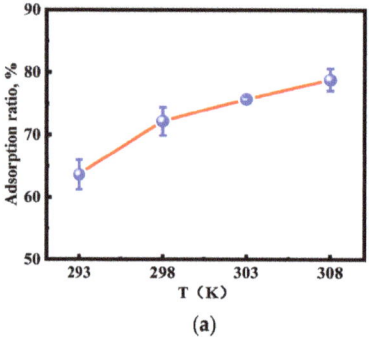 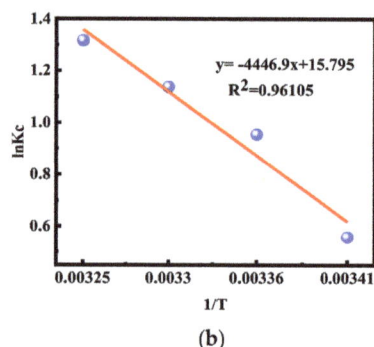

Figure 10. (a) Effect of temperature; (b) plot of ln Kc versus 1/T.

Table 3. The thermodynamic parameters at different temperatures.

T(K)	Kc	ΔG (kJ/moL)	ΔH (kJ/moL)	ΔS (kJ/moL)
293	1.7473	−1.3594		
298	2.5965	−2.3641	36.971	0.13132
303	3.1197	−2.8662		
308	3.7279	−3.3695		

2.7. Selectivity and Practical Application

E-waste containing Au(III) coexists with other metal ions; therefore, the adsorption selectivity of Au(III) was studied to better evaluate the adsorbent's performance. Mg(II), Cu(II), Zn(II), Pb(II), Fe(II), and Ni(II) were chosen as background ions to study the selectivity of [HMIm]$^+$[BF$_4$]$^-$@UiO-66. As shown in Figure 11a, when the concentration ratio of Au(III) to other coexisting ions was 1:1, Au(III) adsorption on the adsorbent reached 98.5%. In contrast, almost no other metal ions were adsorbed. Considering that the concentration of coexisting ions in e-waste leachate is several times higher than that of Au(III), the adsorption of Au(III) with other coexisting ions at a concentration ratio of 1:150 was investigated. The results show that the adsorption of Au(III) remained unaffected by the high concentration of coexisting ions, and that Au(III) could be 100% adsorbed by the adsorbent. The excellent Au(III) adsorption performance of the adsorbent can be attributed to the physicochemical properties of the metal atoms, such as the ionic radius (R), electronegativity (Xm), and covalent index (Xm^{2}r). The Xm^{2}r (5.48) and Xm (2.54) of Au(III) are higher, which allows Au(III) to be preferentially adsorbed [46]. Additionally, according to hard–soft acid–base (HSAB) theory, Au(III) can form strong bonds with N-containing functional groups, which may also contribute to its high adsorption [46,60]. In addition, at a lower pH, other coexisting ions may exist as cations or neutrals. Therefore, the coexisting ions are not adsorbed by [HMIm]$^+$[BF$_4$]$^-$@UiO-66 due to electrostatic repulsion with positively charged [HMIm]$^+$[BF$_4$]$^-$@UiO-66 on the surface [61]. In addition, some anions will be inevitably introduced into the system during the leaching of Au(III) from e-waste. Therefore, the effect of several representative anions (Cl$^-$, SO$_4^{2-}$, PO$_4^{3-}$, and NO$_3^-$) on Au(III) adsorption at different concentrations (0, 0.001, 0.01, and 0.1 mol/L) was investigated. As can be seen in Figure 11b, the adsorption of Au(III) was inhibited to a greater extent by PO$_4^{3-}$ as the anion concentration increased. When PO$_4^{3-}$ was 0.1 mol/L, the adsorption of Au(III) was only 36%. On the contrary, the adsorption of Au(III) was slightly inhibited by SO_4^{2-}. When SO_4^{2-} was 0.1 mol/L, the adsorption of Au(III) was still above 85%, while Cl^- and NO_3^- hardly affected the adsorption of Au(III). In the presence of different Cl^- and NO_3^- concentrations, Au(III) adsorption remained at around 95%. Therefore, leaching agents containing Cl^- and NO_3^- media are preferred in the Au(III) leaching process. When the commonly used aqua regia ablates e-waste to extract Au(III), the adsorption of Au(III) on the composites is not affected. In addition, the aqua regia—based leaching of e-waste is a flexible and low-cost method. At present, it is also a widespread process in the industry [62,63].

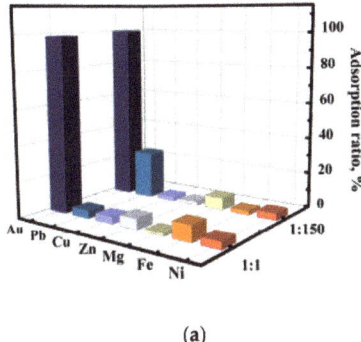 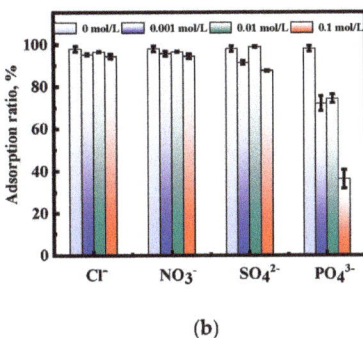

(a) (b)

Figure 11. (a) Effect of coexisting metal ions on Au(III) adsorption ($c_{Au(III) \text{ and coexisting ions}}$ = 100 mg/L, $c_{Au(III)}$ = 10 mg/L and $c_{coexisting ions}$ = 1500 mg/L); (b) effect of acid radical ions on Au(III) adsorption.

The practical application value of $[HMIm]^+[BF_4]^-$@UiO-66 was further evaluated to recover precious metals from discarded CPU motherboard pins. The main metal elements in CPU pins are shown in Figure 12a, and the adsorption rate of each metal ion is shown in Figure 12b. In the figure, it can be seen that the total percentage of the primary metals Ni(II), Cu(II), and Zn(II) in the CPU is more than 99%. In the presence of high concentrations of coexisting ions, it is clear that Ni(II), Cu(II), and Zn(II) are hardly adsorbed. In contrast, up to 96% of Au(III) is adsorbed on the adsorbent at low concentrations. The high selectivity of $[HMIm]^+[BF_4]^-$@UiO-66 for Au(III) in practical applications is consistent with the results of previous selectivity studies, further demonstrating that $[HMIm]^+[BF_4]^-$@UiO-66 has practical application value.

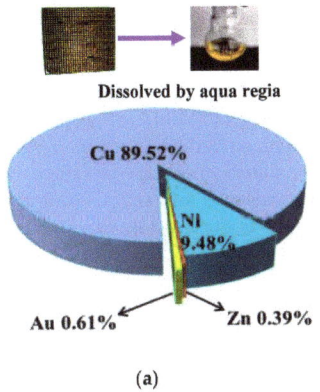 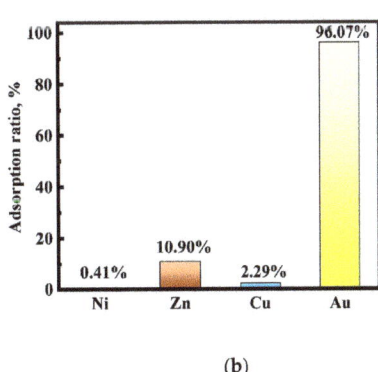

(a) (b)

Figure 12. (a) Percentage of major metal elements; (b) adsorption rate of each metal ion.

2.8. Reusability

In order to understand the recyclability of the adsorbent and to judge whether it has practical value, the reusability of the composite was investigated. The Au(III) adsorption rate of [HMIm]$^+$[BF$_4$]$^-$@UiO-66 in the reusability experiments is shown in Figure 13. After three consecutive cycles, the removal of Au(III) was greater than 95%, and there was no obvious decrease. However, no subsequent experiments were performed due to the large material loss during adsorption resolution. The results show that [HMIm]$^+$[BF$_4$]$^-$@UiO-66 has relatively stable adsorption properties and can be used as an excellent adsorbent for separating Au(III) from aqueous media.

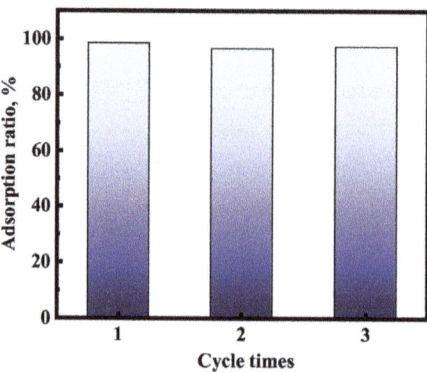

Figure 13. Reusability of [HMIm]$^+$[BF$_4$]$^-$@UiO-66.

3. Discussion

[HMIm]$^+$[BF$_4$]$^-$@UiO-66 was prepared as an adsorbent material for Au(III) recovery from acidic solutions, and it effectively avoided a series of contamination problems associated with the dissolution of ILs in water due to ion exchange. A series of characterizations, including SEM, FTIR, XRD, and N$_2$ adsorption and desorption experiments, confirmed the successful sequestration of ILs, and the maximum adsorption of Au(III) by [HMIm]$^+$[BF$_4$]$^-$@UiO-66 at pH 2.0 and 35 °C was 284.64 mg/g. In the kinetic and thermodynamic studies, the adsorption process was found to be an endothermic, feasible, and spontaneous reaction, while the rate-limiting step of adsorption was found to be a chemical reaction. In the isotherm studies, the adsorption processes at different temperatures were consistent with the Langmuir model. [HMIm]$^+$[BF$_4$]$^-$@UiO-66 showed good selectivity for Au(III) adsorption and successfully recovered Au(III) from e-waste. The effect of the pH on Au(III) adsorption and the XPS results indicated that the Au(III) adsorption mechanism was either an electrostatic, redox, coordination, or complexation mechanism. The results indicate that [HMIm]$^+$[BF$_4$]$^-$@UiO-66 successfully enclosed ILs. As an advanced adsorptive material for gold recovery from e-waste, [HMIm]$^+$[BF$_4$]$^-$@UiO-66 can be reused for three cycles without any significant decrease in the adsorption rate. It has a simple synthesis, appropriate adsorption kinetics and adsorption capacity, and excellent selectivity and regeneration, and it can successfully recover gold in practical applications. The results of the present work show that the material has application value and practicability in industry.

4. Materials and Methods

4.1. Materials and Chemicals

The standard stock solution of Au(III) (1000 mg/L) was obtained from the Shandong Metallurgical Research Institute. The working solutions were prepared daily by diluting the standard stock solution with deionized water. Zirconium chloride (ZrCl$_2$, 99.5%) and terephthalic acid (TPA, 99.0%) were obtained from Shanghai McLean Biochemical Technology Co., Ltd, Shanghai, China. All of the ILs, including 1-ethyl-3-methylimidazole tetrafluoroborate ([EMIm]$^+$[BF$_4$]$^-$, ≥98%), 1-octyl-3-methylimidazole tetrafluoroborate ([OMIm]$^+$[BF$_4$]$^-$, ≥98%), 1-butyl-3-methylimidazole tetrafluoroborate ([BMIm]$^+$[BF$_4$]$^-$, ≥98%), 1-hexyl-3-methylimidazole tetrafluoroborate ([HMIm]$^+$[BF$_4$]$^-$, ≥98%), 1-hexyl-3-methylimidazole hexafluorophosphate ([HMIm]$^+$[BF$_6$]$^-$, 99%), 1-hexyl-3-methylimidazole bis trifluoromethylsulfonimide salt ([HMIm]$^+$[NTf$_2$]$^-$, 99%), and 1-hexyl-3-methylimidazole acetate ([HMIm]$^+$[OAc]$^-$, ≥98%) were provided by the Lanzhou Institute of Chemical Physics, Chinese Academy of Sciences. Other chemicals were of analytical grade and obtained from Sinopharm Chemical Reagent limited corporation. All reagents and solvents were used without additional purification.

4.2. Synthesis of Adsorbents

4.2.1. Synthesis of UiO-66

The synthesis process of UiO-66 was based on a previously reported method [64]. The details are reported in the Supplementary Materials.

4.2.2. Synthesis of IL/UiO-66

Amounts of 1.5 g UiO-66, 1 g ILs, and 3 mL C_2H_5OH were mixed in a glass bottle and stirred at room temperature for 15 h. Then, the composite was filtered, washed with C_2H_5OH, and dried in a vacuum oven overnight at 80 °C.

4.3. Characterization

The remaining contents of Au(III) in water and other metal ions (Mg(II), Cu(II), Zn(II), Pb(II), Fe(III), and Ni (II)) were determined using a flame atomic absorption spectrometer (TAS990, Beijing Purkinje General Instrument Co., Ltd., Beijing, China) (for specific experimental details, see the SI). The concentration of [BF$_4$]$^-$ in the aqueous phase was determined using a high-pressure ion chromatograph (Integrion, Thermo Fisher Scientific, Massachusetts, US) with an ion Dionex IonPac AS 11-HC chromatography column with 20 mM KOH mobile phase at a flow rate of 1.0 mL/min (the sample injection volume was 25 µL, and the column oven temperature was 30 °C). The FT-IR spectra of UiO-66 and [HMIm]$^+$[BF$_4$]$^-$@UiO-66 were investigated in the range of 400–4000 cm^{-1} via Fourier transform infrared spectroscopy (FT-IR) (IRAffinity-15, Shimadzu production Institute, Kyodo, Japan) with KBr pellets. The morphologies of UiO-66 and [HMIm]$^+$[BF$_4$]$^-$@UiO-66 were recorded using a scanning transmission electron microscope (SEM) (STEM, FEI Tecnai G2 TF20, Frequency Electronics, Inc. Hillsboro, US). The zeta potential of [HMIm]$^+$[BF$_4$]$^-$@UiO-66 was measured using a nanometer particle size and zeta potential analyzer (Malvern Nano ZS, Malvern, UK). The powder X-ray diffraction spectrometry (PXRD) patterns of UiO-66 and [HMIm]$^+$[BF$_4$]$^-$@UiO-66 were captured by a D/max-2500 diffractometer (Rigaku, Tokyo, Japan) using CuKα radiation (λ = 1.5418 Å). The Brunauer–Emmett–Teller (BET) surface area and pore size distribution were calculated using N_2 adsorption–desorption methods with an ASAP 2020 V4.00 G instrument (Micromeritics Instrument Crop., Norcross, GA, USA) at 77 K. The X-ray photoelectron spectroscopy (XPS) analysis was conducted using a ThermoFisher scientific spectrometer (Nexsa base, Thermo Fisher Scientific, Massachusetts, US) equipped with a micro-focused monochromatic Al Kα source (hν = 1486.6 eV).

4.4. Batch Adsorption Experiment

The batch adsorption experiments were performed to study the Au(III) adsorption performance of [HMIm]$^+$[BF$_4$]$^-$@UiO-66. Generally, 10 mg [HMIm]$^+$[BF$_4$]$^-$@UiO-66 was added into a 50 mL plastic centrifugal tube containing 10 mL Au(III) solution with different concentrations at pH 2. The solution was shaken in a constant oscillator at 170 rpm for the desired duration, and then the mixture was filtered through a 0.45 µm filter membrane to separate the adsorbent from the aqueous phase.

Adsorption experiments were performed under optimized parameters (oscillation frequency: 170 rpm; t: 6 h; pH: 2; T: 35°C; $V_{Au(III)}$: 10 mL) unless otherwise indicated. The effect of pH was studied in the pH range of 1–9 (adjusted by adding diluted NaOH or HCl solutions). The experiment of the adsorption kinetics was carried out at different adsorption times ranging from 0 to 360 min. In contrast, the experiment of the adsorption isotherms was evaluated at different initial Au(III) concentrations ranging from 0 to 600 mg/L. The thermodynamic experiment was studied by controlling the temperature (from 298 to 308 K). The selectivity of the adsorbent for Au(III) was studied with different concentrations of Au(III) in the hybrid solution prepared by dissolving $HAuCl_4$, $MgCl_2$, $CuCl_2$, $ZnCl_2$, $PbCl_2$, $FeCl_3$, and $NiCl_2$ in DI water. With initial Au(III) concentrations of 10 and 100 mg/L, the mass ratios of Au(III) to coexisting ions were 1:150 and 1:1, respectively. The effect of anions was examined in the presence of Cl^-, SO_4^{2-}, PO_4^{3-}, and NO_3^-, and the concentrations were set to 0, 0.001, 0.01, and 0.1 mol/L.

Liquid–liquid extraction experiments were performed by adding 1 mL $[HMIm]^+[BF_4]^-$ to 10 mL of 150 mg/L Au(III) solution for 6 h with shaking under optimized parameters. After centrifugation (3000 r/min) of the mixed solution for 5 min, the aqueous was taken for measurement.

The regeneration experiment of $[HMIm]^+[BF_4]^-$@UiO-66 was implemented as follows: 20 mg $[HMIm]^+[BF_4]^-$@UiO-66 was mixed to 20 mL Au(III) aqueous solution (60 mg/L) at pH 2. Then, the mixed solution was shaken for 6 h at 35 °C and was separated by high-speed centrifugation (8000 r/min). The supernatant was tested to obtain the remaining Au(III) concentration. The residual solid was immersed for about 12 h with 20 mL 1 mol/L HCl and 5% thiourea solution (the gold was eluted into an acid thiourea solution), rinsed three times with DI water, and executed to the second time of adsorption–desorption cycle. The adsorption capacity and ratio of Au(III) were calculated using the following equations:

$$q = \frac{(c_0 - c_e)}{m} V, \tag{11}$$

$$\text{Adsorption ratio} = \frac{c_0 - c_e}{c_0} \times 100\%, \tag{12}$$

where q (mg/g) is the Au(III) adsorption capacity, c_0 and c_e (mg/L) are the initial and equilibrium concentrations of Au(III) in solution, respectively, m (mg) is the mass of the adsorbent used, and V (mL) is the volume of the Au(III) solution.

4.5. Recovery of Au(III) from Waste CPUs

The waste CPUs we used had an array of pins. The pins were detached from the CPU and immersed in aqua regia solution for 2 h (magnetically stirred for 1 h at room temperature and then 1 h at 75 °C) until the pins were wholly dissolved without residue. The obtained solution was diluted 10 times with DI water. Then, 10 mL of the diluted solution and 10 mg of adsorbent were added into a centrifuge tube, shaken for 6 h, and then filtered through a filter membrane.

Supplementary Materials: The following supporting information can be downloaded at: https://www.mdpi.com/article/10.3390/molecules28052165/s1.

Author Contributions: Conceptualization, W.L., P.Z. and Y.W. (Yanfeng Wang); writing—original draft preparation, X.C. and Y.W. (Yani Wang); writing—review and editing, X.C. and W.L. All authors have read and agreed to the published version of the manuscript.

Funding: This research was funded by the National Natural Science Foundation of China (21606147) and the Academic Promotion Program of Shandong First Medical University (No. 2019QL011; No. 2019LJ003).

Institutional Review Board Statement: Not applicable.

Informed Consent Statement: Not applicable.

Data Availability Statement: Data are contained within the article or Supplementary Materials.

Conflicts of Interest: The authors declare no conflict of interest.

Sample Availability: Samples of [HMIm]⁺[BF₄]⁻@UiO-66 are available from the authors.

References

1. Visser, A.E.; Swatloski, R.P.; Reichert, W.M.; Davis, J.H., Jr.; Rogers, R.D.; Mayton, R.; Sheff, S.; Wierzbicki, A. Task-specific ionic liquids for the extraction of metal ions from aqueous solutions. *Chem. Commun.* **2001**, *1*, 135–136. [CrossRef]
2. Regel-Rosocka, M.; Materna, K. Ionic Liquids for Separation of Metal Ions and Organic Compounds from Aqueous Solutions. In *Ionic Liquids in Separation Technology*; Elsevier: Amsterdam, The Netherlands, 2014; pp. 153–188. [CrossRef]
3. Zheng, Y.; Tong, Y.; Wang, S.; Zhang, H.; Yang, Y. Mechanism of gold (III) extraction using a novel ionic liquid-based aqueous two phase system without additional extractants. *Sep. Purif. Technol.* **2015**, *154*, 123–127. [CrossRef]
4. Wang, N.; Wang, Q.; Geng, Y.; Sun, X.; Wu, D.; Yang, Y. Recovery of Au(III) from Acidic Chloride Media by Homogenous Liquid–Liquid Extraction with UCST-Type Ionic Liquids. *ACS Sustain. Chem. Eng.* **2019**, *7*, 19975–19983. [CrossRef]
5. Boudesocque, S.; Mohamadou, A.; Conreux, A.; Marin, B.; Dupont, L. The recovery and selective extraction of gold and platinum by novel ionic liquids. *Sep. Purif. Technol.* **2019**, *210*, 824–834. [CrossRef]
6. Wang, M.; Wang, Q.; Geng, Y.; Wang, N.; Yang, Y. Gold(III) separation from acidic medium by amine-based ionic liquid. *J. Mol. Liq.* **2020**, *304*, 112735. [CrossRef]
7. Wang, M.; Wang, Q.; Wang, J.; Liu, R.; Zhang, G.; Yang, Y. Homogenous Liquid–Liquid Extraction of Au(III) from Acidic Medium by Ionic Liquid Thermomorphic Systems. *ACS Sustain. Chem. Eng.* **2021**, *9*, 4894–4902. [CrossRef]
8. Liu, X.; Wu, Y.; Wang, Y.; Wei, H.; Guo, J.; Yang, Y. Extraction of Au(iii) from hydrochloric acid media using a novel amide-based ionic liquid. *New J. Chem.* **2022**, *46*, 19824–19833. [CrossRef]
9. Tong, Y.; Yang, H.; Li, J.; Yang, Y. Extraction of Au(III) by ionic liquid from hydrochloric acid medium. *Sep. Purif. Technol.* **2013**, *120*, 367–372. [CrossRef]
10. Freire, M.G.; Neves, C.S.; Marrucho, I.M.; Coutinho, J.P.; Fernandes, A.M. Hydrolysis of Tetrafluoroborate and Hexafluorophosphate Counter Ions in Imidazolium-Based Ionic Liquids. *J. Phys. Chem. A* **2010**, *114*, 3744–3749. [CrossRef]
11. Liu, X.; Yang, F.; Wu, L.; Zhou, Q.; Ren, R.; Lv, Y.K. Ionic liquid-loaded covalent organic frameworks with favorable electrochemical properties as a potential electrode material. *Microporous Mesoporous Mater.* **2022**, *336*, 111906. [CrossRef]
12. Peng, X.; Chen, L.; Liu, S.; Hu, L.; Zhang, J.; Wang, A.; Yu, X.; Yan, Z. Insights into the interfacial interaction mechanisms of p-arsanilic acid adsorption on ionic liquid modified porous cellulose. *J. Environ. Chem. Eng.* **2021**, *9*, 105225. [CrossRef]
13. Dong, Z.; Zhao, L. Surface modification of cellulose microsphere with imidazolium-based ionic liquid as adsorbent: Effect of anion variation on adsorption ability towards Au(III). *Cellulose* **2018**, *25*, 2205–2216. [CrossRef]
14. Yang, N.; Wang, R. Molecular sieve supported ionic liquids as efficient adsorbent for CO_2 capture. *J. Serb. Chem. Soc.* **2015**, *80*, 265–275. [CrossRef]
15. Wu, Y.; Jia, Z.; Bo, C.; Dai, X. Preparation of magnetic β-cyclodextrin ionic liquid composite material with different ionic liquid functional group substitution contents and evaluation of adsorption performance for anionic dyes. *Colloids Surf. A Physicochem. Eng. Asp.* **2021**, *614*, 126147. [CrossRef]
16. Navarro, R.; Saucedo, I.; Lira, M.A.; Guibal, E. Gold(III) Recovery From HCl Solutions using Amberlite XAD-7 Impregnated with an Ionic Liquid (Cyphos IL-101). *Sep. Sci. Technol.* **2010**, *45*, 1950–1962. [CrossRef]
17. Wu, H.; Kudo, T.; Kim, S.-Y.; Miwa, M.; Matsuyama, S. Recovery of cesium ions from seawater using a porous silica-based ionic liquid impregnated adsorbent. *Nucl. Eng. Technol.* **2022**, *54*, 1597–1605. [CrossRef]
18. Campos, K.; Domingo, R.; Vincent, T.; Ruiz, M.; Sastre, A.M.; Guibal, E. Bismuth recovery from acidic solutions using Cyphos IL-101 immobilized in a composite biopolymer matrix. *Water Res.* **2008**, *42*, 4019–4031. [CrossRef]
19. Sadjadi, S.; Koohestani, F. Synthesis and catalytic activity of a novel ionic liquid-functionalized metal–organic framework. *Res. Chem. Intermed.* **2021**, *48*, 291–306. [CrossRef]
20. Ma, Y.; Li, A.; Wang, C. Experimental study on adsorption removal of SO_2 in flue gas by defective UiO-66. *Chem. Eng. J.* **2022**, *455*, 140687. [CrossRef]
21. Dhumal, N.R.; Singh, M.P.; Anderson, J.A.; Kiefer, J.; Kim, H.J. Molecular Interactions of a Cu-Based Metal–Organic Framework with a Confined Imidazolium-Based Ionic Liquid: A Combined Density Functional Theory and Experimental Vibrational Spectroscopy Study. *J. Phys. Chem. C* **2016**, *120*, 3295–3304. [CrossRef]
22. Kinik, F.P.; Uzun, A.; Keskin, S. Ionic Liquid/Metal-Organic Framework Composites: From Synthesis to Applications. *ChemSusChem* **2017**, *10*, 2842–2863. [CrossRef]
23. Ahmad, K.; Nazir, M.A.; Qureshi, A.K.; Hussain, E.; Najam, T.; Javed, M.S.; Shah, S.S.A.; Tufail, M.K.; Hussain, S.; Khan, N.A.; et al. Engineering of Zirconium based metal-organic frameworks (Zr-MOFs) as efficient adsorbents. *Mater. Sci. Eng. B* **2020**, *262*, 114766. [CrossRef]
24. Ahmadijokani, F.; Molavi, H.; Rezakazemi, M.; Tajahmadi, S.; Bahi, A.; Ko, F.; Aminabhavi, T.M.; Li, J.-R.; Arjmand, M. UiO-66 metal–organic frameworks in water treatment: A critical review. *Prog. Mater. Sci.* **2022**, *125*, 100904. [CrossRef]
25. Zhao, M.; Huang, Z.; Wang, S.; Zhang, L.; Wang, C. Experimental and DFT study on the selective adsorption mechanism of Au(III) using amidinothiourea-functionalized UiO-66-NH_2. *Microporous Mesoporous Mater.* **2020**, *294*, 109905. [CrossRef]
26. Wang, Y.; Zhang, N.; Chen, D.; Ma, D.; Liu, G.; Zou, X.; Chen, Y.; Shu, R.; Song, Q.; Lv, W. Facile synthesis of acid-modified UiO-66 to enhance the removal of Cr(VI) from aqueous solutions. *Sci. Total. Environ.* **2019**, *682*, 118–127. [CrossRef]

27. Hu, S.-Z.; Huang, T.; Zhang, N.; Lei, Y.-Z.; Wang, Y. Enhanced removal of lead ions and methyl orange from wastewater using polyethyleneimine grafted UiO-66-NH2 nanoparticles. *Sep. Purif. Technol.* **2022**, *297*, 121470. [CrossRef]
28. Piscopo, C.G.; Polyzoidis, A.; Schwarzer, M.; Loebbecke, S. Stability of UiO-66 under acidic treatment: Opportunities and limitations for post-synthetic modifications. *Microporous Mesoporous Mater.* **2015**, *208*, 30–35. [CrossRef]
29. Ouyang, J.; Chen, J.; Ma, S.; Xing, X.; Zhou, L.; Liu, Z.; Zhang, C. Adsorption removal of sulfamethoxazole from water using UiO-66 and UiO-66-BC composites. *Particuology* **2022**, *62*, 71–78. [CrossRef]
30. Xue, W.; Li, Z.; Huang, H.; Yang, Q.; Liu, D.; Xu, Q.; Zhong, C. Effects of ionic liquid dispersion in metal-organic frameworks and covalent organic frameworks on CO_2 capture: A computational study. *Chem. Eng. Sci.* **2016**, *140*, 1–9. [CrossRef]
31. Li, Z.; Xiao, Y.; Xue, W.; Yang, Q.; Zhong, C. Ionic Liquid/Metal–Organic Framework Composites for H_2S Removal from Natural Gas: A Computational Exploration. *J. Phys. Chem. C* **2015**, *119*, 3674–3683. [CrossRef]
32. Rao, M.D.; Singh, K.K.; Morrison, C.A.; Love, J.B. Challenges and opportunities in the recovery of gold from electronic waste. *RSC Adv.* **2020**, *10*, 4300–4309. [CrossRef]
33. Cui, J.; Zhang, L. Metallurgical recovery of metals from electronic waste: A review. *J. Hazard. Mater.* **2008**, *158*, 228–256. [CrossRef]
34. Doidge, E.D.; Kinsman, L.M.M.; Ji, Y.; Carson, I.; Duffy, A.J.; Kordas, I.A.; Shao, E.; Tasker, P.A.; Ngwenya, B.T.; Morrison, C.A.; et al. Evaluation of Simple Amides in the Selective Recovery of Gold from Secondary Sources by Solvent Extraction. *ACS Sustain. Chem. Eng.* **2019**, *7*, 15019–15029. [CrossRef]
35. Wang, R.; Zhang, C.; Zhao, Y.; Zhou, Y.; Ma, E.; Bai, J.; Wang, J. Recycling gold from printed circuit boards gold-plated layer of waste mobile phones in "mild aqua regia" system. *J. Clean. Prod.* **2021**, *278*, 123597. [CrossRef]
36. Zhang, L.; Zha, X.; Zhang, G.; Gu, J.; Zhang, W.; Huang, Y.; Zhang, J.; Chen, T. Designing a reductive hybrid membrane to selectively capture noble metallic ions during oil/water emulsion separation with further function enhancement. *J. Mater. Chem. A* **2018**, *6*, 10217–10225. [CrossRef]
37. Chen, Y.; Li, Z.; Ding, R.; Liu, T.; Zhao, H.; Zhang, X. Construction of porphyrin and viologen-linked cationic porous organic polymer for efficient and selective gold recovery. *J. Hazard. Mater.* **2022**, *426*, 128073. [CrossRef]
38. Liu, F.; Peng, G.; Li, T.; Yu, G.; Deng, S. Au(III) adsorption and reduction to gold particles on cost-effective tannin acid immobilized dialdehyde corn starch. *Chem. Eng. J.* **2019**, *370*, 228–236. [CrossRef]
39. Fan, R.; Xie, F.; Guan, X.; Zhang, Q.; Luo, Z. Selective adsorption and recovery of Au(III) from three kinds of acidic systems by persimmon residual based bio-sorbent: A method for gold recycling from e-wastes. *Bioresour. Technol.* **2014**, *163*, 167–171. [CrossRef]
40. Cavka, J.H.; Jakobsen, S.; Olsbye, U.; Guillou, N.; Lamberti, C.; Bordiga, S.; Lillerud, K.P. A New Zirconium Inorganic Building Brick Forming Metal Organic Frameworks with Exceptional Stability. *J. Am. Chem. Soc.* **2008**, *130*, 13850–13851. [CrossRef]
41. Katsuta, S.; Watanabe, Y.; Araki, Y.; Kudo, Y. Extraction of Gold(III) from Hydrochloric Acid into Various Ionic Liquids: Relationship between Extraction Efficiency and Aqueous Solubility of Ionic Liquids. *ACS Sustain. Chem. Eng.* **2015**, *4*, 564–571. [CrossRef]
42. Ahmed, I.; Adhikary, K.K.; Lee, Y.-R.; Ho Row, K.; Kang, K.-K.; Ahn, W.-S. Ionic liquid entrapped UiO-66: Efficient adsorbent for Gd3+ capture from water. *Chem. Eng. J.* **2019**, *370*, 792–799. [CrossRef]
43. Kanj, A.B.; Verma, R.; Liu, M.; Helfferich, J.; Wenzel, W.; Heinke, L. Bunching and Immobilization of Ionic Liquids in Nanoporous Metal-Organic Framework. *Nano Lett.* **2019**, *19*, 2114–2120. [CrossRef]
44. Campos, K.; Vincent, T.; Bunio, P.; Trochimczuk, A.; Guibal, E. Gold Recovery from HCl Solutions using Cyphos IL-101 (a Quaternary Phosphonium Ionic Liquid) Immobilized in Biopolymer Capsules. *Solvent Extr. Ion Exch.* **2008**, *26*, 570–601. [CrossRef]
45. Wang, S.; Wang, H.; Wang, S.; Zhang, L.; Fu, L. Highly effective and selective adsorption of Au(III) from aqueous solution by poly(ethylene sulfide) functionalized chitosan: Kinetics, isothermal adsorption and thermodynamics. *Microporous Mesoporous Mater.* **2022**, *341*, 112074. [CrossRef]
46. Chen, Y.; Tang, J.; Wang, S.; Zhang, L. Facile preparation of a remarkable MOF adsorbent for Au(III) selective separation from wastewater: Adsorption, regeneration and mechanism. *J. Mol. Liq.* **2022**, *349*, 118137. [CrossRef]
47. Zhang, M.; Dong, Z.; Hao, F.; Xie, K.; Qi, W.; Zhai, M.; Zhao, L. Ultrahigh and selective adsorption of Au(III) by rich sulfur and nitrogen-bearing cellulose microspheres and their applications in gold recovery from gold slag leaching solution. *Sep. Purif. Technol.* **2021**, *274*, 119016. [CrossRef]
48. Zhou, X.; Mo, X.; Zhu, W.; Xu, W.; Tang, K.; Lei, Y. Selective adsorption of Au(III) with ultra-fast kinetics by a new metal-organic polymer. *J. Mol. Liq.* **2020**, *319*, 114125. [CrossRef]
49. Xiong, C.; Wang, S.; Zhang, L.; Li, Y.; Zhou, Y.; Peng, J. Preparation of 2-Aminothiazole-Functionalized Poly(glycidyl methacrylate) Microspheres and Their Excellent Gold Ion Adsorption Properties. *Polymers* **2018**, *10*, 159. [CrossRef]
50. Lin, G.; Wang, S.; Zhang, L.; Hu, T.; Peng, J.; Cheng, S.; Fu, L.; Srinivasakannan, C. Selective recovery of Au(III) from aqueous solutions using 2-aminothiazole functionalized corn bract as low-cost bioadsorbent. *J. Clean. Prod.* **2018**, *196*, 1007–1015. [CrossRef]
51. Wu, C.; Zhu, X.; Wang, Z.; Yang, J.; Li, Y.; Gu, J. Specific Recovery and In Situ Reduction of Precious Metals from Waste To Create MOF Composites with Immobilized Nanoclusters. *Ind. Eng. Chem. Res.* **2017**, *56*, 13975–13982. [CrossRef]

52. Wang, Z.; Li, X.; Liang, H.; Ning, J.; Zhou, Z.; Li, G. Equilibrium, kinetics and mechanism of Au(3+), Pd(2+) and Ag(+) ions adsorption from aqueous solutions by graphene oxide functionalized persimmon tannin. *Mater. Sci. Eng. C Mater. Biol. Appl.* **2017**, *79*, 227–236. [CrossRef] [PubMed]
53. Geng, Y.; Li, J.; Lu, W.; Wang, N.; Xiang, Z.; Yang, Y. Au(III), Pd(II) and Pt(IV) adsorption on amino-functionalized magnetic sorbents: Behaviors and cycling separation routines. *Chem. Eng. J.* **2020**, *381*, 122627. [CrossRef]
54. Zhao, J.; Wang, C.; Wang, S.; Zhang, L.; Zhang, B. Selective recovery of Au(III) from wastewater by a recyclable magnetic $Ni_{0.6}Fe_{2.4}O_4$ nanoparticels with mercaptothiadiazole: Interaction models and adsorption mechanisms. *J. Clean. Prod.* **2019**, *236*, 117605. [CrossRef]
55. Saadi, R.; Saadi, Z.; Fazaeli, R.; Fard, N.E. Monolayer and multilayer adsorption isotherm models for sorption from aqueous media. *Korean J. Chem. Eng.* **2015**, *32*, 787–799. [CrossRef]
56. Hu, Q.; Zhang, Z. Application of Dubinin–Radushkevich isotherm model at the solid/solution interface: A theoretical analysis. *J. Mol. Liq.* **2019**, *277*, 646–648. [CrossRef]
57. Zhao, M.; Huang, Z.; Wang, S.; Zhang, L. Ultrahigh efficient and selective adsorption of Au(III) from water by novel Chitosan-coated MoS2 biosorbents: Performance and mechanisms. *Chem. Eng. J.* **2020**, *401*, 126006. [CrossRef]
58. Wang, S.; Shen, M.; Qu, J.; Zhuang, X.; Ni, S.-Q.; Wu, X. Synthesis and characterization of mercapto-modified graphene/multi-walled carbon nanotube aerogels and their adsorption of Au(III) from environmental samples. *J. Non-Cryst. Solids* **2020**, *536*, 120008. [CrossRef]
59. Fagbohun, E.O.; Wang, Q.; Spessato, L.; Zheng, Y.; Li, W.; Olatoye, A.G.; Cui, Y. Physicochemical regeneration of industrial spent activated carbons using a green activating agent and their adsorption for methyl orange. *Surf. Interfaces* **2022**, *29*, 101696. [CrossRef]
60. Wu, S.; Wang, F.; Yuan, H.; Zhang, L.; Mao, S.; Liu, X.; Alharbi, N.S.; Rohani, S.; Lu, J. Fabrication of xanthate-modified chitosan/poly(N-isopropylacrylamide) composite hydrogel for the selective adsorption of Cu(II), Pb(II) and Ni(II) metal ions. *Chem. Eng. Res. Des.* **2018**, *139*, 197–210. [CrossRef]
61. Liu, F.; Wang, S.; Chen, S. Adsorption behavior of Au(III) and Pd(II) on persimmon tannin functionalized viscose fiber and the mechanism. *Int. J. Biol. Macromol.* **2020**, *152*, 1242–1251. [CrossRef]
62. Karamanoğlu, P.; Aydın, S. An economic analysis of the recovery of gold from CPU, boards, and connectors using aqua regia. *Desalination Water Treat.* **2015**, *57*, 2570–2575. [CrossRef]
63. Syed, S. Recovery of gold from secondary sources—A review. *Hydrometallurgy* **2012**, *115–116*, 30–51. [CrossRef]
64. Song, J.Y.; Ahmed, I.; Seo, P.W.; Jhung, S.H. UiO-66-Type Metal-Organic Framework with Free Carboxylic Acid: Versatile Adsorbents via H-bond for Both Aqueous and Nonaqueous Phases. *ACS Appl. Mater. Interfaces* **2016**, *8*, 27394–27402. [CrossRef] [PubMed]

Disclaimer/Publisher's Note: The statements, opinions and data contained in all publications are solely those of the individual author(s) and contributor(s) and not of MDPI and/or the editor(s). MDPI and/or the editor(s) disclaim responsibility for any injury to people or property resulting from any ideas, methods, instructions or products referred to in the content.

Article

A Computational Analysis of the Reaction of SO$_2$ with Amino Acid Anions: Implications for Its Chemisorption in Biobased Ionic Liquids

Vanessa Piacentini [1], Andrea Le Donne [2], Stefano Russo [1] and Enrico Bodo [1,*]

[1] Chemistry Department, University of Rome "La Sapienza", 00185 Rome, Italy; vanessa.piacentini@uniroma1.it (V.P.); stefano.russo@uniroma1.it (S.R.)
[2] Department of Chemical, Pharmaceutical and Agricultural Sciences, University of Ferrara, 44121 Ferrara, Italy; andrea.ledonne@unife.it
* Correspondence: enrico.bodo@uniroma1.it

Abstract: We report a series of calculations to elucidate one possible mechanism of SO$_2$ chemisorption in amino acid-based ionic liquids. Such systems have been successfully exploited as CO$_2$ absorbents and, since SO$_2$ is also a by-product of fossil fuels' combustion, their ability in capturing SO$_2$ has been assessed by recent experiments. This work is exclusively focused on evaluating the efficiency of the chemical trapping of SO$_2$ by analyzing its reaction with the amino group of the amino acid. We have found that, overall, SO$_2$ is less reactive than CO$_2$, and that the specific amino acid side chain (either acid or basic) does not play a relevant role. We noticed that bimolecular absorption processes are quite unlikely to take place, a notable difference with CO$_2$. The barriers along the reaction paths are found to be non-negligible, around 7–11 kcal/mol, and the thermodynamic of the reaction appears, from our models, unfavorable.

Keywords: SO$_2$ capture; amino acids; ionic liquids

1. Introduction

SO$_2$ is a colorless and poisonous gas and represents one of the main atmospheric pollutants generated by anthropic activities. Each year, around 150 million tons of SO$_2$ are produced by various human activities, mainly burning fossil fuels, industrial processes, and energy production [1–3]. The anthropic SO$_2$ is mainly produced by fossil fuel processing since SO$_2$ is typically removed from liquid fuels before usage. Atmospheric SO$_2$ can be oxidated to SO$_3$, which reacts with water to form H$_2$SO$_4$, leading to the appearance of corrosive agents in the atmosphere. The ensuing soil and water acidification has been an environmental problem that has fortunately decreased its impact in recent years due to the adoption of measures aimed at a strong reduction of SO$_2$ emissions, especially from gas exhaust due to the automotive compartment [4,5].

A reduction in SO$_2$ pollution is achieved either by removing the gas from fuel prior to combustion or removing the gas after combustion by treatment of the flue gas (FDG, flue gas desulfurization) [6,7].

In FGD, different chemistries can be used to remove SO$_2$, among others the most notable involve the use of limestone, magnesium oxide, and ammonia [8]. Most, if not all, of these absorption processes are irreversible and do not allow for SO$_2$ recycling despite it being a useful chemical in industry. The only reversible absorption processes are those based on the use of amine solutions (in the same way they are employed for CO$_2$ absorption [9]), but the process is far from being optimal due to the loss of amines in the stream gas.

Due to their low vapor pressure and their intrinsic ability to be functionalized for specific purposes [10], ionic liquids (ILs) have been proposed as possible green absorbing agents for CO_2 and SO_2 removal from gasses in industrial settings [8,11,12]. The absorption of SO_2 in ILs can be physical [8,13,14] or chemical [15–18]. Our interest, in the present work, is the chemical absorption process whereby an SO_2 molecule reacts with a specific group of the IL molecular ions and gives rise to a new chemical specie.

In analogy with what has been seen for CO_2 [19,20], we explore the reaction mechanisms between an amino acid (AA) anion and SO_2. Amino acid-based ILs (AAILS) have attracted the interest of a part of the research community because of their intrinsic biocompatibility [21–29] and, in particular, have been implemented as efficient chemisorbing media for CO_2 absorption from flue gas [30–36].

A possible mechanism of the chemical reaction has already been proposed in [37], where the products have been characterized using FTIR spectroscopy. In [38], it has been shown how aqueous solutions of AAILs are efficient absorbent media and how SO_2 can be regenerated after absorption by heating at 100 °C. In [39], a guanidium-based IL has been shown to undergo chemical reaction by means of its NH_2 group to yield an $-NH-SO_2$ structure.

This work is focused on understanding the mechanisms of the reaction between an AA anion and the SO_2 molecule using ab initio calculations. We do not consider the problem of the diffusion of the SO_2 molecule inside the fluid, nor do we include water in the calculation, but we shall assume an anhydrous IL and that the SO_2 is already in the proximity of the reactive terminal of the AA anion, which is the NH_2 group. To maintain generality, we had to adopt a simplified model that produces universal results, possibly independent of the many variables at play in real systems: we treat environmental effects using an approximate, continuum solvent model, and we do not include a cationic partner. While the latter sounds like an oversimplification, it is the only way to maintain the widest generality. While the cation does play a role in the reaction, it appears very difficult to assess it using ab initio modeling, since the binding motif and energy of an ionic couple in the bulk phase would be inevitably different from those that we can obtain in an isolated system. We also point out that the study of the effect of the cationic partner on the reaction mechanisms would require a different computational strategy and a systematic study of its presence upon varying coordinating properties, size, and steric shape. Such study lies outside the scope of the present work, which focuses exclusively on the reactivity of the anions.

Finally, to support our choice, we mention the data reported in [32,40] for the analogous reaction with CO_2, that show how different cations have a limited effect in modifying both the reaction profile and the energetic barriers associated with the proton transfer step, especially when compared to the effect of a change in the AA anion.

2. Methods

The calculations have been performed using the g16 program [41] and the B3LYP functional combined with the 6−311+G(d,p) basis set, and corrected for dispersion interaction using the empirical GD3 approach of Grimme [42]. All structures have been fully optimized without constraints. Each stationary point has been further characterized by a harmonic frequency calculation, and the zero-point-energies (ZPE) have been added to the electronic energy. Thermodynamic functions such as the Gibbs free energy have been computed under room conditions. The uniqueness of the transition state has been verified using a suitable IRC (intrinsic reaction coordinate) calculation for all the structures presented. Electronic density and population analysis has been performed using Janpa [43].

All calculations have been performed in gas phase and in the solvent phase using the SMD, the universal solvation model based on solute electron density [44], with parameters of the acetonitrile solvent that can be used as representative of the dielectric screening acting in these liquids [19]. When possible, the SMD model has been modified to account for the measured dielectric constant of the IL.

The three AA anions that have been used to study the SO_2 absorption process are: glycine (Gly), cysteine (Cys), and lysine (Lys), which have different chemical motifs on their side chains and could be representative of the variety of AA. The generic chemical reaction [37] between SO_2 and the AA anion proceeds as reported in Scheme 1: Initially, there is a non-covalent complex, **R**, between the two molecules. This complex evolves through a transition state, **TS**, where a proton transfer occurs to produce the product, **P**. The specific mechanism depends on the type of AA anion and, in particular, in which way the proton transfer proceeds; for example, it can also take place between the $-NH_2^+-$ and the SO_2^- groups, thus producing, first, a sulfinic acid derivative, which undergoes an isomerization to the carboxylic acid form **P**. In addition, and as we shall see below, a second basic functional group of the AA can be, at least in principle, involved in the proton transfer step.

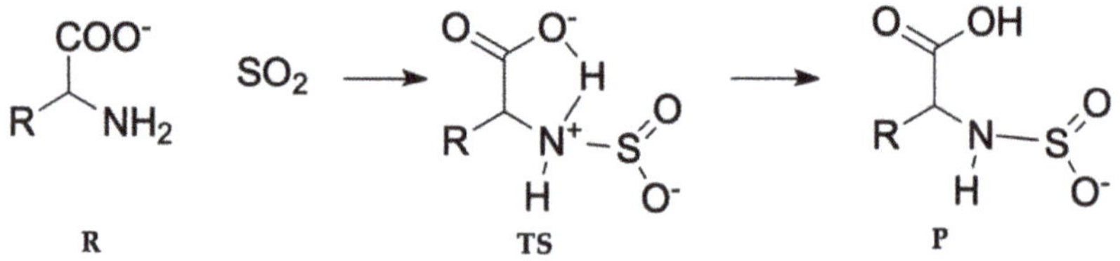

Scheme 1. Generic reaction for the addition of SO_2 to an AA anion.

3. Results

3.1. The Glycinate Reaction

We begin our presentation by looking at the results on the simplest AA anion, i.e., Gly, that serves here as an exemplar case for all aliphatic AA anions. The optimized structures along two possible reaction paths (that differs from the nature of the proton transfer) are reported in Figure 1. The top one involves a transition state with a 5-member ring (**Gly-TS5**) structure, and the bottom one a 4-member ring (**Gly-TS4**). In the former, the proton transfer involves the carboxylate, and in the latter the SO_2^- group. Both transition states require the formation of a positive charge formally located on the tetravalent nitrogen atom and of a negative charge on the oxygens of the $-SO_2$ group.

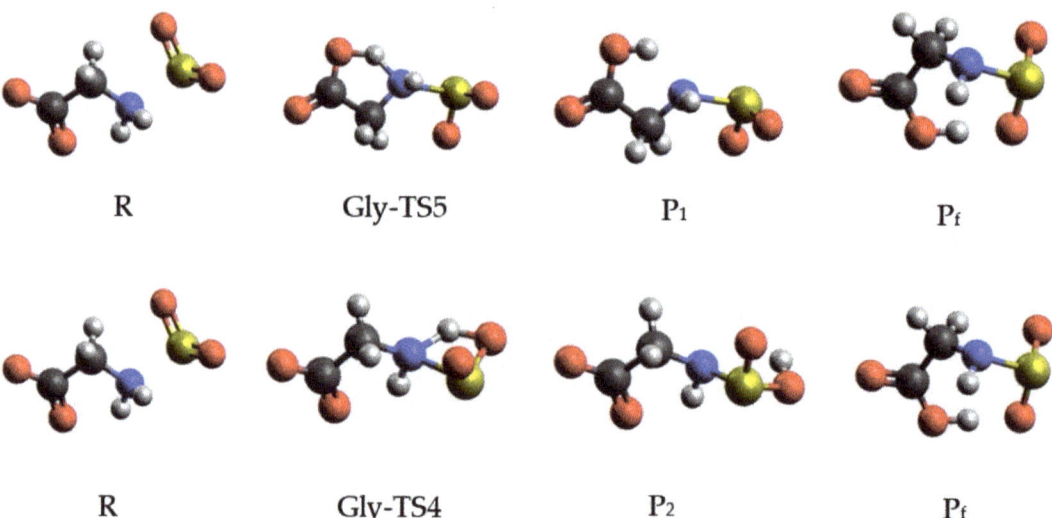

Figure 1. Stationary point geometries along the reaction of $[Gly]^-$ with SO_2. (**Top**): Path with a 5-mamber transition state that sees the transfer of a proton toward the carboxylate. (**Bottom**): Path involving a 4-member ring with proton transfer on the SO_2. The last two structures in the top sequence are conformers, while those of the bottom sequence are isomers.

While for a CO_2 molecule the pre-reaction complex has a strong zwitterionic character, for SO_2, the appearance of a charge separation takes place only when the process has already significantly evolved through the reaction coordinate [19]. Our calculations show that for SO_2, there is almost no trace of zwitterion formation in the pre-reaction complex. This is already a clear sign of a diminished propensity of SO_2 to react with the amino group.

The path involving **Gly-TS5** terminates to the product P_1 that has the proton on the carboxylate, as expected, and the final product P_f is only a more stable conformer of the same structure. The path involving **Gly-PT4** produces a structure that has the proton on the $-SO_2$ group that evolves to the same P_f product of the first path by isomerization. The energetic path is in accordance with the expected acidity of the $-SO_2H$ and $-CO_2H$ groups.

The energies of the **Gly-TS5** path in the gas phase are reported in Figure 2. With respect to the isolated reactants (**IR**), the reaction is highly exothermic with a $\Delta_r H$ of -26.4 kcal/mol and a $\Delta_r G$ of -13.0 kcal/mol. The reduced value of the reaction free energy is due to the unfavorable entropic contribution that emerges in an association reaction. If the initial reaction complex, **R**, is efficiently quenched by the environment, the addition reaction of SO_2 to the AA anion is almost isoergonic, with little or no thermodynamic propensity.

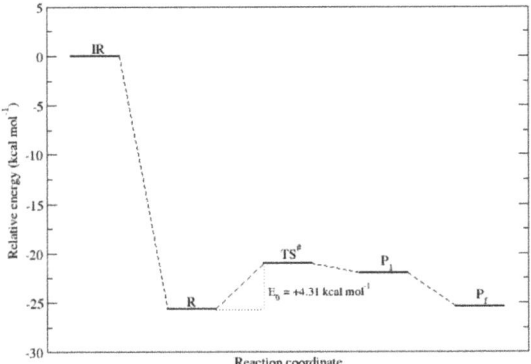

Figure 2. Reaction energies of the **Gly-TS5** path with respect to isolated reactants as computed using ZPE corrected electronic energies.

The energies of the **Gly-TS4** reaction path are reported in Figure 3. This path is hindered by a substantial activation barrier of about 27 kcal/mol that is due to the strain in the 4-member ring that characterizes the transition state. This path is obviously very inefficient, especially in an environment that can quench the initial reactant energy.

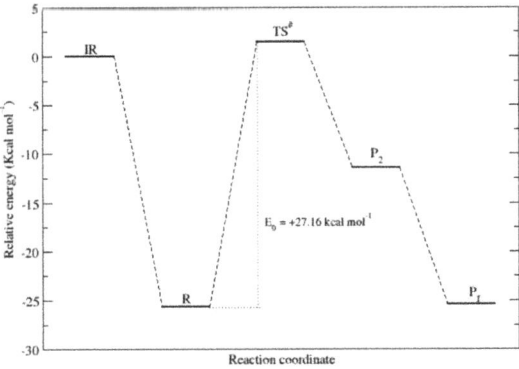

Figure 3. Reaction energies of the **Gly-TS4** path with respect to isolated reactants as computed using ZPE corrected electronic energies.

The same calculations performed in a solvent model with a dielectric constant of 35.8 (acetonitrile) produced the numbers shown in Figure 4, where we report only the **Gly-TS5** path since the **Gly-TS4** path remains hindered by a barrier of 28.8 kcal/mol. The addition of the solvent has the effect of lowering the energy of the isolated reactants and to reduce the overall enthalpic gain to 14.82 kcal/mol and the corresponding $\Delta_r G$ to only -0.5 kcal/mol. The barrier with respect to the quenched reactants **R** is increased to 7.1 kcal/mol, a value that still allows the reaction to proceed quite efficiently at moderate temperatures.

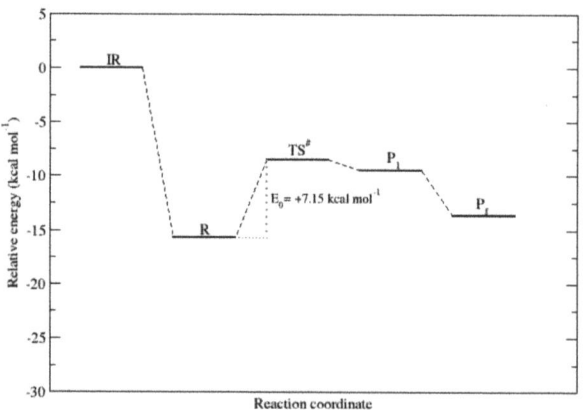

Figure 4. Reaction energies of the **Gly-TS5** path with respect to isolated reactants as computed using ZPE corrected electronic energies in the SMD solvent model.

Notably, the glycinate reaction with SO_2 in a solvent medium becomes slightly endoergic if the pre-reaction complex **R** is sufficiently long-lived to be quenched by the environment. While this reduces the overall efficiency of the incorporation process, it also indicates that the above reaction can be reversible and that the AA anions can serve as a temporary storage of SO_2.

The final product is characterized by an unusual $NH-SO_2^-$ bond that is the sulfur equivalent of the N–C bond in carbamates. The N–S bond is significantly weaker than its N–C counterpart. Its bond distance (from our calculations) is large and is 1.8 Å for all three AA derivatives. The lowest energy dissociation path in vacuo is heterolytic and leads to negative $-NH^-$ and neutral SO_2. The adiabatic bond dissociation energies with respect to this fragmentation are 25.4, 22.2, and 24.0 kcal/mol for Gly, Cys, and Lys, respectively. This is a much lower bond energy when compared to that of the C–N bond in carbamates, which is about 100 kcal/mol [45]. In the sulfinic derivative, both the NH group and the SO_2 are negatively charged with roughly half an electron each, as computed from natural orbital populations. The Wiberg–Mayer bond indices of the S–N bond for the three AA derivatives are 0.85, 0.72, and 0.70 for Gly, Cys, and Lys, respectively, again pointing to a weak single bond. The N–S bond can be described by the localized molecular orbital (CLPO) arising from the combination of two optimally paired hybrids [43], where the contribution of the N hybrid is the dominant one, as shown in Figure 5 for the Gly derivative.

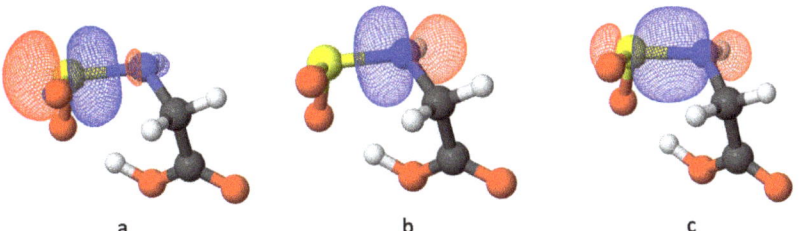

Figure 5. Localized molecular orbitals characterizing the N–S bond in the Gly derivative, P_f. (**a,b**): Optimized atomic hybrids (LHOs) centered on S and N, respectively. (**c**): CLPO responsible for the bond, where the LHO centered on N accounts for 0.64% of the linear combination.

3.2. The Cysteinate Reaction

The reaction with a cysteinate anion proceeds substantially in the same way. A first possible path (**Cys-TS5**) for the reaction with SO_2 requires a 5-member transition state and a proton migration toward the carboxylate group. This path in the gas phase, as in the Gly case, has a low activation barrier of 5.1 kcal/mol, and the $\Delta_r H$ and $\Delta_r G$ are -23.19 and -9.65 kcal/mol, respectively. The same calculation repeated in the SMD model, with the dielectric constant of 9.5 (which is the measured dielectric constant for the [Ch][Cys] IL reported in [46]), yields a $\Delta_r H$ of -14.4 kcal/mol and a $\Delta_r G$ of -1.0 kcal/mol. The reaction paths in vacuo and in solvent are reported in Figure 6, along with the geometries of the stationary points.

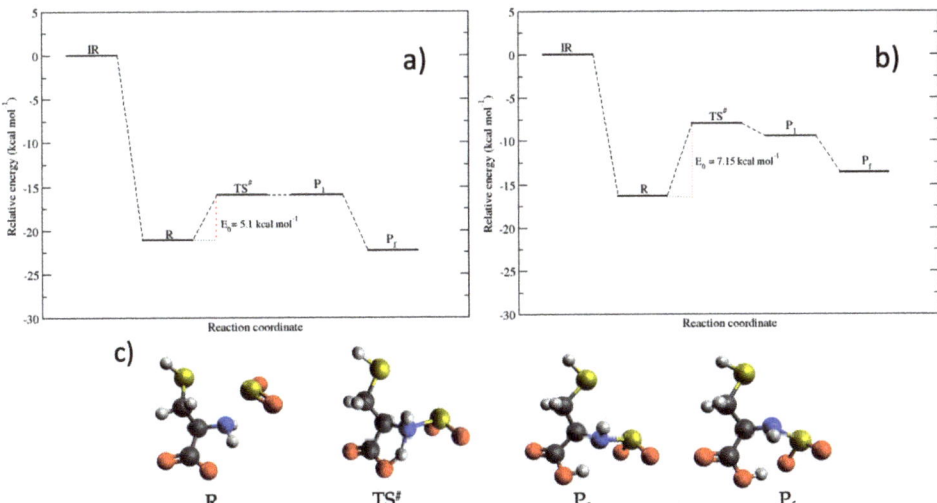

Figure 6. Reaction energies of the **Cys-TS5** path with respect to isolated reactants as computed using ZPE corrected electronic energies. (**a**): Reaction in the gas phase. (**b**): Reaction in the SMD solvent model. (**c**): Optimized geometries in the gas phase of the stationary points along the reaction path.

As we have shown in [47], the cysteinate anion can present itself in an isomeric form, where the proton of the thiol group has migrated onto the carboxylate. The result is that this isomeric form of the AA anion can bind the SO_2 molecule only via the reaction path that requires a proton transfer from the $-NH_2$ group through the highly unstable transition state characterized by a 4-member ring. Needless to say, this path is hindered by a large activation barrier of 27.1 kcal/mol. In addition, the overall reaction free energy becomes positive and equal to 5.8 kcal/mol, thus rendering this reaction channel extremely unlikely to be of any relevance for the pollutant absorption.

We tried to locate a possible reaction path that involved a proton transfer toward the thiolate group, but none of our attempts were successful. In other words, it appears that the –SH group does not play an active role in the SO$_2$ absorption processes.

3.3. The Lysinate Reaction

A third set of calculations were repeated on the Lys anion, where it can be argued that the additional basic NH$_2$ group on the side chain could play a role in the reaction mechanism or even give rise to a bimolecular absorption path, in which both amino groups are converted to sulfinic acids.

The analogous of the **Gly-TS5** and **Cys-TS5** mechanisms is also active for Lys and is reported in terms of energies and structures in Figure 7. The barriers in the gas phase and in the model solvent (with a dielectric constant of 9.8 [46]) are 6.8 and 11.2 kcal/mol, respectively. For the gas phase, the $\Delta_r H$ and $\Delta_r G$ are −26.4 and −11.8 kcal/mol, respectively. These values are reduced to −15.6 and −2.4 kcal/mol in the solvent model.

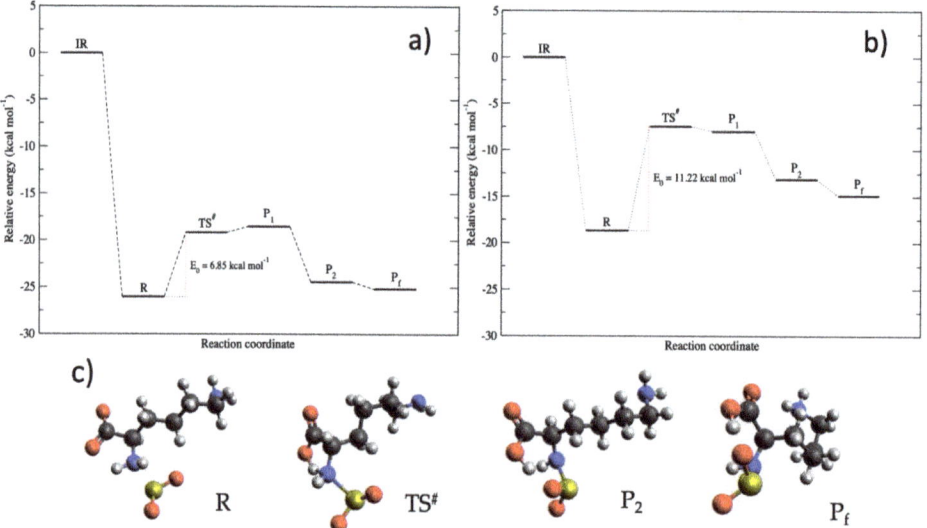

Figure 7. Reaction energies of the **Lys-TS5** path with respect to isolated reactants as computed using zero-point energy-corrected electronic energies. (**a**): Reaction in the gas phase. (**b**): Reaction in the SMD solvent model. (**c**): Optimized geometries in the gas phase of the stationary points along the reaction path. The **P$_1$** product is a high-energy isomer of **P$_2$**.

We have tried to trace a possible reactive path that involved the side chain NH$_2$, but this group seems impervious to the SO$_2$ attack. This is probably due to its higher basicity and to the absence of the carboxylate in β that can activate the reaction. This result, albeit negative, is important because the SO$_2$ molecule appears less reactive than CO$_2$, especially toward the AA side chains. It therefore appears unlikely that a single AA anion would be able to capture more than one SO$_2$ molecule unless its molecular structure is purposedly engineered by placing more than one "activated" amino group. In comparison, we defer the reader to our study in [20], where we had shown how efficient a bimolecular reaction of the same AA can be when CO$_2$ is involved.

4. Conclusions

This study attempted to elucidate a possible reaction mechanism based on the chemisorption of SO$_2$ in biobased ionic liquids, namely, its reaction with the amino group of the AA anion. The model employed here did not account for the problem of SO$_2$ diffusion in the liquid, and instead only focused on the reactive step once the SO_2 had the chance to attack the amino group of the AA anion. Solvation and stabilization due to a dielectric has been nevertheless considered by using an approximate continuum solvent model. This approach, albeit quite simplified, has already been successfully used by us to illustrate the mechanisms at the basis of the quite efficient absorption of CO_2 in these liquids [19,20].

In comparison to CO_2, where a set of reaction mechanisms were found to be essentially barrierless and strongly exoergic with negative $\Delta_r G$, SO_2 was less reactive towards the AA anions. This reduced reactivity substantially prevents SO_2 from taking advantage of the presence of specific functional side chains, and the only mechanism that is viable is a direct addition of it to the amino group next to the carboxylate, a reactive path that should be common to all AA anions.

The mechanism involves passing through a cyclic transition state, where one of the protons initially belonging to NH_2 is transferred to the carboxylate, neutralizing it. The resulting deprotonated sulfinic acid derivative is the final product. We have tried to find other mechanisms that involve the side chain, but we were unable to find any other efficient route. In contrast with the case of CO_2, where the slightness or even absence of kinetic barriers makes the reaction extremely efficient, SO_2 appears less prone to chemisorption due to the appearance of non-negligible barriers and to a much less exoergic thermodynamic of the involved reactions. In general, the barrier to the chemisorption reaction appeared to lie in the 7–11 kcal/mol interval, while the reaction free energies were only slightly negative, with values between -0.5 and -2 kcal/mol. The reason for this behavior lies in the relative weakness of the N–S bond when compared to its analogous in carbamates (the result of the CO_2 reaction). The bond tends to break heterolytically, leading to neutral SO_2, and its dissociation requires only 22–25 kcal/mol. This last aspect, albeit reducing the overall reaction rate, did however indicate a facile reversibility of the absorption reaction, thus pointing to a possible use of these materials as temporary storage for SO_2.

While in the case of CO_2 the overall absorption process, due to the high rate of the absorption reaction, seemed to be dominated by the rate of diffusion of the molecule in the liquid, for SO_2, the chemical step is very likely to be much slower (especially at room temperature) and thermodynamically inefficient. This means that the absorption of SO_2 in these liquids can easily be dominated by physisorption, with chemisorption being only a minor factor contributing to their overall intake capacity.

Author Contributions: Conceptualization and writing, E.B.; investigation, V.P., S.R. and A.L.D. All authors have read and agreed to the published version of the manuscript.

Funding: This research received no external funding.

Data Availability Statement: The data presented in this study are available upon reasonable request from the corresponding author.

Conflicts of Interest: The authors declare no conflict of interest.

References

1. World Health Organization (Ed.) *Air Quality Guidelines: Global Update 2005: Particulate Matter, Ozone, Nitrogen Dioxide, and Sulfur Dioxide*; World Health Organization: Copenhagen, Denmark, 2006; ISBN 978-92-890-2192-0.
2. Hoesly, R.M.; Smith, S.J.; Feng, L.; Klimont, Z.; Janssens-Maenhout, G.; Pitkanen, T.; Seibert, J.J.; Vu, L.; Andres, R.J.; Bolt, R.M.; et al. Historical (1750–2014) Anthropogenic Emissions of Reactive Gases and Aerosols from the Community Emissions Data System (CEDS). *Geosci. Model Dev.* **2018**, *11*, 369–408. [CrossRef]
3. Chen, S.; Li, Y.; Yao, Q. The Health Costs of the Industrial Leap Forward in China: Evidence from the Sulfur Dioxide Emissions of Coal-Fired Power Stations. *China Econ. Rev.* **2018**, *49*, 68–83. [CrossRef]
4. Lawrence, G.B.; Hazlett, P.W.; Fernandez, I.J.; Ouimet, R.; Bailey, S.W.; Shortle, W.C.; Smith, K.T.; Antidormi, M.R. Declining Acidic Deposition Begins Reversal of Forest-Soil Acidification in the Northeastern U.S. and Eastern Canada. *Environ. Sci. Technol.* **2015**, *49*, 13103–13111. [CrossRef] [PubMed]
5. Kahl, J.S.; Stoddard, J.L.; Haeuber, R.; Paulsen, S.G.; Birnbaum, R.; Deviney, F.A.; Webb, J.R.; DeWalle, D.R.; Sharpe, W.; Driscoll, C.T.; et al. Peer Reviewed: Have U.S. Surface Waters Responded to the 1990 Clean Air Act Amendments? *Environ. Sci. Technol.* **2004**, *38*, 484A–490A. [CrossRef]

6. Hansen, B.B.; Kiil, S.; Johnsson, J.E.; Sønder, K.B. Foaming in Wet Flue Gas Desulfurization Plants: The Influence of Particles, Electrolytes, and Buffers. *Ind. Eng. Chem. Res.* **2008**, *47*, 3239–3246. [CrossRef]
7. Mirdrikvand, M.; Moqadam, S.I.; Kharaghani, A.; Roozbehani, B.; Jadidi, N. Optimization of a Pilot-Scale Amine Scrubber to Remove SO_2: Higher Selectivity and Lower Solvent Consumption. *Chem. Eng. Technol.* **2016**, *39*, 246–254. [CrossRef]
8. Ren, S.; Hou, Y.; Zhang, K.; Wu, W. Ionic Liquids: Functionalization and Absorption of SO_2. *Green Energy Environ.* **2018**, *3*, 179–190. [CrossRef]
9. Bates, E.D.; Mayton, R.D.; Ntai, I.; Davis, J.H. CO_2 Capture by a Task-Specific Ionic Liquid. *J. Am. Chem. Soc.* **2002**, *124*, 926–927. [CrossRef]
10. Giernoth, R. Task-Specific Ionic Liquids. *Angew. Chem. Int. Ed.* **2010**, *49*, 2834–2839. [CrossRef]
11. Yan, S.; Han, F.; Hou, Q.; Zhang, S.; Ai, S. Recent Advances in Ionic Liquid-Mediated SO_2 Capture. *Ind. Eng. Chem. Res.* **2019**, *58*, 13804–13818. [CrossRef]
12. Rashid, T.U. Ionic Liquids: Innovative Fluids for Sustainable Gas Separation from Industrial Waste Stream. *J. Mol. Liq.* **2021**, *321*, 114916. [CrossRef]
13. Zeng, S.; Gao, H.; Zhang, X.; Dong, H.; Zhang, X.; Zhang, S. Efficient and Reversible Capture of SO_2 by Pyridinium-Based Ionic Liquids. *Chem. Eng. J.* **2014**, *251*, 248–256. [CrossRef]
14. Jiang, L.; Mei, K.; Chen, K.; Dao, R.; Li, H.; Wang, C. Design and Prediction for Highly Efficient SO_2 Capture from Flue Gas by Imidazolium Ionic Liquids. *Green Energy Environ.* **2022**, *7*, 130–136. [CrossRef]
15. Wang, L.; Zhang, Y.; Liu, Y.; Xie, H.; Xu, Y.; Wei, J. SO_2 Absorption in Pure Ionic Liquids: Solubility and Functionalization. *J. Hazard. Mater.* **2020**, *392*, 122504. [CrossRef]
16. Mao, F.-F.; Zhou, Y.; Zhu, W.; Sang, X.-Y.; Li, Z.-M.; Tao, D.-J. Synthesis of Guanidinium-Based Poly(Ionic Liquids) with Nonporosity for Highly Efficient SO_2 Capture from Flue Gas. *Ind. Eng. Chem. Res.* **2021**, *60*, 5984–5991. [CrossRef]
17. Geng, Z.; Xie, Q.; Fan, Z.; Sun, W.; Zhao, W.; Zhang, J.; Chen, J.; Xu, Y. Investigation of Tertiary Amine-Based PILs for Ideal Efficient SO_2 Capture from CO_2. *J. Environ. Chem. Eng.* **2021**, *9*, 105824. [CrossRef]
18. Hou, Y.; Zhang, Q.; Gao, M.; Ren, S.; Wu, W. Absorption and Conversion of SO_2 in Functional Ionic Liquids: Effect of Water on the Claus Reaction. *ACS Omega* **2022**, *7*, 10413–10419. [CrossRef]
19. Onofri, S.; Adenusi, H.; Le Donne, A.; Bodo, E. CO_2 Capture in Ionic Liquids Based on Amino Acid Anions With Protic Side Chains: A Computational Assessment of Kinetically Efficient Reaction Mechanisms. *ChemistryOpen* **2020**, *9*, 1153–1160. [CrossRef]
20. Onofri, S.; Bodo, E. CO_2 Capture in Biocompatible Amino Acid Ionic Liquids: Exploring the Reaction Mechanisms for Bimolecular Absorption Processes. *J. Phys. Chem. B* **2021**, *125*, 5611–5619. [CrossRef]
21. Hou, X.-D.; Liu, Q.-P.; Smith, T.J.; Li, N.; Zong, M.-H. Evaluation of Toxicity and Biodegradability of Cholinium Amino Acids Ionic Liquids. *PLoS ONE* **2013**, *8*, e59145. [CrossRef]
22. Gontrani, L.; Scarpellini, E.; Caminiti, R.; Campetella, M. Bio Ionic Liquids and Water Mixtures: A Structural Study. *RSC Adv.* **2017**, *7*, 19338–19344. [CrossRef]
23. Gontrani, L. Choline-Amino Acid Ionic Liquids: Past and Recent Achievements about the Structure and Properties of These Really "Green" Chemicals. *Biophys. Rev.* **2018**, *10*, 873–880. [CrossRef]
24. Caparica, R.; Júlio, A.; Baby, A.; Araújo, M.; Fernandes, A.; Costa, J.; Santos de Almeida, T. Choline-Amino Acid Ionic Liquids as Green Functional Excipients to Enhance Drug Solubility. *Pharmaceutics* **2018**, *10*, 288. [CrossRef] [PubMed]
25. Le Donne, A.; Adenusi, H.; Porcelli, F.; Bodo, E. Structural Features of Cholinium Based Protic Ionic Liquids through Molecular Dynamics. *J. Phys. Chem. B* **2019**, *123*, 5568–5576. [CrossRef] [PubMed]
26. Chen, X.; Luo, X.; Li, J.; Qiu, R.; Lin, J. Cooperative CO_2 Absorption by Amino Acid-Based Ionic Liquids with Balanced Dual Sites. *RSC Adv.* **2020**, *10*, 7751–7757. [CrossRef] [PubMed]
27. Bodo, E. Modelling Biocompatible Ionic Liquids Based on Organic Acids and Amino Acids: Challenges for Computational Models and Future Perspectives. *Org. Biomol. Chem.* **2021**, *19*, 4002–4013. [CrossRef] [PubMed]
28. Dhattarwal, H.S.; Kashyap, H.K. Unique and Generic Structural Features of Cholinium Amino Acid-Based Biocompatible Ionic Liquids. *Phys. Chem. Chem. Phys.* **2021**, *23*, 10662–10669. [CrossRef]
29. Bodo, E. Perspectives in the Computational Modeling of New Generation, Biocompatible Ionic Liquids. *J. Phys. Chem. B* **2022**, *126*, 3–13. [CrossRef]
30. Hussain, M.A.; Soujanya, Y.; Sastry, G.N. Evaluating the Efficacy of Amino Acids as CO_2 Capturing Agents: A First Principles Investigation. *Environ. Sci. Technol.* **2011**, *45*, 8582–8588. [CrossRef]
31. Bhattacharyya, S.; Shah, F.U. Ether Functionalized Choline Tethered Amino Acid Ionic Liquids for Enhanced CO_2 Capture. *ACS Sustain. Chem. Eng.* **2016**, *4*, 5441–5449. [CrossRef]
32. Firaha, D.S.; Kirchner, B. Tuning the Carbon Dioxide Absorption in Amino Acid Ionic Liquids. *ChemSusChem* **2016**, *9*, 1591–1599. [CrossRef]
33. Chen, F.-F.; Huang, K.; Zhou, Y.; Tian, Z.-Q.; Zhu, X.; Tao, D.-J.; Jiang, D.; Dai, S. Multi-Molar Absorption of CO_2 by the Activation of Carboxylate Groups in Amino Acid Ionic Liquids. *Angew. Chem. Int. Ed.* **2016**, *55*, 7166–7170. [CrossRef]
34. Chen, K.; Wang, Y.; Yao, J.; Li, H. Equilibrium in Protic Ionic Liquids: The Degree of Proton Transfer and Thermodynamic Properties. *J. Phys. Chem. B* **2018**, *122*, 309–315. [CrossRef]
35. Latini, G.; Signorile, M.; Crocellà, V.; Bocchini, S.; Pirri, C.F.; Bordiga, S. Unraveling the CO_2 Reaction Mechanism in Bio-Based Amino-Acid Ionic Liquids by Operando ATR-IR Spectroscopy. *Catal. Today* **2019**, *336*, 148–160. [CrossRef]

36. Davarpanah, E.; Hernández, S.; Latini, G.; Pirri, C.F.; Bocchini, S. Enhanced CO_2 Absorption in Organic Solutions of Biobased Ionic Liquids. *Adv. Sustain. Syst.* **2020**, *4*, 1900067. [CrossRef]
37. Meng, X.; Wang, J.; Jiang, H.; Shi, X.; Hu, Y. 2-Ethyl-4-Methylimidazolium Alaninate Ionic Liquid: Properties and Mechanism of SO_2 Absorption. *Energy Fuels* **2017**, *31*, 2996–3001. [CrossRef]
38. Wang, H.; Wu, P.; Li, C.; Zhang, J.; Deng, R. Reversible and Efficient Absorption of SO_2 with Natural Amino Acid Aqueous Solutions: Performance and Mechanism. *ACS Sustain. Chem. Eng.* **2022**, *10*, 4451–4461. [CrossRef]
39. Wu, C.; Lü, R.; Gates, I.D. Computational Study on the Absorption Mechanisms of SO_2 by Ionic Liquids. *ChemistrySelect* **2018**, *3*, 4330–4338. [CrossRef]
40. Shaikh, A.R.; Ashraf, M.; AlMayef, T.; Chawla, M.; Poater, A.; Cavallo, L. Amino Acid Ionic Liquids as Potential Candidates for CO_2 Capture: Combined Density Functional Theory and Molecular Dynamics Simulations. *Chem. Phys. Lett.* **2020**, *745*, 137239. [CrossRef]
41. Frisch, M.J.; Trucks, G.W.; Schlegel, H.B.; Scuseria, G.E.; Robb, M.A.; Cheeseman, J.R.; Scalmani, G.; Barone, V.; Petersson, G.A.; Nakatsuji, H.; et al. *Gaussian 16 Rev. C.01*; Gaussian, Inc.: Wallingford, CT, USA, 2016.
42. Grimme, S.; Antony, J.; Ehrlich, S.; Krieg, H. A Consistent and Accurate Ab Initio Parametrization of Density Functional Dispersion Correction (DFT-D) for the 94 Elements H-Pu. *J. Chem. Phys.* **2010**, *132*, 154104. [CrossRef]
43. Nikolaienko, T.Y.; Bulavin, L.A. Localized Orbitals for Optimal Decomposition of Molecular Properties. *Int. J. Quantum Chem.* **2019**, *119*, e25798. [CrossRef]
44. Marenich, A.V.; Cramer, C.J.; Truhlar, D.G. Universal Solvation Model Based on Solute Electron Density and on a Continuum Model of the Solvent Defined by the Bulk Dielectric Constant and Atomic Surface Tensions. *J. Phys. Chem. B* **2009**, *113*, 6378–6396. [CrossRef]
45. Kaur, R.P.; Singh, H. Effect of Cyclization on Bond Dissociation Enthalpies, Acidities and Proton Affinities of Carbamate Molecules: A Theoretical Study. *Results Chem.* **2019**, *1*, 100003. [CrossRef]
46. Bennett, E.L.; Song, C.; Huang, Y.; Xiao, J. Measured Relative Complex Permittivities for Multiple Series of Ionic Liquids. *J. Mol. Liq.* **2019**, *294*, 111571. [CrossRef]
47. Le Donne, A.; Bodo, E. Isomerization Patterns and Proton Transfer in Ionic Liquids Constituents as Probed by Ab-Initio Computation. *J. Mol. Liq.* **2018**, *249*, 1075–1082. [CrossRef]

Article

An Ionic-Liquid-Imprinted Nanocomposite Adsorbent: Simulation, Kinetics and Thermodynamic Studies of Triclosan Endocrine Disturbing Water Contaminant Removal

Imran Ali [1,2,*], Gunel T. Imanova [3], Hassan M. Albishri [2], Wael Hamad Alshitari [4], Marcello Locatelli [5], Mohammad Nahid Siddiqui [6] and Ahmed M. Hameed [7]

- [1] Department of Chemistry, Jamia Millia Islamia (Central University), New Delhi 110025, India
- [2] Department of Chemistry, King Abdulaziz University, Jeddah 21589, Saudi Arabia
- [3] Department of Physical, Mathematical and Technical Sciences, Institute of Radiation Problems, Azerbaijan National Academy of Sciences, AZ 1143 Baku, Azerbaijan
- [4] Department of Chemistry, College of Science, University of Jeddah, Jeddah 21589, Saudi Arabia
- [5] Department of Pharmacy, University "G. d'Annunzio" of Chieti-Pescara, Build B, Level 2, Via dei Vestini, 31, 66100 Chieti, Italy
- [6] Department of Chemistry and IRC Membranes and Water Security, King Fahd University of Petroleum and Minerals (KFUPM), Dhahran 31261, Saudi Arabia
- [7] Department of Chemistry, Faculty of Applied Sciences, Umm Al-Qura University, Makkah 21955, Saudi Arabia
- * Correspondence: drimran.chiral@gmail.com or drimran_ali@yahoo.com

Citation: Ali, I.; Imanova, G.T.; Albishri, H.M.; Alshitari, W.H.; Locatelli, M.; Siddiqui, M.N.; Hameed, A.M. An Ionic-Liquid-Imprinted Nanocomposite Adsorbent: Simulation, Kinetics and Thermodynamic Studies of Triclosan Endocrine Disturbing Water Contaminant Removal. *Molecules* **2022**, *27*, 5358. https://doi.org/10.3390/molecules27175358

Academic Editors: Slavica Ražić, Aleksandra Cvetanović Kljakić and Enrico Bodo

Received: 15 July 2022
Accepted: 10 August 2022
Published: 23 August 2022

Publisher's Note: MDPI stays neutral with regard to jurisdictional claims in published maps and institutional affiliations.

Copyright: © 2022 by the authors. Licensee MDPI, Basel, Switzerland. This article is an open access article distributed under the terms and conditions of the Creative Commons Attribution (CC BY) license (https://creativecommons.org/licenses/by/4.0/).

Abstract: The presence of triclosan in water is toxic to human beings, hazardous to the environment and creates side effects and problems because this is an endocrine-disturbing water pollutant. Therefore, there is a great need for the separation of this notorious water pollutant at an effective, economic and eco-friendly level. The interface sorption was achieved on synthesized ionic liquid-based nanocomposites. An N-methyl butyl imidazolium bromide ionic liquid copper oxide nanocomposite was prepared using green methods and characterized by using proper spectroscopic methods. The nanocomposite was used to remove triclosan in water with the best conditions of time 30 min, concentration 100 µg/L, pH 8.0, dose 1.0 g/L and temperature 25 °C, with 90.2 µg/g removal capacity. The results obeyed Langmuir, Temkin and D-Rs isotherms with a first-order kinetic and liquid-film-diffusion kinetic model. The positive entropy value was 0.47 kJ/mol K, while the negative value of enthalpy was −0.11 kJ/mol. The negative values of free energy were −53.18, −74.17 and −76.14 kJ/mol at 20, 25 and 30 °C. These values confirmed exothermic and spontaneous sorption of triclosan. The combined effects of 3D parameters were also discussed. The supramolecular model was developed by simulation and chemical studies and suggested electrovalent bonding between triclosan and N-methyl butyl imidazolium bromide ionic liquid. Finally, this method is assumed as valuable for the elimination of triclosan in water.

Keywords: ionic liquid nanocomposite; water treatment; endocrine-disturbing triclosan; simulation; thermodynamics; kinetics

1. Introduction

Nowadays, new-generation water pollutants are getting more attention for their removal. These are personal-care products, medicinal and pharmaceutical residues. Among many residues, triclosan has been found as a water pollutant in some places in the world [1]. Further, this compound has been reported in some aquatic animals [2,3]. Basically, triclosan is used as an antimicrobial agent with a wide range of activities. It is used as oral medicine in some countries [4]. It is mixed with many personal products during their preparation. Most commonly, triclosan is mixed with sanitizers, soaps, skin creams, etc. [5]. It is also used in houses, hotels and hospitals to maintain hygiene [6–11]. In addition, it is also used as an additive in polymer production, such as polyethylene and polyolefin polymers. Consequently, there are great chances of water contamination by triclosan. It is a pollutant of high concern because of its endocrine-disturbing nature. A good value of log K_{ow} of 4.8 octanol–water partition coefficient made it capable to be bioaccumulative in fatty tissues, which is responsible for toxicity. In addition, the hormonal activity of these pollutants is studied in vitro and in vivo (animals) and showed serious effects [12–15]. These effects may affect human beings. The most notorious effect is that triclosan may enter the human body through the skin [16,17], which may be a dangerous sign for health. Briefly, triclosan is a serious and toxic pollutant and its detailed health hazards are discussed by Olaniyan et al. [1]. Chemically, triclosan is known as 5-Chloro-2-(2,4-dichlorophenoxy)phenol (Figure 1) with $C_{12}H_7Cl_3O_2$ molecular formula and 289.54 g as molecular mass.

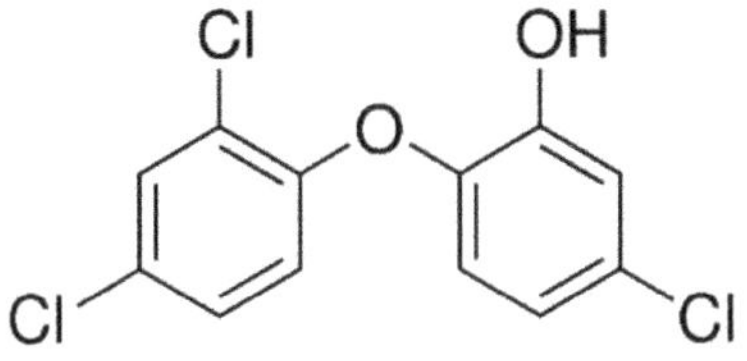

Figure 1. Structure of triclosan.

Due to the above discussion, it is very clear that there is an urgent need for fast, economic, reproducible and reliable methods for the removal of triclosan in water. Among many materials, nanomaterials are gaining utility in a wide range of applications, including water treatment [18–22]. Some papers are available on the removal of triclosan. Brose et al. [23] defined triclosan removal in influents, effluents and biosolids. Wu et al. [24] defined the subtraction of triclosan using clay and sandy soils with maximum sorption at a high concentration of 0.05 mg/L. In addition, the maximum sorption was at low pH, which is not the pH of natural water. Moreover, the authors reported maximum sorption in days, which is a high time for fast and economic processes. Tonga et al. [25] described the removal of triclosan using biosolid-derived biochar. The authors reported high sorption at low pH. Khori et al. [26] reported the removal of triclosan by using activated carbon prepared from waste biomass. Although the equilibrium achieved was within 20 min, the adsorption capacity was low, i.e., 80.77%. Fard and Barkdoll [27] reported the removal of triclosan on magnetic nanoparticles. The authors reported maximum removal in 47 min. Yi et al. [28] reported the removal of triclosan using core-shell Fe_3O_4@COFs nanoparticles. The removal was good but the adsorbent is costly and it is difficult to use on a large scale. Triwiswaraa et al. [29] reported the removal of triclosan by using char obtained from palm kernel. The authors reported maximum sorption from 6 to 12 h, which is a high time period for a fast and economic process. Jiang et al. [30] reported the removal of triclosan on a silica–zeolite sorbent. The authors reported the removal of triclosan at a high concentration of about 2.0 mico mole. It is clear from the above-cited literature that no one can be used to remove this pollutant on a large scale economically. The reason is that almost all methods took a longer time, which is not feasible in natural water treatment on a large scale. Moreover, mostly, methods are applicable at high concentrations of triclosan while it is present at the microgram level in the water. Taking these facts into consideration, a new N-methyl butyl imidazolium bromide ionic-liquid-based copper oxide nanocomposite was prepared using a facile and green method. The nanocomposite was characterized and used for the removal of triclosan. The sorption mechanism was developed by simulation and chemical methods. The results of the findings are discussed in this article.

2. Experimental

Details of the chemicals, reagents, instruments used and experimental protocol are given in the Supplementary Information. However, some important information is provided herein.

2.1. Preparation of N-Methyl Butyl Imidazolium Bromide CuO Nanocomposite

Copper nanoparticles were prepared using green methods by utilizing the appropriate amount of copper acetate and *Acacia arabica* leave extract. Thus, 50 g leaves of *Acacia arabica* was dried in shade, ground in the form of water paste and heated in 1000 mL distilled water at 80 °C for 2 h followed by cooling at room temperature and filtration through Whatman filter paper number 1. Next, 250 mL solution of copper acetate of 0.1 M concentration was prepared and the leaf extract (500 mL) was mixed slowly with constant stirring. The pH of the solution was adjusted to 10 by using dilute sodium hydroxide solution. The solution was stirred for 5 h and kept standing undisturbed at room temperature. The solid part of the solution was centrifuged and washed with water following its activation in a furnace at 500 °C for 5 h. The so-formed nanoparticles of CuO were used to prepare the composite with N-methyl butyl imidazolium bromide. Further, 0.5 g CuO nanoparticles was suspended in 100 mL of water and after that, a solution of N-methyl butyl imidazolium bromide in 50 mL of the concentration of 1.0 g/L was added to a suspension of CuO nanoparticles. The suspension was stirred 2–3 h with constant heating. A blackish product was obtained which was washed with hot water and ethanol. Finally, the product was dried in an oven at 100 °C (yield = 99.5%) and used for further sorption studies.

2.2. Characterization of N-Methyl Butyl Imidazolium Bromide Nanocomposite

The obtained N-methyl butyl imidazolium bromide ionic liquid copper oxide nanocomposite was characterized by FT-IR, XRD, SEM and TEM spectroscopic techniques. These methods were utilized as per the typical protocol.

2.3. Sorption Study

The uptake capability of N-methyl butyl imidazolium bromide ionic liquid copper oxide nanocomposite was determined by the typical measures of batch mode. Several factors were optimized and these were the amount of triclosan, sorbent dose, medium pH, time of contact and temperature. The variables used were 20–120 µg/L triclosan; 0.1–1.5 g/L for N-methyl butyl imidazolium bromide copper oxide nanocomposite dosage; 3–12 medium pHs; time of 5–40 min with a temperature of 20–30 °C. The lasting triclosan was detected by HPLC as described in the Supplementary Information. The different models are also given in Supporting Information to analyze the results. The equilibrium capacity of sorption was fixed by the following equation.

$$Q_e \ (\mu g/g) = (C_i - C_e) \times v/m \quad (1)$$

where C_i and C_e are starting and equilibrium amounts of triclosan in µg/L whilst m is the weight of N-methyl butyl imidazolium bromide copper oxide nanocomposite in g/L. The solution volume (v) was in milliliters. Triclosan dismissal efficacy was measured by the given equation.

$$\text{Elimination (\%)} = [(C_i - C_t)/C_0] \times 100 \quad (2)$$

2.4. Kinetic Study

The kinetic study of triclosan removal by N-methyl butyl imidazolium bromide copper oxide nanocomposite was fixed at 5 to 40 min by taking a known quantity of N-methyl butyl imidazolium bromide copper oxide nanocomposite with triclosan solution in a water bath which was temperature controlled. The experiments were performed until the equilibrium was achieved. The amount of triclosan was determined at dissimilar time periods. The trials were completed in a similar method as for the sorption ones.

2.5. Thermodynamics Study

This study of triclosan was performed by using 20, 25 and 30 °C temperatures. The adjusted quantities of triclosan were utilized under these conditions.

2.6. Analysis of Triclosan by HPLC

The residual quantities of the triclosan were estimated by the HPLC procedure as given in the Supplementary Information (SI). The mobile phase used was 2.5 pH acetic acid buffer-acetonitrile (30:70) with 1.0 mL/min flow rate. The column used was Sunniest RP Aqua C_{28} (25 cm × 4.6 mm id) and detection was achieved at 280 nm. The identification of triclosan was determined by equating the retention time of the standard with sample ones. The retention time of triclosan was 11.35 min.

2.7. Simulation Study

The simulation docking of triclosan uptake was performed as per the typical method given in Supplementary Information.

3. Results and Discussion
3.1. Characterization

N-methyl butyl imidazolium bromide copper oxide nanocomposite was characterized by FT-IR, which shows the presence of Cu-O in CuO. No peak was seen in the 600 to 610 region; confirming the non-availability of Cu_2O [31]. The results are in agreement with Wu et al. [32] findings. The crystallinity of the prepared N-methyl butyl imidazolium bromide copper oxide nanoparticles was ascertained by XRD studies and confirmed monoclinic hexagonal crystalline CuO. The crystal size was calculated by the following formula.

$$D = K\lambda/d\cos\theta$$

where, d, θ, K and D are full width at half maximum, reflection angle and constant, usually taken as 0.94. λ = 0.154 nm for Cu-Kα, β. The particle sizes calculated were from 50 to 98.5 nm.

The morphology of the prepared N-methyl butyl imidazolium bromide copper oxide nanoparticles was ascertained by SEM and the two photographs at 10,000 and 2,00,000 magnifications, showing the crystalline and rough surface. The shapes of the particles were irregular and spherical. Another study for characterization was TEM and the picture is shown in Figure 2, confirming the round and irregular shape of nanoparticles with sizes ranging from 50 to 99.5 nm. These results are in agreement with the XRD results. Based on the above argument, the following N-methyl butyl imidazolium bromide copper oxide nanoparticle structure (Figure 3) was developed, in which four molecules of N-methyl butyl imidazolium bromide are found on the exterior of CuO. This structure has a positive charge and a very good sorbent for negatively charged species.

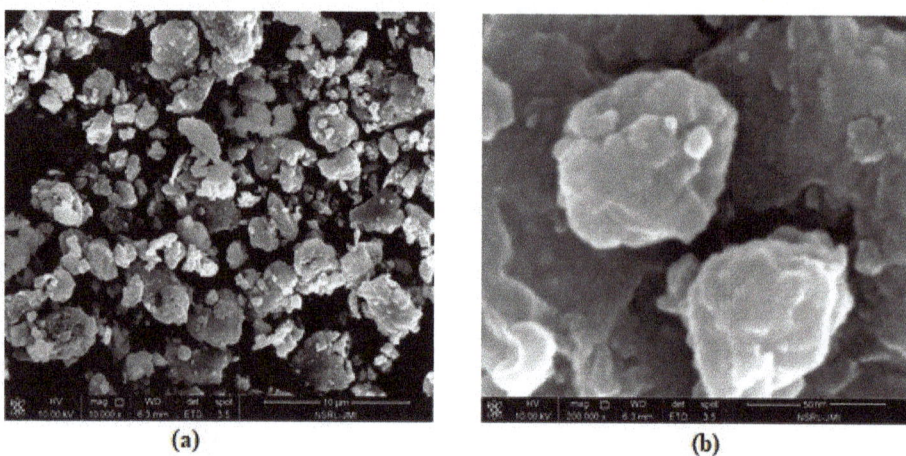

Figure 2. TEM images of spectrum of N-methyl butyl imidazolium bromide copper oxide nanocomposite: (**a**) at 10,000 and (**b**) 2,00,000 magnifications (ranging from 50 to 99.5 nm).

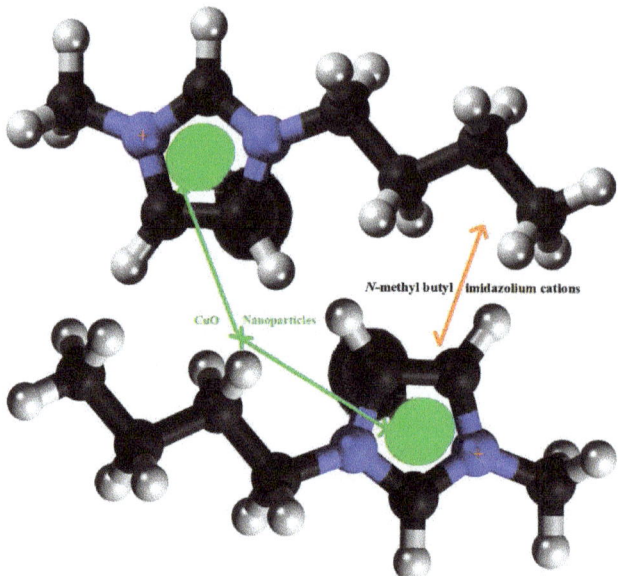

Figure 3. The structure of N-methyl butyl imidazolium cation copper oxide nanocomposite.

3.2. Sorption Study

The uptake of triclosan was optimized by numerous parameters. The improved parameters were 20–120 μg/L triclosan; 0.1–1.5 g/L for N-methyl butyl imidazolium bromide copper oxide nanocomposite dosage; 3–12 medium pHs; time of 5–40 min with a temperature of 20–30 °C. The results of all these sets of experiments are given in the following sections.

3.2.1. Concentration of Triclosan

The concentration of triclosan was optimized by taking 10 to 130 μg/L. The other variables controlled were contact time of 30 min, pH 8.0, dose 1.0 and 25 °C temperature.

The results of these sets of experiments findings are graphed in Figure 4a and it is clear from this figure that the triclosan removal incased from 10 to 100 g/L concentration while further augment in the concentration could not result in more triclosan uptake. The sorption of triclosan were 9, 18.7, 28, 36, 45, 53.4, 63, 71.5, 80.5, 90.2, 90.3, 90.3 and 90.3 µg/g at 10, 20, 30, 40, 50, 60, 70, 80, 90, 100, 110, 120 and 130 µg/L concentrations. The percentage calculations were conducted and the percentage removal at equilibrium was 90.2. These data are a confirmation of the fact that 100 µg/L concentration was the optimized one at equilibrium. Therefore, 100 µg/L concentration of triclosan was considered as the best one in this set of experiments.

3.2.2. Contact Time

The contact time for triclosan removal was optimized by taking 5 to 40 min. The other variables controlled were starting concentration 100 µg/L, pH 8.0, dose 1.0 and 25 °C temperature. The results of these sets of experiments findings are graphed in Figure 4b and it is clear from this figure that the triclosan removal increased from 5 to 30 min time while further augment in the contact time could not result in more triclosan uptake. The sorption of triclosan were 13.6, 27.9, 42.5, 57.15, 72.90, 90.2, 91.3 and 91.3 µg/g at 5, 10, 15, 20, 25, 30, 35, and 40 min. The percentage calculations were done and the percentage removals were 13.6, 27.9, 42.5, 57.15, 72.90, 90.2, 91.3 and 91.3. These data are a confirmation of the fact that 30 min of contact time was the optimized one at equilibrium. Therefore, 30 min contact time for triclosan was considered the best one in this set of experiments.

3.2.3. pH of the Solution

The solution pH for triclosan removal was optimized by taking 3 to 12 units. The other variables controlled were starting concentration 100 µg/L, contact time 30 min, dose 1.0 and 25 °C temperature. The results of these sets of experiment findings are graphed in Figure 4c and it is clear from this figure that the triclosan removal increased from 3 to 8 pH while further augment in the pH could not result in more triclosan uptake. The sorptions of triclosan were 22.3, 32.4, 41.5, 57.4, 83.6, 91.2, 93.2, 93.3, 94.5 and 94.4 µg/g at 3, 4, 5, 6, 7, 8, 9, 10, 11 and 12 pH. The percentage calculations were done and the percentage removals were 22.3, 32.4, 41.5, 57.4, 83.6, 91.2, 93.2, 93.3, 94.5 and 94.4. These data are a confirmation of the fact that 8.0 pH was the optimized one at equilibrium. Therefore, 8.0 pH for triclosan was considered the best one in this set of experiments. It is important to emphasize that many natural water resources have 7–8 pH and, therefore, it may be concluded that the given method may be highly useful to natural water systems.

3.2.4. Dose of the Nanocomposite

The nanocomposite dose for triclosan removal was optimized by taking 0.1, 0.25, 0.50, 0.75, 1.0, 1.25 and 1.5 g/L concentration. The other variables controlled were starting concentration 100 µg/L, contact time 30 min, pH 8.0 and 25 °C temperature. The results of these sets of experiment findings are graphed in Figure 4d and it is clear from this figure that the triclosan removal increased from 0.1 to 1.0 g/L while further augment in dose could not result in more triclosan uptake. The sorptions of triclosan were 35.6, 55.4, 72.8, 81.7, 91.2, 91.2 and 92.4 µg/g at 0.1, 0.25, 0.50, 0.75, 1.0, 1.25 and 1.50 g/L. The percentage calculations were done and the percentage removals were 35.6, 55.4, 72.8, 81.7, 91.2, 91.2 and 92.4. These data are a confirmation of the fact that 1.0 g/L dose was the optimized one at equilibrium. Therefore, 1.0 g/L dose for triclosan was considered as the best one in this set of experiments.

3.2.5. Temperature of the Solution

The experimental temperature for triclosan removal was optimized by taking 20, 25 and 30 °C. The other variables controlled were starting concentration 100 µg/L, contact time 30 min, 1.0 g/L dose and pH 7.0. The results of these sets of experiment findings are graphed in Figure 4e and it is clear from this figure that the triclosan removal decreased from 20 to 30 °C while the experiments that further increased or decreased temperature were not conducted because these are not the temperatures of natural water resources. The sorptions of triclosan were 9.0, 18.7, 28.5, 37.0, 46.4, 54.7, 64.2, 72.8, 82.2, 91.5, 92.3 and 92.3 µg/g at 20 °C while these values at 25 °C were 9.0, 18.7, 28.0, 36, 45, 53.4, 63, 71.5, 80.5, 90.2, 90.3, and 90.3 µg/g. The values at 30 °C were 9.0, 18.7, 26.8, 35, 43.6, 51.5, 60.4, 70.1, 79.5, 88.5, 88.3 and 88.2 µg/g. These data are a confirmation of the fact that the sorption was in the order of 20 > 25 > 30 °C, confirming the exothermic sorption of triclosan on the reported nanocomposite.

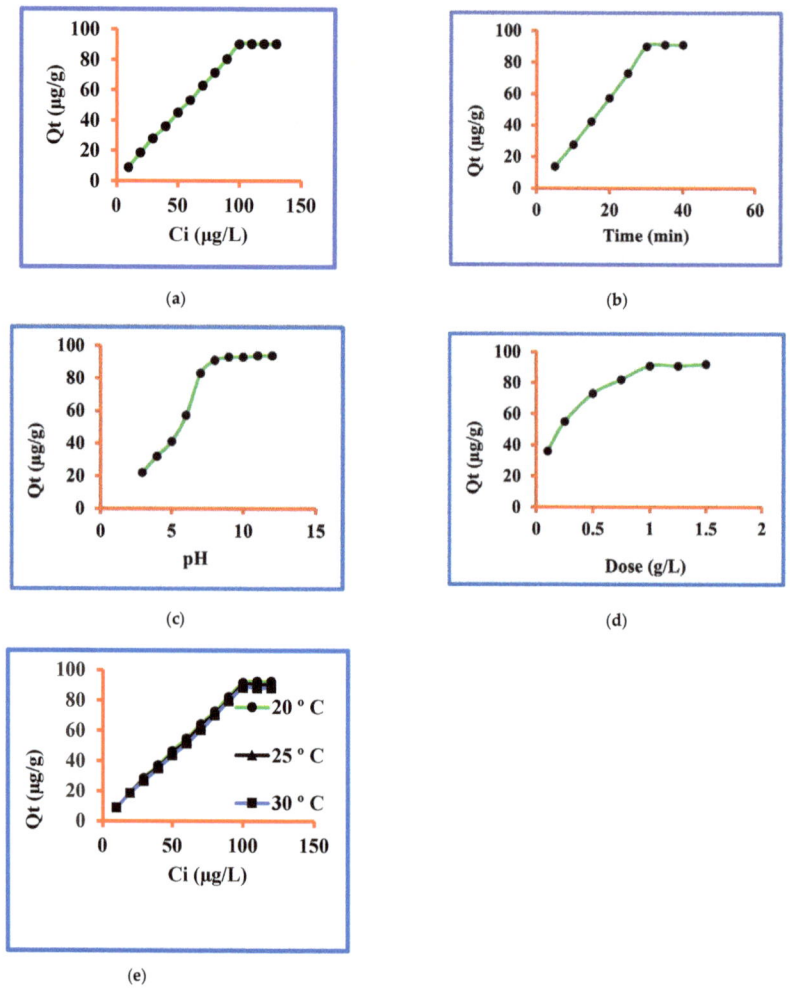

Figure 4. Sorption of triclosan (**a**) amounts, (**b**) time, (**c**) pH, (**d**) dose and (**e**) temperature.

3.3. Combined Effect of the 3D Parameters

pH is one of the most important factors for determining the applicability of a method in the treatment of water in real-life problems. Therefore, efforts were made to study the combined effect of pH with initial concentration, time, dose and temperature. The results are given in Figure 5. A look at Figure 5a indicates that the best removal was 90.2% at 100 µg/L concentration and pH 8. A further increase in both concentration and pH could not augment the removal of triclosan. Similarly, a look at Figure 5b indicates that the best removal was 90.0% at 30 min and pH 8. A further increase in both contact time and pH could not augment the removal of triclosan. A look at Figure 5c indicates that the best removal was 91.0% at a 1.0 g/L dose and pH 8. A further increase in both dose and pH could not augment the removal of triclosan. In addition, the effects of %removal vs. temperature vs. pH are shown in Figure 5d, which indicates that the best removal was 91.5% at 20 °C temperature and pH 8. A further increase in pH could not augment the sorption of triclosan. It is interesting to mention that the adsorption decreased at high temperatures. Therefore, it was decided to select 25 °C temperature throughout this study because this is the temperature of most of the water resources. Finally, 25 °C temperature was selected to carry out all the experiments and the maximum sorption of triclosan was 91.0% at pH 8. The experiments were conducted five times ($n = 5$) and the percentage errors were calculated. The percentage error values were in a range of 2.5 to 4.5. These values clearly showed the approval of the experiments.

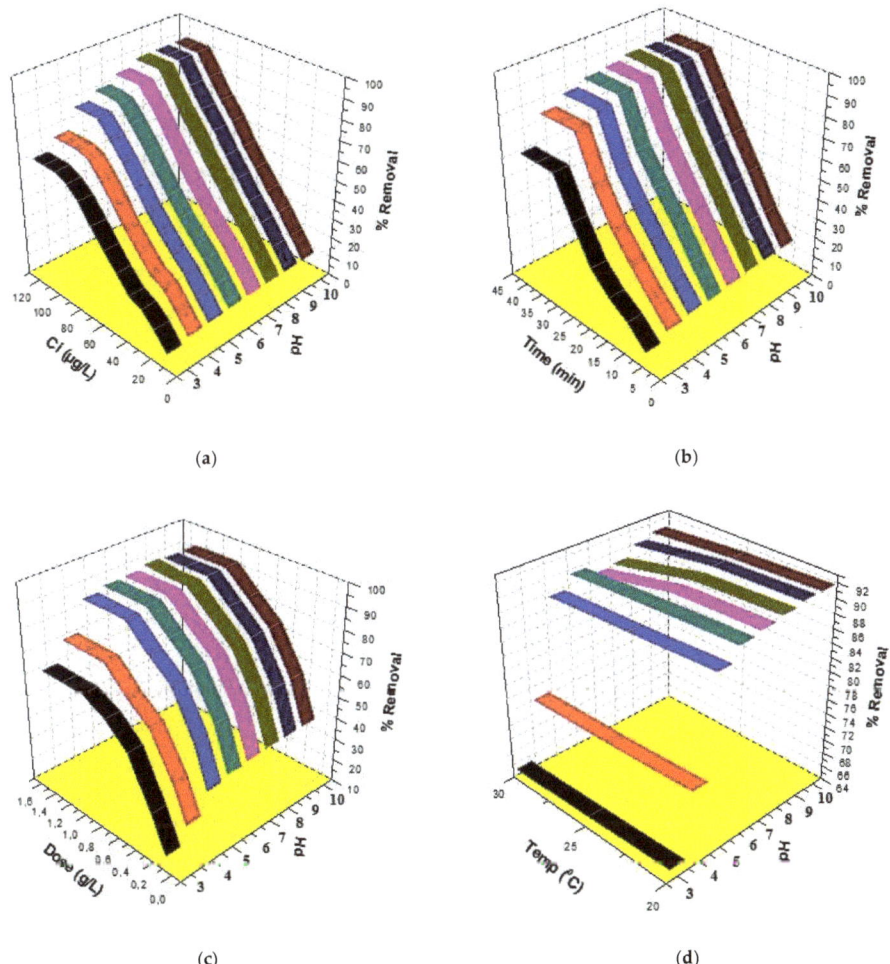

Figure 5. Three-dimensional plots for triclosan removal: (**a**) %removal vs. Ci vs. pH, (**b**) %removal vs. time vs. pH, (**c**) %removal vs. dose vs. pH and (**d**) %removal vs. temperature vs. pH.

3.4. Modeling

The data obtained after experiments were used to run the models and the most significant isotherms used were Langmuir, Freundlich, Temkin and Dubinin–Radushkevich. The findings after running these models are given below.

3.4.1. Langmuir

The equation for the Langmuir model is given in the Supplementary Information (Table S1) and this equation was used to model the experimental data. The Langmuir model parameters were determined and expressed by b (L/µg) and X_m (µg/g). The magnitudes of these parameters were exploited by the intercept and slope of the plot drawn between $1/Q_t$ and $1/C_t$ (Figure 6a). These values are presented in Table 1, which confirmed 0.087, 0.015 and 0.008 µg/g values of X_{max} at 20, 25 and 30 °C, respectively, while the values of b were 188.68, 625 and 833.33 L/µg at these temperatures. The magnitude of regressing constants was 0.929, 0.967 and 0.901.

Table 1. The adsorption isotherm parameters of triclosan.

Isotherms	Temperatures		
	20.0 °C	25.0 °C	30.0 °C
Langmuir			
X_{max} (µg/g)	0.087	0.015	0.008
b (L/µg)	188.68	625	833.33
R^2	0.929	0.967	0.901
Freundlich			
k_F (µg/g)	17.64	10.11	5.08
n (µg/L)	1.38	1.09	0.89
R^2	0.949	0.968	0.885
Temkin			
K_T (L/µg)	5.55	2.99	1.86
B_T (kJ/mol)	3.35	2.70	2.23
R^2	0.949	0.968	0.885
Dubinin–Radushkevich			
Q_m (µg/g)	128.83	158.48	177.83
K_{ad} (mol²/kJ²)	1.22	3.20	5.38
E (kJ/mol)	0.53	0.42	0.36
R^2	0.901	0.954	0.849

X_{max} and b = Langmuir; k_F and n = Freundlich constants; k_T and B_T = Temkin constants; Qm, Kad and E = Dubinin–Radushkevich constants and R^2 = Regression coefficient.

3.4.2. Freundlich

The equation for the Freundlich model is given in the Supplementary Information (Table S1) and this equation was used to model the experimental data. The Freundlich model parameters were determined and expressed n (µg/L) and k_F (µg/g) (Figure 6b). The magnitudes of these parameters were exploited by intercept and slope of the plot drawn between logQt and logCt. These values are presented in Table 1, which confirmed 17.64, 10.11 and 5.08 µg/g values of k_F at 20, 25 and 30 °C, respectively, while the values of n were 1.38, 1.09 and 0.89 L/µg at these temperatures. The magnitude of regressing constants was 0.949, 0.968 and 0.885. The outcome of the experimental data by Freundlich and Langmuir models was studied. A comparison can be seen in Table 1 and it is clear that the regression constants for Langmuir are close to unity in comparison to the Freundlich. It means that the experimental data followed the criterion of the Langmuir model rather than Freundlich one. Finally, it was concluded that the sorption process proceeded through the Langmuir model.

3.4.3. Temkin

The equation of the Temkin model is given in the Supplementary Information (Table S1) and this equation was used to model the experimental data. The Temkin model parameters were determined and expressed by K_T (L/μg) B_T and (kJ/mol) constants. The magnitudes of these parameters were exploited by slope and intercept of the plot drawn between logQt and logCt (Figure 6c). These values are presented in Table 1, which confirmed 5.55, 2.99 and 1.86 L/μg of K_T at 20, 25 and 30 °C, respectively, while 3.35, 2.70 and 2.23 were the values of B_T at 20, 25 and 30 °C, respectively. The magnitude of regressing constants was 0.949, 0.968 and 0.885 at K_T at 20, 25 and 30 °C, respectively. These values are approaching unity, which is the confirmation of the validity of the data by the Temkin model.

3.4.4. Dubinin–Radushkevich

The equation for the Dubinin–Radushkevich model is given in the Supplementary Information (Table S1) and this equation was used to model the experimental data. The Dubinin–Radushkevich model parameters were determined and expressed by Q_m (μg/g), K_{ad} (mol²/kJ²) and E (kJ/mol). The values of Q_m (μg/g) and E (kJ/mol) were determined by the intercept and slope of the plot drawn between the graph of $1/Q_t$ and $1/C_t$ (Figure 6d). The degrees of K_{ad} (mol²/kJ²) were computed by exploiting an equation as given below.

$$\ln Q_e = \ln Q_m - K_{ad} \cdot E^2 \quad (3)$$

where Q_e, Q_m and E have the usual meaning.

The values of different constants are presented in Table 1, which confirmed 128.83, 158.48 and 177.83 μg/g of Q_m at 20, 25 and 30 °C, respectively, while 1.22, 3.20 and 5.83 mol²/kJ² were the values of K_{ad} at 20, 25 and 30 °C, respectively. The degrees of E were 0.53, 0.42 and 0.36 kJ/mol at 20, 25 and 30 °C, respectively. The magnitude of regressing constants was 0.901, 0.954 and 0.849 at K_T at 20, 25 and 30 °C, respectively. These values are approaching unity, which is the confirmation of the validity of the data by the Dubinin–Radushkevich model.

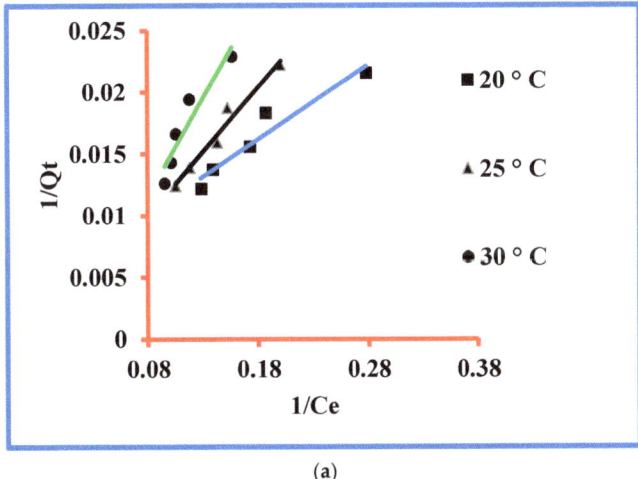

(a)

Figure 6. Cont.

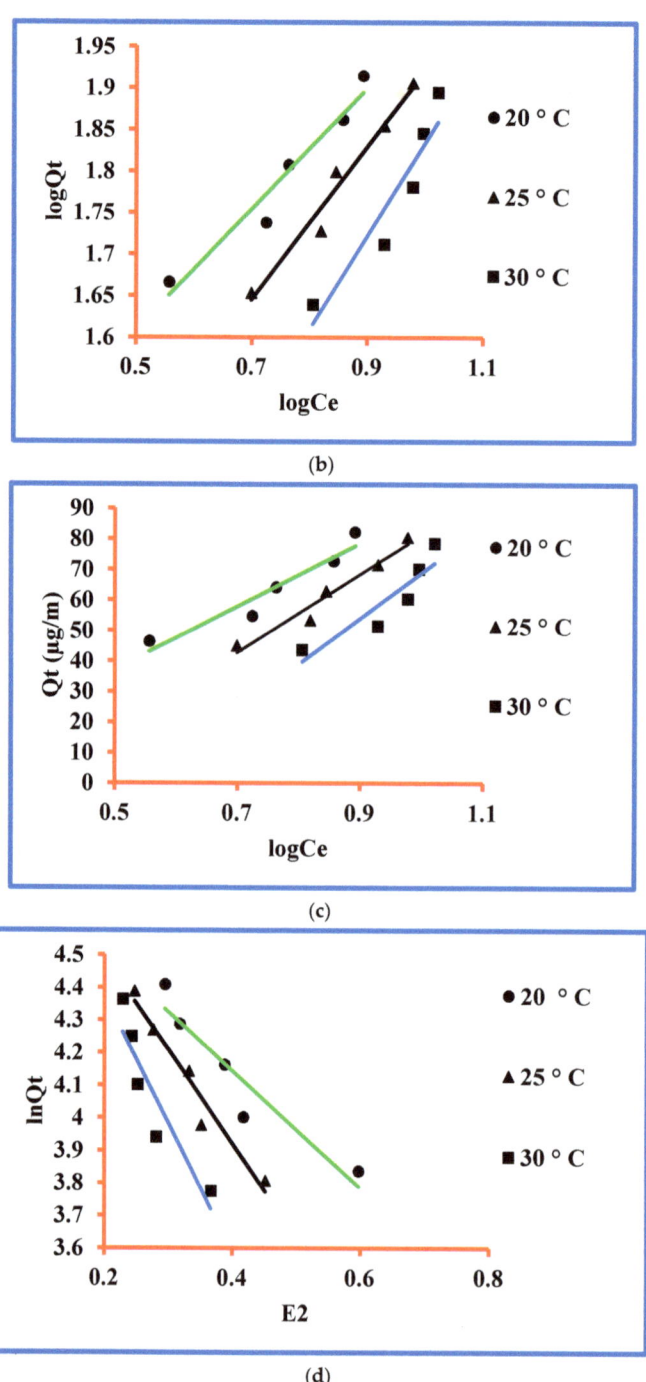

Figure 6. Models for triclosan: (**a**) Langmuir, (**b**) Freundlich, (**c**) Temkin and (**d**) Dubinin–Radushkevich.

3.5. Kinetic Study

First, second and Elovich's kinetic models were exploited for computing the kinetics of the triclosan sorption process. The equations used for these models are given in the SI (Table S2). First, second and Elovich's kinetic models' graphs are plotted between $\log(Q_e-Q_t)$, t vs. Qt and t, lnt vs. Qt, respectively. The slopes and intercepts of these plots were exploited to compute the degrees of the various constants. The values of these different constants are presented in Table 2. Taking the example of first order, the value of the rate constant was 0.073 1/min with 90.2 experimental and 124.11 theoretical values of the equilibrium concentration of triclosan. These values varied from each other by 33.91, confirming their closeness. The value of the regression constant was 0.995, approaching unity. Taking the example of second order, the values of rate constant were 3.26×10^{-6} g/µg min with 90.2 experimental value and 909.09 theoretical value. These values varied from each other by 818.89%, confirming their non-closeness. The value of h was 2.70 µg/g min, confirming the good value of the initial rate of the sorption. The value of the regression constant was 0.994. The data of the first- and second-pseudo-order reaction models were matched. It was observed that the experimental and theoretical degrees of the sorbed triclosan amount were in agreement in the first-pseudo-order reaction, rather than in the second-pseudo-order reaction model. Further, the degrees of the regression constants of triclosan were slightly higher in the case of first order in comparison to the second-order reaction. Consequently, it was decided that the investigational data obeyed the first-pseudo-kinetic reaction. The degrees of adsorption (α) and desorption (β) rates were computed by Elovich's kinetic model and the computed values are presented in Table 2; it is clear that α was 9.37 µg/g min while β was 0.28 g/µg, confirming faster sorption than desorption. The degree of the regressing coefficient was 0.939, approaching unity. These data confirmed the applicability of Elovich's kinetic model.

Table 2. Kinetic parameters of triclosan.

Kinetic Models and Parameters	Numerical Values
First-second-order	
k_1 (1/min)	0.073
The experimental Qe (µg/g)	90.2
The theoretical Qe (µg/g)	124.11
R^2	0.995
Pseudo-second-order	
k_2 (g/µg min)	3.26×10^{-6}
The experimental Qe (µg/g)	90.2
The theoretical Qe (µg/g)	909.09
h (µg/g min)	2.70
R^2	0.994
Elovich	
α (µg/g min)	9.37
β (g/µg)	0.028
R^2	0.939

3.6. Thermodynamics Study

Triclosan sorption thermodynamics of the adsorption phenomenon was ascertained by the thermodynamics equations, which are given in the Supplementary Information. The thermodynamics were estimated in terms of entropy, enthalpy and free energy. The degrees of entropy and enthalpy were 0.47, −0.11 and −5.43 × 10^{-2}, respectively. The degrees of free energy at 20, 25 and 30 °C were −53.18, −74.17 and −76.14, respectively. The negative values of enthalpy, free energy and entropy are an indication of impulsive sorption and exothermic process.

3.7. Adsorption Mechanism

In scientific discoveries, the mechanism of any phenomenon is very important and it is also true with this manuscript. The mechanism of triclosan sorption on N-methyl butyl imidazolium bromide copper oxide nanocomposite was studied by exploiting the liquid-film diffusion kinetic model and the intra-particle diffusion kinetic models (Figure 7a,b). The equations for the liquid-film diffusion kinetic model and the intra-particle diffusion kinetic models are given in the Supplementary Information (Table S2) and these equations were used to model the experimental data. The liquid-film diffusion kinetic model parameters were determined and expressed by the rate constant (1/min), intercept and regression constant. The magnitude of the rate constant was exploited by the slope of the plot drawn between t and $\ln(1-F)$. This value is presented in Table 3, which confirmed 0.073 1/min. The magnitude of regressing constant was 0.946 with 0.32 as the intercept value. This information shows that the graph is approaching its origin. Contrarily, the intra-particle diffusion kinetic model parameters were determined and expressed by the rate constant (k_{ipd1}), intercept and regression constant. The magnitude of this constant was exploited by the slope of the plot drawn between Q_t and $t^{0.5}$ (Figure 7a,b). The values are presented in Table 3, which confirmed 23.37 k_{ipd1}. The magnitude of regressing constant was 0.976 with -43.72 as the intercept value. This information shows that the graph is not going through the origin. By comparing both models, it was found that the graph of the liquid-film diffusion kinetic model is passing closer to the origin in comparison to the intra-particle diffusion kinetic model. Moreover, the value of the regression constant was greater in the former model in comparison to the latter. These findings are clear proof of the applicability of the liquid-film diffusion kinetic model in the present sorption process.

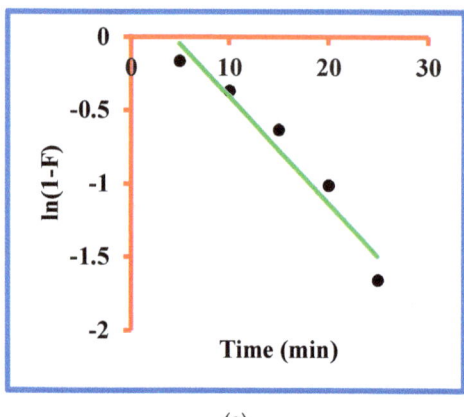 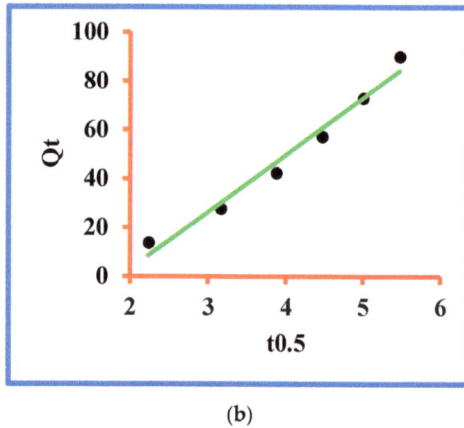

Figure 7. Mechanism of the sorption: (**a**) liquid-film diffusion and (**b**) intra-particular diffusion models.

Table 3. The values of the uptake mechanism parameters of triclosan.

The Kinetic Models and Parameters	Numerical Values
The intraparticle diffusion kinetic model	
k_{id} (μg/g min$^{0.5}$)	23.37
Intercept	-43.72
R^2	0.976
Liquid film diffusion kinetic model	
k_{fd} (1/min)	0.073
Intercept	0.32
R^2	0.946

3.8. Supramolecular Mechanism of Uptake

In addition to using the sorption mechanism model, it is also interesting if the sorption mechanism is developed and this is possible by a simulation study and triclosan and N-methyl butyl imidazolium bromide ionic liquid chemistry. The procedure for simulation between triclosan and N-methyl butyl imidazolium bromide is given in the Supplementary Information. The outcomes of the binding were calculated in terms of bonding and residues involved. The model developed after simulation between triclosan and N-methyl butyl imidazolium bromide and by chemical methods is shown in Figure 8. This figure indicates bonding between the hydroxyl group of triclosan and imidazole ring of N-methyl butyl imidazolium bromide. The maximum adsorption was observed at pH 8.0 and it can be explained at the supramolecular level by considering the model presented in Figure 8. The pKa value of N-methyl butyl imidazolium bromide is 7.0–7.4 and it means that N-methyl butyl imidazolium bromide ionizes at pH 7.0–7.4, leading to the formation of a cation of N-methyl butyl imidazolium bromide. On the other hand, the pKa value of 7.9–8.1 means that triclosan ionizes at 7.9–8.1 pH, leading to the formation of the anion of triclosan. Finally, the cation and anion interacted by electrovalent bonding and formed a complex structure, as shown in Figure 8. Further, it is important to mention that the pKa value of triclosan is controlling sorption because the pKa value of triclosan is higher than the pKa value of N-methyl butyl imidazolium bromide. The residues involved in the formation of the complex structure are the phenoxide group of triclosan and imidazolium ring of N-methyl butyl imidazolium bromide. These interpretations are in a good arrangement for supporting the adsorption behavior of triclosan. Finally, the uptake mechanism at the supramolecular scale is developed and established.

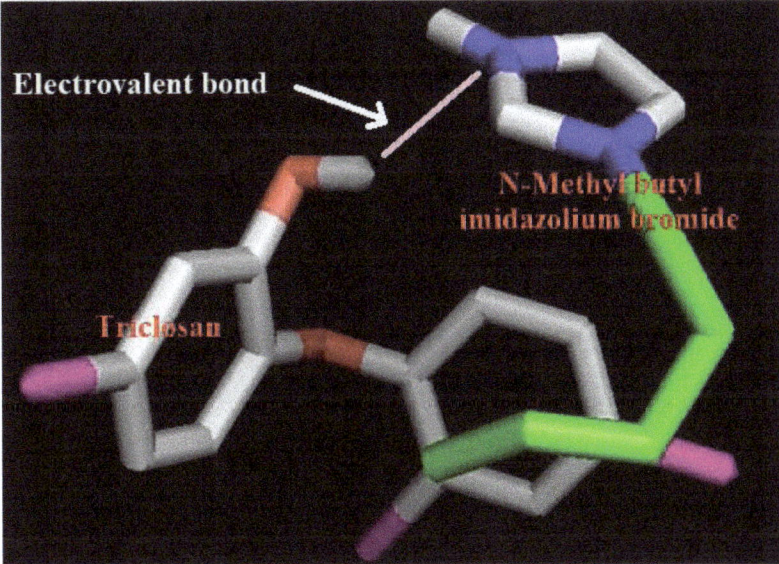

Figure 8. Interacting model of triclosan with N-methyl butyl imidazolium bromide.

3.9. Regeneration of Sorbent

The regeneration of any sorbent is one of the important issues in developing economic water treatment methods. The regeneration was tried with water of different pHs, buffers of different pHs, acids, bases and organic solvents. The maximum desorption (98.6%) was obtained with acetone and the efficiency of the regenerated sorbent was not good, as it could remove further triclosan to 73%. On the other hand, the ethanol could desorb 96.8%

triclosan and the removal capacity of the sorbent was good, i.e., 90.0%, which decreased to 85.6% after five cycles of regeneration.

4. Conclusions

Copper oxide nanoparticles were synthesized using green methods and the nanocomposite was prepared by using N-methyl butyl imidazolium bromide ionic liquid. The nanocomposite was characterized by using proper spectroscopic methods. The nanocomposite was used to remove triclosan in water and the percentage removal was 90.2% at pH 8.0, which is an asset for using this method in natural water conditions. In addition, 30 min is the fast time to apply the method in natural situations. The conditions developed for this method were time 30 min, concentration 100 µg/L, pH 8.0, dose 1.0 g/L and 25 °C temperature. The results were verified by Langmuir isotherm. In addition, Temkin and and D-Rs models were followed well by the data. The sorption mechanism was through the liquid-film diffusion kinetic model. The enthalpy and free energy values were negative while the entropy of the value was positive, which confirmed the natural sorption phenomenon at all the working temperatures. In addition, these results also confirmed the exothermic elimination of triclosan. The supramolecular simulation study confirmed electrostatic interaction between triclosan and N-methyl butyl imidazolium bromide ionic liquid. This method is green in nature and economical in behavior, giving 90% elimination of triclosan. Finally, this method is considered worthwhile for the elimination of triclosan in water.

Supplementary Materials: The following supporting information can be downloaded at: https://www.mdpi.com/article/10.3390/molecules27175358/s1, Table S1: Langmuir, Freundlich, Temkin, Dubinin–Radushkevich adsorption models*, Table S2. Models implemented to describe the adsorption kinetics and mecahnism*.

Author Contributions: I.A.: Conceptualization, methodology, supervision, writing—review and editing; G.T.I.: software, validation; H.M.A.: formal analysis, investigation; W.H.A.: data curation, visualization; M.L.: writing—review and editing; M.N.S.: writing—review and editing. A.M.H.: software, writing—review and editing. All authors have read and agreed to the published version of the manuscript.

Funding: The authors gratefully acknowledge the support by the King Abdulaziz University, Jeddah, Saudi Arabia. The authors would also like to acknowledge the support provided by the Deanship of Scientific Research (DSR) at King Fahd University of Petroleum & Minerals (KFUPM), Dhahran, Saudi Arabia, for funding this work through project number DF191025. Also, the authors would like to thank the Deanship of Scientific Research at Umm Al-Qura University for supporting this work by Grant Code: (22UQU4280401DSR02).

Institutional Review Board Statement: Not applicable.

Informed Consent Statement: Not applicable.

Data Availability Statement: Data is contained within the article.

Conflicts of Interest: The authors declare no conflict of interest.

References

1. Olaniyan, L.W.B.; Mkwetshana, N.; Okoh, A.I. Triclosan in water, implications for human and environmental health. *SpringerPlus* **2016**, *5*, 1639. [CrossRef] [PubMed]
2. Das Sarkar, S.; Nag, S.K.; Kumari, K.; Saha, K.; Bandyopadhyay, S.; Aftabuddin, M.; Das, B.K. Occurrence and safety evaluation of antimicrobial compounds triclosan and triclocarban in water and fishes of the multitrophic niche of River Torsa, India. *Arch. Environ. Contam. Toxicol.* **2020**, *79*, 488–499. [CrossRef] [PubMed]
3. Vimalkumar, K.; Arun, E.; Kumar, S.K.; Poopal, R.K.; Nikhil, N.P.; Subramanian, A.; Babu-Rajendran, R. Occurrence of triclocarban and benzotriazole ultraviolet stabilizers in water, sediment, and fish from Indian rivers. *Sci. Total Environ.* **2018**, *625*, 1351–1360. [CrossRef] [PubMed]
4. MacIsaa, J.K.; Gerona, R.R.; Blanc, P.D.; Apatira, L.; Friesen, M.W.; Cop-polino, M.; Janssen, S. Health care worker exposures to the antibacterial agent triclosan. *J. Occup. Environ. Med.* **2014**, *56*, 834–839. [CrossRef] [PubMed]

5. Schweizer, H.P. Triclosan: A widely used biocide and its link to antibiotics. *FEMS Microbiol. Lett.* **2001**, *202*, 1–7. [CrossRef]
6. Kolpin, D.W.; Furlong, E.T.; Meyer, M.T.; Thurman, E.M.; Zaugg, S.D.; Barber, L.B.; Buxton, H.T. Pharmaceuticals, hormones, and other organic wastewater contaminants in U.S. streams, 1999–2000: A national reconnaissance. *Environ. Sci. Technol.* **2000**, *36*, 1202–1211. [CrossRef]
7. Reiss, R.; Mackay, N.; Habig, C.; Griffin, J. An ecological risk assessment for triclosan in lotic systems following discharge from wastewater treatment plants in the United States. *Environ. Toxicol. Chem.* **2002**, *21*, 2483–2492. [CrossRef]
8. Loraine, G.A.; Pettigrove, M.E. Seasonal variations in concentrations of pharmaceuticals and personal care products in drinking water and reclaimed wastewater in southern California. *Environ. Sci. Technol.* **2006**, *40*, 687–695. [CrossRef]
9. Li, X.; Ying, G.G.; Su, H.C.; Yang, X.B.; Wang, L. Simultaneous determination and assessment of 4-nonylphenol, bisphenol A and triclosan in tap water, bottled water and baby bottles. *Environ. Int.* **2010**, *36*, 557–562. [CrossRef]
10. Helbing, C.C.; van Aggelen, G.; Veldhoen, N. Triclosan affects thyroid hormone-dependent metamorphosis in anurans. *Toxicol. Sci.* **2011**, *119*, 417–418. [CrossRef]
11. Park, H.G.; Yeo, M.K. The toxicity of triclosan, bisphenol a, bisphenol a diglycidyl ether to the regeneration of cnidarian, Hydra magnipapillata. *Mol. Cell Toxicol.* **2012**, *8*, 209–216. [CrossRef]
12. Crofton, K.M.; Paul, K.B.; De Vito, M.J.; Joan, M.; Hedge, J.M. Short-term in vivo exposure to the water contaminant triclosan: Evidence for disruption of thyroxine. Environ. *Toxicol. Pharmacol.* **2007**, *24*, 194–197. [CrossRef] [PubMed]
13. Huang, H.; Du, G.; Zhang, W.; Hu, J.; Wu, D.I.; Song, L.; Xia, Y.; Wang, X. The in vitro estrogenic activities of triclosan and triclocarban. *J. Appl. Toxicol.* **2014**, *34*, 1060–1067. [CrossRef] [PubMed]
14. Stoker, T.E.; Gibson, E.K.; Zorrilla, L.M. Triclosan exposure modulates estrogen-dependent responses in the female Wistar rat. *Toxicol. Sci.* **2010**, *117*, 45–53. [CrossRef] [PubMed]
15. Ishibashi, H.; Matsumura, N.; Hirano, M.; Matsuoka, M.; Shiratsuchi, H.; Ishibashi, Y.; Takao, Y.; Arizono, K. Effects of triclosan on the early life stages and reproduction of medaka Oryzias latipes and induction of hepatic vitellogenin. *Aquat. Toxicol.* **2004**, *67*, 167–179. [CrossRef]
16. Fang, J.L.; Vanlandingham, M.; da Costa, G.G.; Beland, F.A. Absorption and metabolism of triclosan after application to the skin of B6C3F1 mice. *Environ. Toxicol.* **2014**, *31*, 609–623. [CrossRef]
17. Queckenberg, C.; Meins, J.; Wachall, B.; Doroshyenko, O.; Tomalik-Scharte, D.; Bastian, B.; Abdel-Tawab, M.; Fuhr, U. Absorption, pharmacokinetics, and safety of triclosan after dermal administration. *Antimicrob. Agents Chemother.* **2010**, *54*, 570–572. [CrossRef]
18. Ali, I.; Alharbi, O.M.L.; ALOthman, Z.A.; Badjah, A.Y. Kinetics, thermodynamics and modeling of amido black dye photo-degradation in water using Co/TiO_2 nanoparticles. *Photochem. Photobiol.* **2018**, *94*, 935–941. [CrossRef]
19. Ali, I.; Alharbi, O.M.L.; ALOthman, Z.A.; Alwarthan, A.; Al-Mohaimeed, A.M. Preparation of a carboxymethylcellulose-iron composite for the uptake of atorvastatin in water. *Int. J. Biol. Macromol.* **2019**, *132*, 244–253. [CrossRef]
20. Ali, I.; Burakov, A.E.; Melezhik, A.V.; Babkin, A.V.; Burakova, I.V.; Neskomornaya, E.A.; Galunin, E.V.; Tkachev, A.G.; Kuznetsov, D.V. Kinetics, thermodynamics and mechanism of copper and zinc metal ions removal in water on newly synthesized polyhydroquinone/graphene nanocomposite material. *Chem. Select* **2019**, *4*, 12708–12718.
21. Ali, I.; Basheer, A.A.; Kucherova, A.; Memetov, N.; Pasko, T.; Ovchinnikov, K.; Pershin, V.; Kuznetsov, D.; Galunin, E.; Grachev, V.; et al. Advances in carbon nanomaterials as lubricants modifiers. *J. Mol. Liq.* **2019**, *279*, 251–266. [CrossRef]
22. Ali, I. Nano anti-cancer drugs: Pros and cons and future perspectives. *Curr. Cancer Drugs Target.* **2011**, *11*, 131–134. [CrossRef] [PubMed]
23. Brose, D.A.; Kumar, K.; Liao, A.; Hundal, L.S.; Tian, G.; Cox, A.; Zhang, H.; Podczerwinski, E.W. A reduction in triclosan and triclocarban in water resource recovery facilities' influent, effluent, and biosolids following the U.S. Food and Drug Administration's 2013 proposed rulemaking on antibacterial products. *Water Environ. Res.* **2019**, *91*, 715–721. [CrossRef]
24. Wu, C.; Spongberg, A.L.; Witter, J.D. Adsorption and degradation of triclosan and triclocarban in soils and biosolids-amended soils. *J. Agric. Food Chem.* **2009**, *57*, 4900–4905. [CrossRef] [PubMed]
25. Tohgu, Y.; Mayeri, B.K.; McNamara, P.J. Triclosan adsorption using wastewater biosolids-derived biochar. *Environ. Sci. Water Res. Technol.* **2016**, *2*, 761–768. [CrossRef]
26. Khori, N.K.E.M.; Hadibarata, T.; Elshikh, M.S.; Al-Ghamdi, A.A.; Yusop, S.Z. Triclosan removal by adsorption using activated carbon derived from waste biomass: Isotherms and kinetic studies. *J. Chin. Chem. Soc.* **2018**, *65*, 951–959. [CrossRef]
27. Fard, M.A.; Barkdoll, B. Using recyclable magnetic carbon nanotube to remove micropollutants from aqueous solutions. *J. Mol. Liq.* **2018**, *249*, 193–202. [CrossRef]
28. Li, Y.; Zhang, H.; Chen, Y.; Huang, L.; Lin, Z.; Cai, Z. Core-shell structured magnetic covalent organic framework nanocomposites for triclosan and triclocarban adsorption. *ACS Appl. Mater Interfac.* **2019**, *11*, 22492–22500. [CrossRef]
29. Triwiswaraa, M.; Leeb, C.G.; Moonc, J.K.; Park, S.J. Adsorption of triclosan from aqueous solution onto char derived from palm kernel shell. *Desal. Water Treat.* **2020**, *177*, 71–79. [CrossRef]
30. Jiang, N.; Shang, R.; Heijman, S.G.J.; Rietveld, L.C. Adsorption of triclosan, trichlorophenol and phenol by high-silica zeolites: Adsorption efficiencies and mechanisms. *Sep. Purif. Technol.* **2020**, *235*, 116–152. [CrossRef]
31. Zheng, L.; Liu, X. Solution phase synthesis of CuO hierarchical nanosheets at near neutral PH and near room temperature. *Mater. Lett.* **2007**, *61*, 2222–2226. [CrossRef]
32. Wu, W.; Li, W.; Han, B.; Zhang, Z.; Jiang, T.; Liu, Z. A green and effective method to synthesize ionic liquids: Supercritical CO_2 route. *Green Chem.* **2005**, *7*, 701–704. [CrossRef]

Article

The Formulation and Evaluation of Deep Eutectic Vehicles for the Topical Delivery of Azelaic Acid for Acne Treatment

Dhari K. Luhaibi [1], Hiba H. Mohammed Ali [2], Israa Al-Ani [1,*], Naeem Shalan [1], Faisal Al-Akayleh [3], Mayyas Al-Remawi [3], Jehad Nasereddin [4], Nidal A. Qinna [3], Isi Al-Adham [3] and Mai Khanfar [5]

[1] Faculty of Pharmacy, Pharmacological and Diagnostic Research Center, Al-Ahliyya Amman University, Amman 19328, Jordan; dluhaibi@amman.edu.jo (D.K.L.); n.shalan@ammanu.edu.jo (N.S.)
[2] Department of Pharmaceutics, College of Pharmacy, University of Sulaimani, Sulaimani 46001, Kurdistan Region, Iraq; hiba.mohammed@univsul.edu.iq
[3] Faculty of Pharmacy and Medical Sciences, University of Petra, Amman 11196, Jordan; falakyleh@uop.edu.jo (F.A.-A.); malremawi@uop.edu.jo (M.A.-R.); ialadham@uop.edu.jo (N.A.Q.); nqanna@uop.edu.jo (I.A.)
[4] Faculty of Pharmacy, Department of Pharmaceutical Sciences, Zarqa University, Zarqa 13110, Jordan; jnasereddin@zu.edu.jo
[5] Department of Pharmaceutical Technology, Faculty of Pharmacy, Jordan University of Science and Technology, Irbid 22110, Jordan; mskhanfar@just.edu.jo
* Correspondence: ialani@ammanu.edu.jo; Tel.: +962-790419013

Abstract: The current work was aimed at the development of a topical drug delivery system for azelaic acid (AzA) for acne treatment. The systems tested for this purpose were deep eutectic systems (DESs) prepared from choline chloride (CC), malonic acid (MA), and PEG 400. Three CC to MA and eight different MA: CC: PEG400 ratios were tested. The physical appearance of the tested formulations ranged from solid and liquid to semisolid. Only those that showed liquid formulations of suitable viscosity were considered for further investigations. A eutectic mixture made from MA: CC: PEG400 1:1:6 (MCP 116) showed the best characteristics in terms of viscosity, contact angle, spreadability, partition coefficient, and in vitro diffusion. Moreover, the MCP116 showed close rheological properties to the commercially available market lead acne treatment product (Skinorin®). In addition, the formula showed synergistic antibacterial activity between the MA moiety of the DES and the AzA. In vitro diffusion studies using polyamide membranes demonstrated superior diffusion of MCP116 over the pure drug and the commercial product. No signs of skin irritation and edema were observed when MCP116 was applied to rabbit skin. Additionally, the MCP116 was found to be, physically and chemically, highly stable at 4, 25, and 40 °C for a one-month stability study.

Keywords: Azelaic acid; ionic liquid; deep eutectic mixture; choline chloride; malonic acid

Citation: Luhaibi, D.K.; Ali, H.H.M.; Al-Ani, I.; Shalan, N.; Al-Akayleh, F.; Al-Remawi, M.; Nasereddin, J.; Qinna, N.A.; Al-Adham, I.; Khanfar, M. The Formulation and Evaluation of Deep Eutectic Vehicles for the Topical Delivery of Azelaic Acid for Acne Treatment. *Molecules* 2023, 28, 6927. https://doi.org/10.3390/molecules28196927

Academic Editors: Enrico Bodo, Aleksandra Cvetanović Kljakić and Slavica Ražić

Received: 14 August 2023
Revised: 19 September 2023
Accepted: 22 September 2023
Published: 4 October 2023

Copyright: © 2023 by the authors. Licensee MDPI, Basel, Switzerland. This article is an open access article distributed under the terms and conditions of the Creative Commons Attribution (CC BY) license (https:// creativecommons.org/licenses/by/ 4.0/).

1. Introduction

Acne (also called acne vulgaris (AV)) is a common dermatological disorder that most frequently affects adolescents, yet people from all age groups are candidates to be affected at least once at some point in life [1,2]. The visible nature of acne, symptoms, and sequelae all contribute physically and psychosocially to the overall burden of the disease, as do the costs required for management [3]. Sebum production, accumulation of dead skin cells in follicles, and hormonal factors are the endogenous factors that contribute to acne formation; however, the major cause of acne pathogenesis is the microorganisms [4]. Among these microorganisms, Gram-positive *Propionibacterium acnes* (*P. acnes*) has been the primary cause, whereas microorganisms such as *Staphylococcus aureus* (*S. aureus*) and *Staphylococcus epidermidis* (*S. epidermidis*) are known to increase the severity of this disease [5,6]. The widespread and prolonged use of antibiotics introduces a potential added burden through resulting antimicrobial resistance [7]. Amongst the most commonly used antibiotics, Azelaic acid (AzA, Figure 1) is a naturally occurring saturated dicarboxylic acid that has therapeutic relevance in the management of several dermatological conditions. AzA is reported to possess keratolytic and comedolytic properties, and is reported to be effective in the management of hyperpigmentation disorders like lentigo maligna and melasma [7–9]. AzA is also reported to possess antibacterial, specifically bactericidal, activity against *Cutibacterium acnes* and *Staphylococcus epidermidis*, and is therefore commonly used in the management of acne vulgaris [10–12].

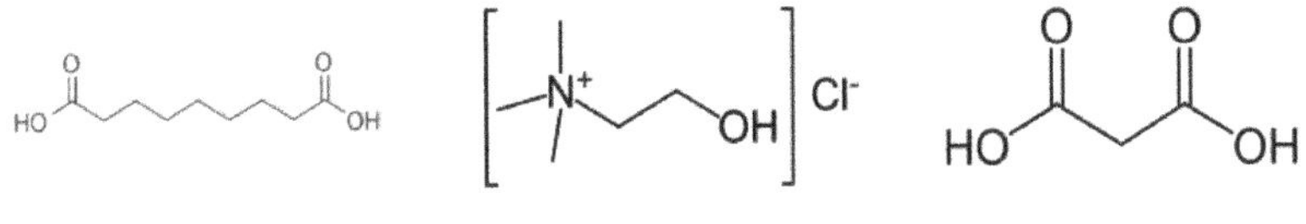

Figure 1. Chemical structure of Azelaic acid, Malonic acid, and Choline chloride.

AzA is regarded as a problematic drug to formulate; it is sparingly soluble in water, and is reported to possess limited skin permeability [13]. Several formulation strategies have been investigated to facilitate the intradermal delivery of AzA, including microemulsions [14], ethosomes [15], liposomes [16], foam formulations [17], hydrogels [18], and chitosan-loaded nanoparticles [19], among others.

Of the various strategies to enhance drug solubility and skin penetrability, Ionic Liquids (ILs) and deep eutectic systems (DESs) have recently gained considerable interest as green solvents with unique tunable physiochemical properties, most interestingly their superior solvation properties, wide liquid ranges, safety, non-toxicity, non-flammability, non-volatility, thermal stability, sustainability, biodegradability, skin penetration enhancement, antibacterial activity, and low cost [20–22]. Plenty of recently published research showed the potential antibacterial activity of some DESs [23]. It was reported that choline chloride (CC, Figure 1) with urea or glucose DESs did not inhibit bacterial growth, while the DES made from CC and malonic acid (MA, Figure 1) had an inhibitory effect, indicating its potential application as an antibacterial agent [24] (Marcel et al., 2022). Recently, DES made from choline and geranic acid has been shown to exhibit excellent antimicrobial activity against a broad variety of pathogens [25]. Further, these DESs have been shown to exhibit deep penetration into the skin, thus suggesting their ability to treat pathogens residing in the skin [26].

From the literature, DESs based on CC and dicarboxylic acids (such as MA) are one of the most commonly used representatives of these fluids that at certain molar ratios are room-temperature liquid low-melting mixtures [24,27]. One of the critical drawbacks of DESs is their relatively higher viscosity compared to molecular solvents [28]. The high viscosity of these DESs limits their pharmaceutical applicability as solubilizers and drug vehicles. To overcome their high viscosity, one efficient way is to use an appropriate co-solvent such as propylene glycol, polyethylene glycol, or glycerol. Moreover, the suggested components of the DES utilized in the current study have been reported as suitable vehicles for topical applications and are generally recognized as safe (GRAS) when used in accordance with good manufacturing practices [29]. To the best of our knowledge, the pharmaceutical application of a DES made from MA and CC as a solubilizing and drug delivery vehicle for AzA was not reported before. Therefore, the present work describes an investigation into the feasibility of the use of a DES formulation consisting of MA, CC, and PEG 400 as a vehicle for the intradermal delivery of AzA to treat acne. The prepared DES will be evaluated in terms of viscosity, spreadability, pH, partition coefficient, contact angle, and solubilization power.

2. Results and Discussion

2.1. Rheological, Solubility, and Contact Angle Studies

The acceptability and efficacy of topically applied products require that they have optimal mechanical properties, adequate rheological behavior, spreadability, and appropriate skin adhesion. Rheological evaluation is mainly to ensure the suitability of product manufacturing, uniformity, extrudability, stability, and suitability with respect to rubbing on the human skin. The rheological study (Figure 2) was conducted at temperatures that could represent product storage in the refrigerator (5 to approximately 10 °C), room temperature (22 to 25 °C), and skin temperature (32 °C). For temperatures in the range from 8 to 20 °C, the order of viscosity was MC21 > MC12 > MC11. The extremely high viscosity of MC21 hinders its application as a suitable solvent and drug carrier system, and it was therefore excluded from further investigations [30]. At room temperature (22 to 25 °C) and skin temperature (32 °C), formulation MC11 showed significantly lower viscosity readings than MC12. To further decrease the viscosity, PEG 400 was added as a co-solvent at different ratios to MC11 and MC12 and was compared with a reference commercial topical product for acne treatment (Skinoren®, 20% AzA, Leo laboratoreies Ltd., Maidenhead, UK). PEG 400 addition resulted in a significant reduction in the viscosity readings of the DESs. The MCP116 showed the lowest viscosity reading among the tested DESs (Table 1). MCP116 viscosity value was closer to that of the commercial product (340 ± 7.1 MPa and 380 ± 5.1 MPa for MCP116 and Skinoren®, respectively). PEG400 has hydrophilic polymer chains with the availability of many terminal hydroxyl groups that may possibly improve the hydrophilicity of biomaterials. In the current study, the addition of PEG 400 to MC11 and MC12 resulted in an increase in the partition coefficient parallel with a decrease in the contact angle on the hydrophobic glass surface. The increased hydrophobicity with increased PEG addition suggests possible interruptions of MA and CC interactions. In addition, PEG400 is less polar as compared to ionic liquids and DESs [31]. Therefore, the PEG addition resulted in a significant decrease in the DES contact angles, with MCP116 being the lowest. The low contact angle of MCP116 indicates its good hydrophobicity. Although AzA contains hydrogen bond donors and acceptors, AzA (pKa = 4.6) in the ionized form has a small diffusivity through the stratum corneum [32]. The pH of the DESs (2.60 to 3.13, acidic pH) renders AzA mostly in the unionized form. Therefore, AzA alone and AzA in the DESs partitioning properties were evaluated. AzA's apparent solubility in octanol was (45 ± 4) mg/mL at 22 °C. The estimated Octanol/water log p value of AzA was approximately 2 ± 0.02. The Octanol/water log p value of AzA is comparable to 1.42 ± 0.06 [33]. Increasing the ratio of PEG with resulted in an increase in log p with MCP116 being the highest. Such an increase in partitioning coefficient values could be due to the existence of the drug in the unionized form in the DESs system. Such results could be due to the availability of many (OH) groups for hydrogen bonding with the AzA and the decrease in the viscosity of the DESs with PEG addition. A decrease in viscosity, a reduction in contact angle, and an elevated partitioning coefficient collectively promote the spreadability of the formula, with MCP116 achieving the highest value, as indicated in Table 1. Additionally, the incorporation of PEG400 in the DESs was observed to have no impact on the pH. This outcome is to be expected, as PEG represents a nonionizable polymer, devoid of proton-donating or proton-accepting attributes. Most importantly, the results presented in Table 1 showed an increase in drug solubility in the DES formulations compared to that of pure water. Among the tested DESs, MCP116 showed the highest solubility, with about 80 times the solubility of the drug in water (AzA solubility in water is 2.4 mg/mL) [13]. The high-required concentration (20%) of AzA to guarantee its availability in the skin increases the incidence of side effects, most likely skin irritation. Moreover, such a high required drug concentration increases the need for a drug carrier system with high loading capacity, acceptable skin tolerability, and good penetration enhancement ability. DESs were shown to be a possible solution to overcome these challenges. The pH values of the formulations in the range of 2.6 to 3.11 are generally lower than the typical pH range of healthy skin, which falls between 4 and 6. Lower pH values can make formulations more acidic, which could potentially lead to skin irritation, dryness, or disrupt the skin's natural pH balance. However, the suitability of these pH values depends on various factors, including the specific ingredients in the formulation, the intended use (e.g., as a spot treatment or all-over application), and individual skin sensitivities. Since the current formulation is intended for short-term and localized use (acne treatment), a slightly lower pH might be acceptable and justifiable as it can help solubilize AzA, and enhance its effectiveness and intradermal penetration. Furthermore, it is worth noting that the components comprising the formulation, namely CC, MA, and PEG400, are generally considered "green" components. They have a well-documented history in skin and cosmetic formulations, demonstrating acceptable levels of skin irritation and tolerability. These attributes can further support the justification for the pH range chosen in the formulation. Because of the high viscosity of PEG-free DESs (MC11, MC12, and MC21) solubility of AzA was not possible to perform.

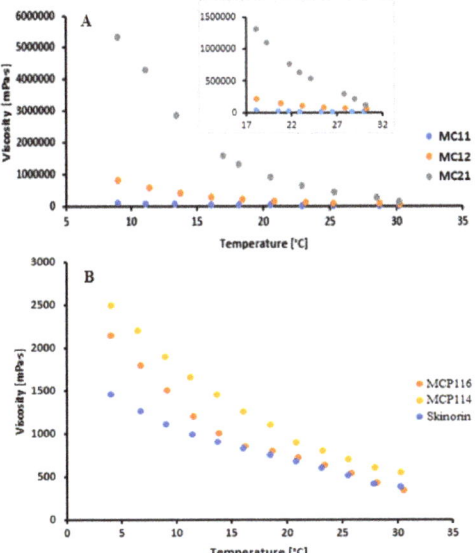

Figure 2. Rheogram of the binary deep eutectic system from malonic acid and choline chloride at different molar ratios (**A**) and malonic acid and choline chloride with PEG400 (**B**) at temperatures from 8 °C to 32 °C.

Table 1. The solubility of AZA, pH of the DESs, partition coefficient of AzA in the DESs, viscosity, spreadability and contact angle of the DESs.

DESs	AzA Solubility mg/g IL Mean ± SD	pH ± SD	Log p	Viscosity (mPa.s) ± SD at 30 °C	Spreadability Cm (Mean ± SD)	Contact Angle (Θ)
MC11	NM	NM	NM	15,872 ± 20.1	2.91 ± 0.13	NM
MC12	NM	NM	NM	51,950 ± 25.1	2.91 ± 0.13	NM
MC21	NM	NM	NM	403,000 ± 25.1	2.91 ± 0.13	NM
MCP111	156.5 ± 7.1	3.01 ± 0.01	2.12	1504 ± 12.1	5.2 ± 0.01	75 ± 3.4
MCP112	172.25 ± 8.2	3.11 ± 0.02	2.40	1076 ± 7.2	8.0 ± 0.12	70 ± 2.9
MCP114	178.0 ± 6.8	3.00 ± 0.01	2.65	721 ± 5.1	9.2 ± 0.21	65 ± 2.4
MCP116	194.1 ± 10.1	3.13 ± 0.01	2.91	340 ± 7.1	11.4 ± 0.31	53 ± 2.1
MCP121	92.0 ± 7.6	2.60 ± 0.01	2.12	5120 ± 19.1	3.01 ± 0.13	79 ± 3.4
MCP122	130.5 ± 8.9	2.61 ± 0.02	2.21	4440 ± 10.2	5.5 ± 0.16	76 ± 2.9
MCP124	145.5 ± 8.1	2.62 ± 0.01	2.21	822 ± 9.2	7.7 ± 0.12	72 ± 2.4
MCP126	152.6 ± 10.1	2.62 ± 0.015	2.45	646 ± 7.1	8.3 ± 0.21	69 ± 2.1
Skinorin®	-	-	-	380 ± 5.1	7.3 ± 0.11	-
Azelaic acid	≈0.24 g/100 g	-	2.01	-	-	-

NM: the reading was not possible to perform due to the high viscosity.

2.2. Attenuated Total Reflectance—Fourier Transformed Infra-Red (ATR–FTIR)

FTIR studies were successfully implemented for the elucidation of the possible interactions between DES components [33]. The ATR-FTIR results of MA, CC, AzA, PEG, and the optimal candidate formulation MCP116 are shown in Figure 3. AzA exhibited characteristic peaks at 1706 cm^{-1}, likely corresponding to the C=O stretching peak. The peak centered around 3000 cm^{-1} likely corresponds to the carboxylic acid dimer structure of AzA. The peaks observed between 2800 and 2950 cm^{-1} are among the specific peaks of the aliphatic chains of AzA [34]. CC exhibited characteristic peaks at 3391 cm^{-1}, and 1086 cm^{-1}, likely corresponding to the -OH stretching, and C-N stretching peaking, respectively. PEG exhibited a characteristic -OH stretching peak at 3374 cm^{-1}. MA exhibited a characteristic, aggregated peak centered on 1695 cm^{-1}, likely corresponding to C=O stretching; it also exhibited the carbonyl dimer ring centered on 3000 cm^{-1}. The spectra of both the blank (placebo DES) and the drug-loaded DES (MCP116) appear nearly identical; the carbonyl dimer is seen shifted to 2900 cm^{-1}, with the peak appearing broader in the drug-loaded MCP116 formulation. Furthermore, there is a clear shifting of the C=O stretching peak MA, indicative of hydrogen bonding. Both the peaks corresponding to a carbonyl functional groups and the -OH stretching peak are seen to be more intense (relative to % transmittance) in the drug-loaded formulation than in the placebo, but the characteristic peaks of AzA were not visible, likely due to the penetration depth limit of the ATR crystal.

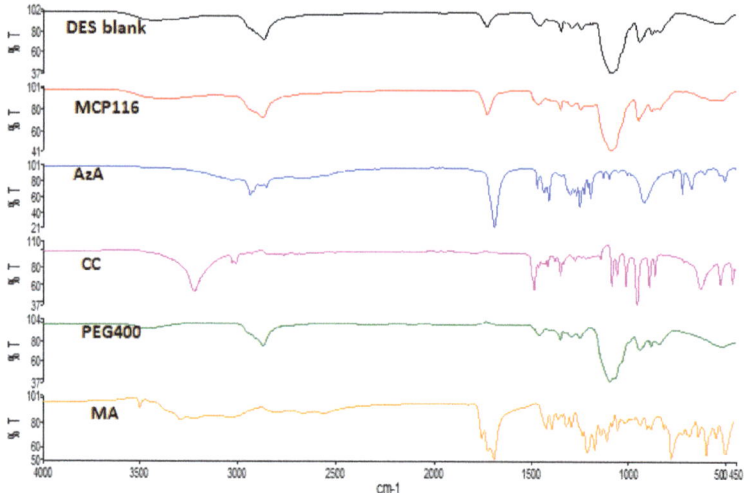

Figure 3. FTIR spectra of MA, CC, AzA, PEG, and the optimal candidate formulation MCP116.

2.3. Differential Scanning Calorimetry (DSC) Study

The solid materials analyzed (Figure 4) exhibited typical thermal behavior and melting endotherms at 105 °C, 151 °C and 325 °C for raw AzA, MA and CC, respectively (Figure 4). Such results are consistent with those reported in the literature [35,36]. Concerning PEG, a substance that remains in a liquid state at room temperature, thermal analysis was initiated at −40 °C when it was in a solid phase. The thermogram revealed an exothermic peak at −30 °C, signifying solidification, and an endothermic peak at 5 °C, indicating melting. Furthermore, Figure 4 illustrates the thermogram of MCP116 in combination with AzA. Notably, this thermogram displays a minor peak at −20 °C, another minor peak at 150 °C, and a broad peak at 260 °C. The findings indicated the absence of the characteristic peaks associated with MA and AzA, replaced by a broad peak at 260 °C, indicative of a lowered melting point for CC. These results substantiate the formation of a DES between MA and CC, the complete miscibility of PEG400, and the full solubilization of AzA within the resultant DES.

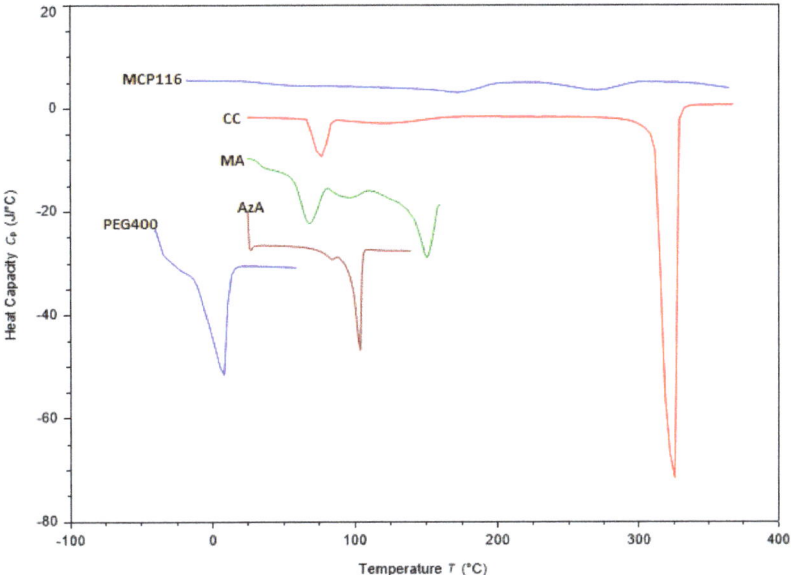

Figure 4. DSC thermogram of choline chloride (CC), malonic acid (MA), Azelaic acid (AzA, PEG400, and the eutectic formulation MCP116, with AzA.

2.4. High-Performance Liquid Chromatography (HPLC)

A singular, sharply defined peak corresponding to AzA, displaying remarkable resolution, emerged at a retention time (RT) of 4.278 min, as depicted in Figure 5. The analytical method exhibited robust linearity, substantiated by a high correlation coefficient ($R^2 = 0.9997$), and demonstrated good selectivity by virtue of the absence of any discernible interference from the excipients with the model drug.

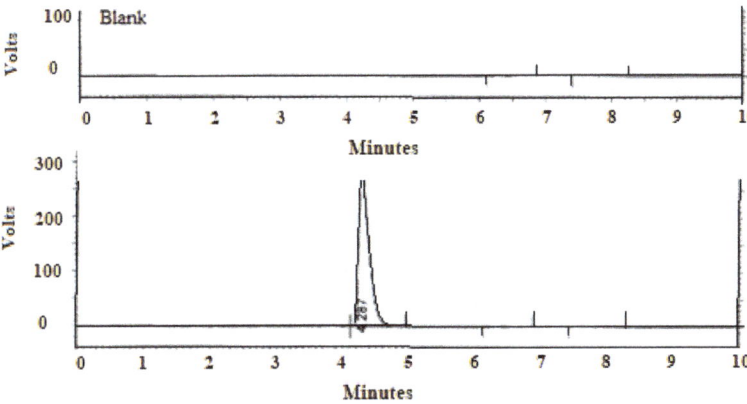

Figure 5. Chromatogram of blank formulation and formulation with Azelaic acid.

2.5. Diffusion Study

Based on the results regarding partitioning coefficient, contact angle, spreadability, and rheology, MCP116 was chosen as the candidate DESs and MCP114 was tested for comparative purposes. To assess the permeation characteristics, equivalent quantities of AzA were investigated from both formulations, alongside a pure drug solution and a commercially available product, using a Franz diffusion cell. Previous studies established the successful use of polyamide membranes as an in vitro model for SC [37,38]. Therefore, the diffusion of AzA and the selected formulations through polyamide membranes was studied. The concentration of AzA in the donor phase was the same in all diffusion studies. As anticipated, due to its higher lipophobicity, the MCP116 formulation exhibited greater diffusion compared to the parent drug.

The findings depicted in Figure 6 indicate that all formulations exhibited comparable flux during the initial 5 h. However, the flux demonstrated a consistent linear augmentation, characterized by a high R^2 value of 0.99, reaching values of 23.5 ± 4.1, 16.7 ± 2.0, and 7.5 ± 1.01 mg.cm^2.h for MCP116, MCP114, and Skinorin®, respectively, after 24 h.

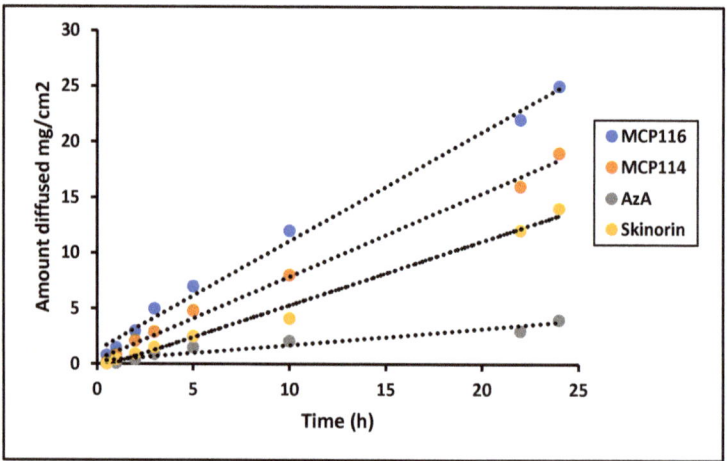

Figure 6. Diffusion study of Formulation MCP116, MCP114, AzA suspension and commercial product (Skinorin) through polyamide membrane, temperature 37 °C.

The cumulative amount released (as an average of three readings) was equal to approximately 100 ± 7%, 75 ± 8%, and 16.5 ± 3%, from MC116, MC114, and Skinorin®, respectively. These results showed the superiority of MC116 over MC114 and the reference product Skinorin®. The pure drug showed only low diffusion through the tested membrane.

2.6. Microbiological Study

Recently many reports indicated the potential application of organic acid-based MA: CC DESs as an antibacterial agent due to their inhibitory effect [23]. A recent study investigated the DES composed of a 2:1 ratio of MA and CC along with its individual component, MA, for their antifungal properties against notable fungal species, including *Aspergillus niger*, *Lentinus tigrinus*, *Candida cylindracea*, and *Cyprinus carpio fish*. The study's findings concluded that the MA present in the DES exhibited lower toxicity when compared to its isolated counterpart, MA. [39]. AzA inhibits protein synthesis inside bacterial cells [40]. In the current study, the low pH offered by the DES (pH 3.13 ± 0.01), renders AzA mostly in the unionized form; thus, enhancing its uptake through the bacterial cell wall as also proposed by a study by Al-Marabeh et al. [41]. Table 2 showed an inhibitory effect on the growth of tested microorganisms with an inhibition zone of 28.62 ± 0.85 mm, 13.03 ± 0.90 mm, and 21.50 ± 0.81 mm for MCP116 with AzA, MCP116 without AzA (blank DES), and Skinorine®, respectively. Such results indicate that the drug alone showed an inhibitory effect higher than the blank DES. MCP116 with AzA showed a significantly higher inhibitory effect against *P. acnes*, mostly due to the additive effect between the drug and MA of the DES. While the effectiveness of the MCP116 with AzA compared to the marketed product was statistically significant ($p < 0.05$), no statistical difference was detected between the inhibition zones of MCP116 without AzA and the commercial product ($p > 0.05$). The relatively small inhibition zone diameter observed for MCP116 in the absence of AzA is likely attributable to MA, a phenomenon previously documented in several references [42,43]. This effect could be a result of the additive interaction between MA and AzA. Such formulation (i.e., MCP116 with AzA) exhibited higher efficacy than the marketed product, since they yielded a larger inhibition zone diameter, which might be explained by the fact that MCP116 with AzA scattered faster in the medium due to its high spreadability and diffusion characteristics.

Table 2. Inhibition zone diameters of the DES and Skinorine® against *C. acnes* at the end of 72 h ($n = 3$; * $p < 0.05$ for MCP116 vs. Skinorine®; $p < 0.05$ for MCP116, with AzA vs. MCP116, without AzA).

Formulation	Inhibition Zone (mm) ± SD
MCP116, with AzA	28.62 ± 0.85 *
MCP116, without AzA	09.03 ± 0.90
Skinorine®	21.50 ± 0.81

The assessment of skin irritation or corrosive potential was based on the Draize Dermal Irritation Scoring model, following a methodology consistent with prior findings [44]. The results are summarized in Table 3, and skin responses were consistently observed throughout all stages of the dermal testing process.

Table 3. Dermal responses observed in individual rabbits for the formulation MCP116.

Rabbit	Erythema				
	Evaluation after removal of the test substance				
	0 min	60 min	24 h	48 h	72 h
1 (Initial)	0	0	0	0	0
2 (Confirmatory)	0	0	0	0	0
Rabbit	Edema				
	Evaluation after removal of test substance				
	0 min	60 min	24 h	48 h	72 h
1 (Initial)	0	0	0	0	0
2 (Confirmatory)	0	0	0	0	0

The results obtained (zero erythema and zero edema) indicate that the application of the formula on rabbit skin, serving as a model for human skin, did not exhibit any signs of irritation for a duration of 72 h. This suggests that both the DES and the formula can be safely applied to the skin without a significant risk of irritation.

2.7. Stability Results

The physical appearances of all formulations at all tested storage conditions were unchanged in terms of phase separation and transparency, demonstrating that the AzA in the DES system is thermodynamic stable. Moreover, the drug did not show any sign of precipitation. After storage at 25 °C for 30 days, the AzA content was 98.2 ± 2.2% and the rheological behavior of the formulation did not change, indicating that the tested AzA-loaded in DES formulations were considered stable.

In summary, this formulation represents a pioneering approach to AzA delivery for acne treatment, utilizing DES. It offers several advantages, including safety, environmental friendliness (as it is free of organic solvents), and simplicity in preparation. Anti-acne studies demonstrated that the DES formulation exhibited inhibition zones comparable to those of the commercially available cream and previously reported formulations [15,34,45].

3. Materials and Methods

3.1. Materials

Azelaic acid was purchased from UFC Biotechnology, USA, Choline Chloride 99.2% and Malonic Acid 99.1% (Xi'an Gawen Biotechnology company (Xi'an, China), Sodium dihydrogen phosphate, Potassium dihydrogen phosphate and Phosphoric acid (Merck, Germany), Formic acid (HPLC grade), Hydrochloric acid 33% (HPLC grade), Polyethylene glycol 400 and Dimethyl sulfoxide (TEDIA, Fairfield, OH, USA), Acetonitrile, 99% (VWR chemicals, BDH®, Radnor, PA, USA), Dialysis tubing (Medical International Ltd., London, UK).

3.2. Preparation of Deep Eutectic Systems (DESs)

DES formulations consisting of different ratios of MA and CC, either with or without the addition of PEG 400, are presented in Table 4. One gram of each eutectic system was prepared by mixing the components in a glass bottle with continuous stirring at room temperature. The temperature was gradually increased from 25 °C to 100 °C, and stirring was continued until a clear liquid was achieved. Each mixture was then degassed using an Ultrasonic Cleaner (Elma Sonicator, Sign, Germany). Finally, each eutectic formulation was equilibrated in a shaking water bath (SB-12L Shaking Water Baths) for 12 h at 190 rpm and 25 °C.

Table 4. Combinations of the composition of DESs.

Code	Components	Ratio (MA:CC: PEG (MCP) Respectively)
MC11	MA: CC	1:1
MC12	MA: CC	1:2
MC21	MA: CC	2:1
MCP111	MA:CC: PEG400	1:1:1
MCP112	MA:CC: PEG400	1:1:2
MCP114	MA:CC: PEG400	1:1:4
MCP116	MA:CC: PEG400	1:1:6
MCP121	MA:CC: PEG400	1:2:1
MCP122	MA:CC: PEG400	1:2:2
MCP124	MA:CC: PEG400	1:2:4
MCP126	MA:CC: PEG400	1:2:6

3.3. pH Measurement of DESs

The pH of each of the prepared DESs was measured using a pH meter (Jenway 3510 pH meter). A total of 1 mL from each eutectic system was transferred into a beaker then diluted with distilled water up to 20 mL, and the pH was measured using a pH meter (Jenway 3510 Standard Digital pH Meter Kit; 230 VAC/UK). Each measurement was conducted in triplicate, and is reported as the mean ± standard deviation.

3.4. Solubility Study

The solubility of AzA in the DESs was measured by adding 200 mg of AzA to 2 gm of each of the DESs in small glass bottles. Each glass was then sealed and transferred to a vortex mixture for 15 min, and then equilibrated in a shaking water bath for 24 h at 190 RPM and 25 °C. Each sample was then centrifuged (Stuart SCF1 Mini Centrifuge Spinner) at $14,000 \times g$ RPM for 5 min. The supernatant was assayed for AzA present. The results are reported as the percent dissolved relative to the original 200 mg placed in each container. Solubility measurements were conducted for samples that showed good viscosity compared to water. Each measurement was conducted in triplicate, and is reported as the mean ± standard deviation.

3.5. High-Performance Liquid Chromatography (HPLC)

HPLC was used to quantify the presence of AzA in the DESs. A Supelcosil LC-18-DB (5 µm, 15 cm × 4.6 mm) column was used. The mobile phase consisted of a 75% Phosphate-Buffered Saline (PBS) at pH 3.5, and a 25% acetonitrile mixture at a flow rate of 1 mL/min. The injection volume was 20 µL, the column temperature was set to 40 °C, and the UV detector was set to 206 nm. The total run time was 10 min. The quantification of AzA was performed against a standard calibration curve that was previously prepared in the region of 22–280 µg/mL.

3.6. Attenuated Total Reflectance—Fourier Transformed Infra-Red (ATR-FTIR) Study

ATR-FTIR spectra of AzA, CC, MA, and DES MCP116 (both with and without the presence of the drug), were carried out using (Perkin Elmer Spectrum Two UATR FT-IR spectrometer, Waltham, MA, USA). Scan resolution was set to 4 cm^{-1}, with 32 samples per scan. Spectra were acquired over a range of 4000–450 cm^{-1}.

3.7. Differential Scanning Calorimetry (DSC) Study

DSC was performed on AzA, MA, CC, PEG 400, and MCP116. Each test was performed by weighing an approximate amount between 5 and 8 mg of solid samples and an amount of 20–30 mg of liquid samples into an aluminum DSC crucible. Analysis was performed using the Mettler Toledo DSC 1 Star system. The temperature limit for the samples was between −60 and 320 °C, depending on the sample. All measurements were performed at a heating rate of 10 °C/min.

3.8. Rheology Study

The viscosity of different prepared DESs was studied using a rheometer (Physica MCR 302, Anton Paar, Graz, Austria) that offers different geometries (concentric cylinder, cone-and-plate, and parallel-plate) with Cp 50 double gap concentric cylinder measurement system to determine the viscosity. After the calibration of the instrument, each DES was loaded between the concentric cylinders at a volume of approximately 5–10 mL. Measurement conditions were: shear rate (0.1–100), temperature set (−10–30 °C), cone angle 1, and zero-gap 0.1.

3.9. Measurement of Partition Coefficient

The partition coefficient of AzA was measured in the selected DES formulations. For measurement, 20 mL octanol was placed in a flask, to which 0.5 g of the sample to be tested was accurately weighed and added, followed by the addition of 30 mL of deionized water. The mixtures were then placed in a shaking water bath and mixed at 190 RPM and room temperature for 24 h. Each mixture was then transferred to a separatory funnel and allowed to equilibrate. The octanol was then carefully decanted using the separatory funnel. A total of 2 mL of each phase was then diluted with the mobile phase and assayed for drug content the HPLC method. The partition coefficients of the tested samples are reported as the ratio of octanol content to water content, respectively.

3.10. Spreadability

The spreadability of the DESs was investigated by placing 1 g of the DES preparation to the center of a 20 cm × 20 cm glass plate. It was then covered with an identical slide and waited for a minute. The diameter of the spread area (cm) of three triplicates was measured as the mean ± SD [46].

3.11. In Vitro Drug Release with Franz Diffusion Cells

AzA diffusion was investigated using standard Franz diffusion cells (SES GmbH-Analyse Systeme/Fridhofstr 7–9D 55234 Bechenheim/Germany), using a polyamide membrane, 0.2 µm pore size (Sartolon Polyamide, Germany) as the diffusion barrier with a cross-sectional area of 1.7 cm^2. The acceptor compartment consisted of 12 mL Phosphate-

Buffered Saline (PBS) pH 7.4 with 30% ethanol as a cosolvent to maintain sink condition (Arpa et al., 2022). The receptor cell was stirred at 550 rpm by a magnetic stirrer, and maintained at 37 °C by circulating water with a thermostatic pump. Pure AzA (200 mg) and a suitable amount of selected DES formulations, and a commercially available AzA formulation (equivalent to 200 mg AzA) were added to the donor compartment. A drug solution in PBS pH 7.4 with 30% ethanol was also tested as a control. Aliquots of 0.5 mL were withdrawn from the acceptor compartment and replaced with equal volumes of the respective PBS pH 7.4 with 30% ethanol for 24 h. The withdrawn samples were analyzed using an HPLC method.

3.12. Contact Angle Measurements

The contact angles of a drop of the DES were measured on a polyethylene plastic surface using the contact angle goniometer (OCA 15 EC, Data Physics instruments GmbH, Filderstadt, Germany) running SCA20 Software (https://sca20.software.informer.com/) for OCA and PCA. For each measurement 500 μg Hamilton syringe was filled with the sample and anchored on the device. The dosing volume of each drop was 4 μg, with a dosing rate of 1 μg/s. High-resolution images of each drop were captured by a fixed camera. Analysis of images was carried out by two baselines, a curve, and the tangent angle drawn by the software.

3.13. Microbiological Study

The microbiological activity of the selected formula (200 mg AzA in MCP116) against Propionibacterium acnes was tested. The sensitivity test to the formula was assessed by the disc diffusion method. AzA was taken as a positive control and the blank formula as a negative control. The concentration of the bacteria was 108 CFU, the temperature of incubation was 37 °C, and the type of media was blood agar and anaerobic jar. Four disks were put in each Petri dish soaked with the diluted sample of formula MCP116 (2 disks), blank formula MCP116 (without AzA), and AzA in the solvent. Dilutions were made by taking 400 mg of both MCP116 and blank MCP116, which were diluted in 0.5 mL DMSO, and 66.5 mg pure AzA was dissolved in 0.5 DMSO. A total of 20 μL of each sample was pipetted and put on the sterilized disk and the disk was put in the specified place on the Petri dish. The Petri dish was then put in the incubator for 72 h in anaerobic conditions.

3.14. Skin Irritation/Corrosive Potential Test

Healthy young albino female rabbits (2250 ± 150 g) were housed and acclimatized at the Laboratory Animal Research Unit of the University of Petra Pharmaceutical Center. All animal studies were conducted following the University of Petra Institutional Guidelines on Animal Use. Rabbits were individually housed in rabbit racks (X-type, Techniplast, Italy) and maintained under controlled conditions of temperature (20 ± 3 °C), humidity (50 ± 15%), and photoperiod cycles (12 light/12 h dark) with a conventional laboratory diet and unrestricted supply of drinking water. Assessment of skin irritation/corrosive potential and the reversibility of dermal effects of a topical preparation of AzA. The presented test was conducted according to the OECD Guideline for Testing of Chemicals, adopting Guideline 404 for Acute Dermal Irritation/Corrosion.

A dose of 0.5 mL of both the formula and vehicle was applied to an area of approximately 6 cm^2 of the skin and covered with a gauze patch. According to the guideline an initial test, using one animal was conducted by applying three patches sequentially on the rabbit at different sites. Briefly, the first patch was removed after three min, then when no serious skin reaction was observed, a second patch was applied and removed after one hour. Thereafter, a third patch was applied and left for four hours only. Animals were examined immediately after patch removal for signs of erythema and edema, and dermal reactions were scored.

A confirmatory test was conducted after the initial test, since no corrosive effects were observed. Thus, the negative response was confirmed using another animal for an exposure period of 4 h. Then, the dermal response was evaluated immediately after the removal of the patch, after 60 min, and then at 24, 48, and 72 h from patch removal. Draize's dermal irritation scoring model was adopted for dermal assessment, as per OECD recommendation shown in Table 5.

Table 5. Draize's dermal irritation scoring model adopted for dermal assessment, as per OECD recommendation.

Erythema and Eschar Formation	Value	Edema Formation	Value
No erythema	0	No edema	0
Very slight erythema (barely perceptible)	1	Very slight edema (barely perceptible)	1
Well-defined erythema	2	Slight edema (edges of area well defined by definite raising)	2
Moderate to severe erythema	3	Moderate edema (raised approximately 1 mm)	3

3.15. Stability Study of the Selected AzA Formula

A preliminary stability study was conducted on MCP116 to determine the physicochemical stability of storage in different environments. A fresh sample of MCP116 was prepared as outlined previously, and the amount was divided into nine tubes; three tubes were sealed and stored in a fridge at 4 °C. Another three tubes were sealed and stored in an oven at 40 °C, and the last three tubes were sealed and stored at room temperature conditions at 20–25 °C. Each sample was wrapped with aluminum foil to mitigate against possible photosensitivity. The samples were tested at regular intervals (10 days, 20 days, and 30 days), investigating for changes in both organoleptic properties and drug content.

4. Conclusions

The choline-based DESs are recently gaining great focus for many researchers in the field of transdermal, topical and oral drug delivery. The work outlined herein is an application of choline-based DESs as a topical delivery system of azelaic acid. The results obtained indicate the suitability of the application of DESs technology as a solubilizing and drug-delivery vehicle of azelaic acid for acne treatment. The method of preparation is simple with high loading capacity, good stability, and higher permeability than the commercial product. The DES itself showed antibacterial activity and synergetic antibacterial effect with AzA. Such synergistic effect due to DES may allow for dose reduction and so the study provides an added value for DES application as solubilizer and drug delivery vehicle. The high reported viscosity of the DES was simply modified by the addition of PEG 400, which showed excellent miscibility with the tested DES system. The DES and the PEG aid in the enhancement of drug diffusion through the tested membrane.

Author Contributions: Conceptualization, F.A.-A. and I.A.-A. (Israa Al-Ani); methodology, D.K.L., M.A.-R. and J.N.; software, M.A.-R.; validation, M.A.-R., D.K.L. and I.A.-A. (Isi Al-Adham); formal analysis, H.H.M.A., I.A.-A. (Israa Al-Ani) and N.S.; investigation, H.H.M.A., N.S., F.A.-A., N.A.Q., I.A.-A. (Isi Al-Adham) and M.K.; resources, N.S. and N.A.Q.; data curation, D.K.L.; writing—original draft preparation, I.A.-A. (Israa Al-Ani), writing—review and editing, F.A.-A.; visualization, M.K.; supervision, J.N.; project administration, F.A.-A. All authors have read and agreed to the published version of the manuscript.

Funding: This research received no external funding.

Institutional Review Board Statement: Not applicable.

Informed Consent Statement: Not applicable.

Data Availability Statement: No data is confidential.

Acknowledgments: The authors would like to thank University of Petra and the PDRC at Al-Ahliyya Amman University for their kind contribution.

Conflicts of Interest: The authors declare no conflict of interest.

References

1. Zaenglein, A.L.; Pathy, A.L.; Schlosser, B.J.; Alikhan, A.; Baldwin, H.E.; Berson, D.S.; Bowe, W.P.; Graber, E.M.; Harper, J.C.; Kang, S.; et al. Guidelines of Care for the Management of Acne Vulgaris. *J. Am. Acad. Dermatol.* **2016**, *74*, 945–973.e33. [CrossRef]
2. Shannon, J.F. Why Do Humans Get Acne? A Hypothesis. *Med. Hypoth.* **2020**, *134*, 109412. [CrossRef]
3. Layton, A.M.; Thiboutot, D.; Tan, J. Reviewing the global burden of acne: How could we improve care to reduce the burden? *Br. J. Dermatol.* **2021**, *184*, 219–225. [CrossRef]
4. Kumar, B.; Pathak, R.; Mary, P.B.; Jha, D.; Sardana, K.; Gautam, H.K. New insights into acne pathogenesis: Exploring the role of acne-associated microbial populations. *Dermatol. Sin.* **2016**, *34*, 67–73. [CrossRef]
5. Cogen, A.L.; Nizet, V.; Gallo, R.L. Skin microbiota: A source of disease or defense? *Br. J. Dermatol.* **2008**, *158*, 442–455. [CrossRef]
6. Goodarzi, A.; Mozafarpoor, S.; Bodaghabadi, M.; Mohamadi, M. The potential of probiotics for treating acne vulgaris: A review of literature on acne and microbiota. *Dermatol. Ther.* **2020**, *33*, e13279. [CrossRef] [PubMed]
7. Schulte, B.C.; Wu, W.; Rosen, T. Azelaic Acid: Evidence-based Update on Mechanism of Action and Clinical Application. *J. Drugs Dermatol.* **2015**, *14*, 964–968. [PubMed]
8. Searle, T.; Ali, F.R.; Al-Niaimi, F. The versatility of azelaic acid in dermatology. *J. Dermatol. Treat.* **2020**, *33*, 722–732. [CrossRef] [PubMed]
9. Breathnach, A.S. Pharmacological Properties of Azelaic Acid. *Clin. Drug Investig.* **1995**, *10*, 27–33. [CrossRef]
10. Savage, L.J.; Layton, A.M. Treating acne vulgaris: Systemic, local and combination therapy. *Expert Rev. Clin. Pharmacol.* **2010**, *3*, 563–580. [CrossRef] [PubMed]
11. Dongdong, Z.; Jin, Y.; Yang, T.; Yang, Q.; Wu, B.; Chen, Y.; Luo, Z.; Liang, L.; Liu, Y.; Xu, A.; et al. Antiproliferative and Immunoregulatory Effects of Azelaic Acid Against Acute Myeloid Leukemia via the Activation of Notch Signaling Pathway. *Front. Pharmacol.* **2019**, *10*, 1396. [CrossRef] [PubMed]
12. Xie, M.; Ma, L.; Xu, T.; Pan, Y.; Wang, Q.; Wei, Y.; Shu, Y. Potential Regulatory Roles of MicroRNAs and Long Noncoding RNAs in Anticancer Therapies. *Mol. Ther. Nucleic Acids* **2018**, *13*, 233–243. [CrossRef]
13. Tomić, I.; Juretić, M.; Jug, M.; Pepić, I.; Cetina Čižmek, B.; Filipović-Grčić, J. Preparation of in situ hydrogels loaded with azelaic acid nanocrystals and their dermal application performance study. *Int. J. Pharm.* **2019**, *563*, 249–258. [CrossRef] [PubMed]
14. Hung, W.-H.; Chen, P.-K.; Fang, C.-W.; Lin, Y.-C.; Wu, P.-C. Preparation and Evaluation of Azelaic Acid Topical Microemulsion Formulation: In Vitro and In Vivo Study. *Pharmaceutics* **2021**, *13*, 410. [CrossRef] [PubMed]
15. Apriani, E.F.; Rosana, Y.; Iskandarsyah, I. Formulation, characterization, and in vitro testing of azelaic acid ethosome-based cream against Propionibacterium acnes for the treatment of acne. *J. Adv. Pharm. Technol. Res.* **2019**, *10*, 75. [CrossRef] [PubMed]
16. Akl, E.M. Liposomal azelaic acid 20% cream vs hydroquinone 4% cream as adjuvant to oral tranexamic acid in melasma: A comparative study. *J. Dermatol. Treat.* **2021**, *33*, 2008–2013. [CrossRef] [PubMed]
17. Hashim, P.W.; Chen, T.; Harper, J.C.; Kircik, L.H. The Efficacy and Safety of Azelaic Acid 15% Foam in the Treatment of Facial Acne Vulgaris. *J. Drugs. Dermatol.* **2018**, *17*, 641–645.
18. Bisht, A.; Hemrajani, C.; Rathore, C.; Dhiman, T.; Rolta, R.; Upadhyay, N.; Nidhi, P.; Gupta, G.; Dua, K.; Chellappan, D.K.; et al. Hydrogel composite containing azelaic acid and tea tree essential oil as a therapeutic strategy for *Propionibacterium* and testosterone-induced acne. *Drug Deliv. Transl. Res.* **2021**, *12*, 2501–2517. [CrossRef]
19. Tarassoli, Z.; Najjar, R.; Amani, A. Formulation and optimization of lemon balm extract loaded azelaic acid-chitosan nanoparticles for antibacterial applications. *J. Drug Deliv. Sci. Technol.* **2021**, *65*, 102687. [CrossRef]
20. Al-Akayleh, F.; Mohammed Ali, H.H.; Ghareeb, M.M.; Al-Remawi, M. Therapeutic deep eutectic system of capric acid and menthol: Characterization and pharmaceutical application. *J. Drug Deliv. Sci. Technol.* **2019**, *53*, 101159. [CrossRef]
21. Alkhawaja, B.; Al-Akayleh, F.; Al-Khateeb, A.; Nasereddin, J.; Ghanim, B.Y.; Bolhuis, A.; Jaber, N.; Al-Remawi, M.; Qinna, N.A. Deep Eutectic Liquids as a Topical Vehicle for Tadalafil: Characterisation and Potential Wound Healing and Antimicrobial Activity. *Molecules* **2023**, *28*, 2402. [CrossRef] [PubMed]
22. Al-Akayleh, F.; Adwan, S.; Khanfar, M.; Idkaidek, N.; Al-Remawi, M. A Novel Eutectic-Based Transdermal Delivery System for Risperidone. *AAPS Pharm. Sci. Tech.* **2020**, *22*, 4. [CrossRef] [PubMed]
23. Al-Akayleh, F.; Khalid, R.M.; Hawash, D.; Al-Kaissi, E.; Al-Adham, I.S.I.; Al-Muhtaseb, N.; Jaber, N.; Al-Remawi, M.; Collier, P.J. Antimicrobial potential of natural deep eutectic solvents. *Lett. Appl. Microbiol.* **2022**, *75*, 607–615. [CrossRef] [PubMed]
24. Marchel, M.; Cieśliński, H.; Boczkaj, G. Thermal Instability of Choline Chloride-Based Deep Eutectic Solvents and Its Influence on Their Toxicity–Important Limitations of DESs as Sustainable Materials. *Ind. Eng. Chem. Res.* **2022**, *61*, 11288–11300. [CrossRef]
25. Ko, J.; Mandal, A.; Dhawan, S.; Shevachman, M.; Mitragotri, S.; Joshi, N. Clinical translation of choline and geranic acid deep eutectic solvent. *Bioeng. Transl. Med.* **2020**, *6*, e10191. [CrossRef] [PubMed]
26. Riaz, M.; Akhlaq, M.; Naz, S.; Uroos, M. An overview of biomedical applications of choline geranate (CAGE): A major breakthrough in drug delivery. *RSC Adv.* **2022**, *12*, 25977–25991. [CrossRef]

27. Yuan, C.; Wang, J.; Zhang, X.; Liang, Y.; Cheng, X.; Zhu, X. High pressure-induced glass transition and stability of choline chloride/malonic acidic deep eutectic solvents with different molar ratios. *J. Mol. Liq.* **2022**, *364*, 120055. [CrossRef]
28. Smith, E.L.; Abbott, A.P.; Ryder, K.S. Deep Eutectic Solvents (DESs) and Their Applications. *Chem. Rev.* **2014**, *114*, 11060–11082. [CrossRef]
29. Sarmad, S.; Xie, Y.; Mikkola, J.-P.; Ji, X. Screening of deep eutectic solvents (DESs) as green CO_2 sorbents: From solubility to viscosity. *New J. Chem.* **2017**, *41*, 290–301. [CrossRef]
30. Abdkarimi, F.; Haghtalab, A. Solubility measurement and thermodynamic modeling of sertraline hydrochloride and clopidogrel bisulfate in deep eutectic solvent of choline chloride and malonic acid. *J. Mol. Liq.* **2021**, *344*, 117940. [CrossRef]
31. Soni, J.; Sahiba, N.; Sethiya, A.; Agarwal, S. Polyethylene glycol: A promising approach for sustainable organic synthesis. *J. Mol. Liq.* **2020**, *315*, 113766. [CrossRef]
32. Li, N.; Wu, X.; Jia, W.; Zhang, M.C.; Tan, F.; Zhang, J. Effect of ionization and vehicle on skin absorption and penetration of azelaic acid. *Drug Dev. Ind. Pharm.* **2012**, *38*, 985–994. [CrossRef] [PubMed]
33. Ali, H.H.; Ghareeb, M.M.; Al-Remawi, M.; Al-Akayleh, F.T. New insight into single phase formation of capric acid/menthol eutectic mixtures by Fourier-transform infrared spectroscopy and differential scanning calorimetry. *Trop. J. Pharm. Res.* **2020**, *19*, 361–369. [CrossRef]
34. Arpa, M.D.; Seçen, İ.M.; Erim, Ü.C.; Hoş, A.; Üstündağ Okur, N. Azelaic acid loaded chitosan and HPMC based hydrogels for treatment of acne: Formulation, characterization, in vitro-ex vivo evaluation. *Pharm. Dev. Technol.* **2022**, *27*, 268–281. [CrossRef] [PubMed]
35. Manosroi, J.; Apriyani, M.G.; Foe, K.; Manosroi, A. Enhancement of the release of azelaic acid through the synthetic membranes by inclusion complex formation with hydroxypropyl-beta-cyclodextrin. *Int. J. Pharm. Res.* **2005**, *293*, 235–240. [CrossRef]
36. Morrison, H.G.; Sun, C.C.; Neervannan, S. Characterization of thermal behavior of deep eutectic solvents and their potential as drug solubilization vehicles. *Int. J. Pharm.* **2009**, *378*, 136–139. [CrossRef]
37. Ng, S.F.; Rouse, J.; Sanderson, D.; Eccleston, G. A Comparative Study of Transmembrane Diffusion and Permeation of Ibuprofen across Synthetic Membranes Using Franz Diffusion Cells. *Pharmaceutics* **2010**, *2*, 209–223. [CrossRef]
38. Fallacara, A.; Marchetti, F.; Pozzoli, M.; Citernesi, U.R.; Manfredini, S.; Vertuani, A.S. Formulation and Characterization of Native and Crosslinked Hyaluronic Acid Microspheres for Dermal Delivery of Sodium Ascorbyl Phosphate: A Comparative Study. *Pharmaceutics* **2018**, *10*, 254. [CrossRef]
39. Kumar Jangir, A.; Patel, D.; More, R.; Parmar, A.; Kuperkar, K. New insight into experimental and computational studies of Choline chloride-based 'green'ternary deep eutectic solvent (TDES). *J. Mol. Struct.* **2019**, *1181*, 295–299. [CrossRef]
40. Holland, K.; Bojar, R. Antimicrobial effects of azelaic acid. *J. Dermatol. Treat.* **1993**, *4*, S8–S11. [CrossRef]
41. Al-Marabeh, S.; Khalil, E.; Khanfar, M.; Al-Bakri, A.G.; Alzweiri, M. A prodrug approach to enhance azelaic acid percutaneous availability. *Pharm. Dev. Technol.* **2017**, *22*, 578–586. [CrossRef] [PubMed]
42. Zainal-Abidin, M.H.; Hayyan, M.; Ngoh, G.C.; Wong, W.F.; Looi, C.Y. Emerging frontiers of deep eutectic solvents in drug discovery and drug delivery systems. *J. Control. Release* **2019**, *316*, 168–195. [CrossRef] [PubMed]
43. Siddiqui, R.; Makhlouf, Z.; Akbar, N.; Khamis, M.; Ibrahim, T.; Khan, A.S.; Khan, N.A. Antiamoebic properties of salicylic acid-based deep eutectic solvents for the development of contact lens disinfecting solutions against Acanthamoeba. *Mol. Biochem. Parasitol.* **2022**, *250*, 111493. [CrossRef]
44. Ubaydee, A.H.N.; Issa, R.; Hajleh, M.N.A.; Ghanim, B.Y.; Al-Akayleh, F.; Qinna, N.A. The effect of *Medicago sativa* extract and light on skin hypopigmentation disorders in C57/BL6 mice. *J. Cosmet. Dermatol.* **2022**, *21*, 6270–6280. [CrossRef] [PubMed]
45. Ghasemiyeh, P.; Mohammadi-Samani, S.; Noorizadeh, K.; Zadmehr, O.; Rasekh, S.; Mohammadi-Samani, S.; Dehghan, D. Novel topical drug delivery systems in acne management: Molecular mechanisms and role of targeted delivery systems for better therapeutic outcomes. *J. Drug Deliv. Sci. Technol.* **2022**, *74*, 103595. [CrossRef]
46. Ugandar, R.E.; Deivi, K.S. Formulation and evaluation of natural palm oil based vanishing cream. *Int. J. Pharm. Sci. Res.* **2013**, *4*, 3375.

Disclaimer/Publisher's Note: The statements, opinions and data contained in all publications are solely those of the individual author(s) and contributor(s) and not of MDPI and/or the editor(s). MDPI and/or the editor(s) disclaim responsibility for any injury to people or property resulting from any ideas, methods, instructions or products referred to in the content.

Review

Deep Eutectic Solvents as Catalysts for Cyclic Carbonates Synthesis from CO_2 and Epoxides

Dorota Mańka and Agnieszka Siewniak *

Department of Chemical Organic Technology and Petrochemistry, Faculty of Chemistry, Silesian University of Technology, Krzywoustego 4, 44-100 Gliwice, Poland
* Correspondence: agnieszka.siewniak@polsl.pl

Abstract: In recent years, the chemical industry has put emphasis on designing or modifying chemical processes that would increasingly meet the requirements of the adopted proecological sustainable development strategy and the principles of green chemistry. The development of cyclic carbonate synthesis from CO_2 and epoxides undoubtedly follows this trend. First, it represents a significant improvement over the older glycol phosgenation method. Second, it uses renewable and naturally abundant carbon dioxide as a raw material. Third, the process is most often solvent-free. However, due to the low reactivity of carbon dioxide, the process of synthesising cyclic carbonates requires the use of a catalyst. The efforts of researchers are mainly focused on the search for new, effective catalysts that will enable this reaction to be carried out under mild conditions with high efficiency and selectivity. Recently, deep eutectic solvents (DES) have become the subject of interest as potential effective, cheap, and biodegradable catalysts for this process. The work presents an up-to-date overview of the method of cyclic carbonate synthesis from CO_2 and epoxides with the use of DES as catalysts.

Keywords: cyclic carbonates; deep eutectic solvent; carbon dioxide; epoxide; cycloaddition of CO_2

1. Introduction

Cyclic carbonates, e.g., propylene and ethylene carbonates, are important compounds in the chemical industry, and have attracted a lot of interest over the last decades (Figure 1) [1]. This is mostly due to their unique property sets: thermal stability, low toxicity and easy biodegradability. Therefore, they have a broad spectrum of applications, mainly as solvents in various products [2,3], e.g., in lithium-ion batteries, cleaning and degreasing agents [3], industrial lubricants [4] and fuel additives [5]. They are also used to produce monomers [4] and polymers such as polycarbonates [2], polyester resins, and polyurethanes [3]. Cyclic carbonates have applications in the production of pharmaceuticals, agrochemicals [2], and other fine chemicals [4]. They are also used as intermediates in many chemical processes, e.g., to produce linear dialkyl carbonates [6], and can undergo chemical reactions such as hydrogenation, transesterification and substitution [3,7].

Cyclic carbonates can be synthesized in many ways, for example from diols and phosgene, urea or carbon oxides [4]. There are also reports of methods using halo-alcohols [6] or olefins with CO_2 [8]. Currently, processes utilizing carbon dioxide as feedstock are gaining the most interest. This is mainly due to the increasing impact of the environmental awareness related to the greenhouse effect, which is contributed to by the large amounts of CO_2 released into the atmosphere [9]. It has become essential to capture and utilize already produced carbon dioxide. This abundant compound can be a valuable source of raw material for further syntheses in the chemical industry.

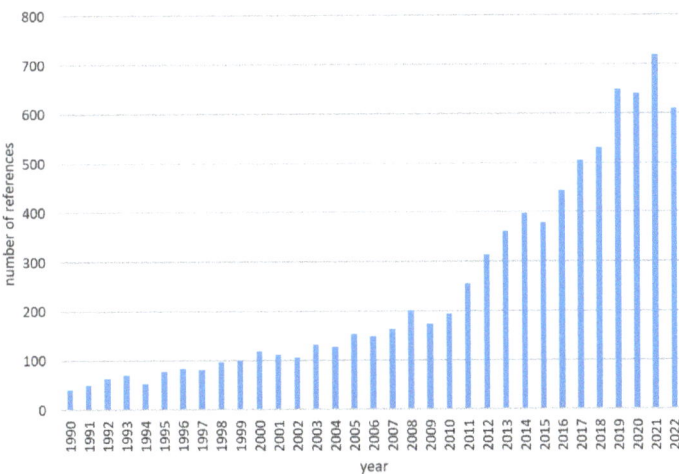

Figure 1. Number of publications on cyclic carbonates in the last 30 years (CAS SciFinder database).

One of the most attractive approaches leading to the obtaining of cyclic carbonates is the catalytic reaction of CO_2 with epoxides. This reaction has an atom economy of 100% and is a greener and safer alternative to the toxic phosgene route [10]. As the reaction of carbon dioxide and epoxide can only occur in the presence of a catalyst [11], many different catalytic systems have been developed for this synthesis. The most common catalysts are quaternary onium salts (including ammonium, phosphonium and imidazolium halides [10,12–14]), Lewis bases [10,15,16], N-heterocyclic carbenes [17], ionic liquids (ILs) [18–21], deep eutectic solvents (DES) [22–24], metal oxides (e.g., MgO, Nb_2O_5) [10,25], metal halides [10] and metal-organic complexes [10,26], including metal-organic frameworks [27].

Using deep eutectic solvents as catalysts for the cycloaddition of CO_2 to epoxides is particularly interesting due to their characteristics and high catalytic activity [28]. They have properties similar to ionic liquids, such as being a liquid at room temperature, low volatility, tuneability, non-flammability, chemical and thermal stability, and being a good solvent. However, they have more advantages than ILs, such as lower price (resulting from cheaper substrates and also ease of preparation and storage), lower toxicity, and also good biodegradability and biocompatibility, which lead to fewer problems with waste disposal [29,30]. Deep eutectic solvents are defined as mixtures with a melting point significantly lower than the melting points of their constituents. They usually consist of two components—a hydrogen bond donor (HBD) and a hydrogen bond acceptor (HBA), although ternary mixtures are also possible [31,32]. They can often be easily prepared from natural and abundant components, such as dicarboxylic acids, urea, choline chloride (ChCl), polyols, sugars, or amino acids [28,33–37]. The application of DESs in the synthesis of cyclic carbonates is also beneficial due to the better solubility of CO_2 in them compared to IL [38] and enhancement of the reaction rate by hydrogen bonds [39], which virtually create this kind of solvent [33]. In addition, homogeneous catalysts are often difficult to separate after the reaction, which is a significant drawback, especially in large-scale production, while deep eutectic solvents are often easy to separate and recycle and can be used in many consecutive reaction runs [40].

This article provides an overview of the up-to-date progress in DES-based catalytic systems for the preparation of cyclic carbonates from CO_2 and epoxides. In this work, we present the most important, interesting and noted examples of cyclic carbonate synthesis to show the development of this issue over the last dozen or so years, and to present the potential of this method with its advantages and problems. We also indicate the main directions of development of the synthesis of cyclic carbonates involving DESs.

2. Deep Eutectic Solvents as Catalysts for Cyclic Carbonates Synthesis from Epoxides and CO_2

2.1. General Information

Deep eutectic solvents generally consist of two compounds—a hydrogen bond donor and a hydrogen bond acceptor [31]. They interact with each other through hydrogen bonds and also through electrostatic and van der Waals forces. However, hydrogen bonds are the dominant interactions and they have the main influence on the properties of DESs [30]. Based on the character of the components of which DESs are comprised, several types of DES can be distinguished [29,31,41,42]: type I: quaternary salt + metal halide; type II: quaternary salt + metal halide hydrate; type III: quaternary salt + HBD; type IV: metal halide hydrate + HBD; and type V: non-ionic molecular HBAs + HBDs.

2.1.1. DES Synthesis

The preparation of DESs is simple. Normally, it requires both compounds (HBA and HBD) to be mixed in an appropriate molar ratio, then heated to about 60–110 °C and stirred until a homogenous liquid is formed. Usually, it takes about an hour. The components are also often dried beforehand due to the influence of water on the properties of DESs, which will be discussed in more detail in Section 2.1.3. of this review [22,28,36,43].

2.1.2. Mechanism of Cyclic Carbonate Synthesis from CO_2 and Epoxides in the Presence of DES

The mechanism of CO_2 cycloaddition to epoxides in the presence of a deep eutectic solvent as a catalyst, using the example of a choline chloride (ChCl)-based DES, is depicted in Figure 2 [36]. A hydrogen bond is created between the hydrogen bond donor and the oxygen of the epoxide, which activates the molecule and promotes the opening of its ring. This is continued by the attack of a nucleophile, which leads to the formation of an alkoxide intermediate. Then the CO_2 molecule is incorporated, the ring closes back, forming the cyclic carbonate, and the catalyst is regenerated. Due to the formation of hydrogen bonds, the HBD here helps to stabilize the intermediates involved in the reaction.

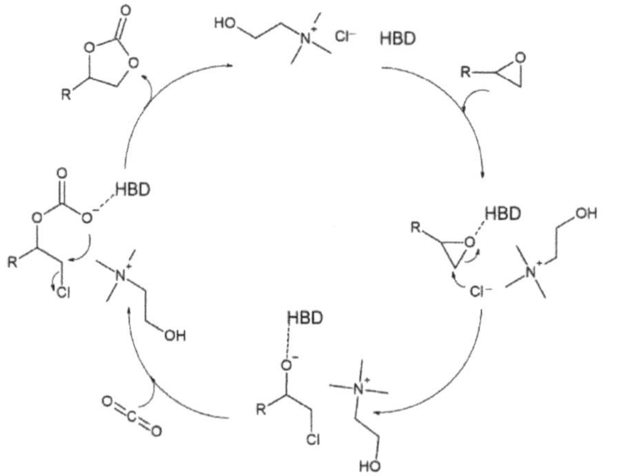

Figure 2. Proposed mechanism for cyclic carbonate synthesis catalyzed by a ChCl-based deep eutectic solvent [36].

2.1.3. The Influence of Water

Deep eutectic solvents are usually hygroscopic substances, and they usually cannot be dried completely. Investigating the influence of water on DESs is important because it can interact with each component of such a solvent. The addition of water can break the hydrogen bonds existing between DES's substrates. For example, water added to a ChCl:urea mixture reduces the strength of the hydrogen bonds between these two components and forms new ones between itself and urea. Even a very small amount of water added to DES can change its physical properties and the mixture becomes ternary [44].

2.1.4. The Influence of Hydroxyl Groups, pK_a and Acidity of DES

The coupling of HBDs with organic salts is a common method used for developing new catalytic systems for the synthesis of cyclic carbonates. One of the most popular donor classes are compounds that have one or more hydroxyl groups, e.g., alcohols, polyalcohols, glycols or fluorinated alcohols and compounds with phenols and polyphenols. In the 2019 study carried out by Yingcharoen et al. [45], the question of whether the number of hydroxyl groups in HBD influences their performance in the synthesis of cyclic carbonates under ambient conditions was investigated. They did not observe a correlation between the number of hydroxyl groups and HBD activity. However, it transpired that in some cases the catalytic activity can be tuned by the presence of additional hydroxyl groups as long as their presence does not influence the pK_a of the donor. The researchers also showed that an important correlation occurs between the catalytic activity of HBD and the pK_a value of its most acidic proton. The activity of HBD increases with the increase in pK_a in the range of 3 to 11, while pK_a values over 12 have the opposite effect. Hydroxyl donors with a pK_a between 9 and 11 give the best results in obtaining cyclic carbonates from epoxides and CO_2.

The influence of DES acidity on its activity in the synthesis of cyclic carbonates was investigated by Wang et al. [32]. It was demonstrated that deep eutectic solvents with higher acidity show stronger catalytic activity in the reaction of CO_2 with epoxides, leading to the formation of cyclic carbonates. The description of these studies has been elaborated in Section 2.2. of this paper.

2.1.5. The Influence of Temperature, Pressure and Reaction Time

An increase in temperature and pressure within a certain range usually has a beneficial effect on the reaction of CO_2 with epoxide. In a paper from 2017, the influence of temperature and pressure used during the reaction of cycloaddition of CO_2 to propylene oxide (PO) epoxides in the presence of DES composed of L-proline (L-Pro) and propanedioic acid (PA), was investigated [33]. It turned out that increasing the reaction temperature has a positive effect on the yield, but only in a certain temperature range. When the temperature exceeded 150 °C, the yield of the product decreased. This dependency is probably due to the intensification of side reactions that occur under such conditions. A similar phenomenon was observed in [28]. However, in the work of Liu et al., a decrease in product yield was observed above 120 °C [46].

As for the pressure, in the reaction of CO_2 and PO catalyzed by L-Pro/PA, increasing it up to 1.2 MPa had a positive effect on the yield of propylene carbonate (PC) [33]. However, a further rise in pressure caused a drop in cyclic carbonate yield. Similar results were observed in other works [28,42,46], although the pressure limit was different and depended on the reaction system. For the reaction involving DES composed of choline chloride and PEG, the product yield occurred after exceeding the pressure of 0.8 MPa [28]. It is worth noting that the CO_2 solubility in DES increases with higher pressure, which results in a better reaction outcome, but only up to a certain value. The further increase causes the dilution of epoxide with carbon dioxide, which has a negative effect on the reaction.

The reaction time depends mainly on the reaction system and the reaction conditions used. Usually, increasing the reaction time has a positive effect on the outcome. In the reaction of CO_2 and PO in the presence of a L-Pro/PA catalyst, the product yield increased when the time was extended to 5 h, but it started declining when the time was longer [33]. Similarly to using a too-high temperature, carrying out the reaction for too long causes a rise in side reactions.

During the process of synthesizing cyclic carbonates from CO_2 and epoxides, side reactions such as isomerization and dimerization of epoxides and hydrolysis leading to the formation of appropriate glycols may occur [36,46–48].

2.1.6. The Influence of Substrates

Many of the developed DES were tested in the synthesis of a wide variety of cyclic carbonates from CO_2 and various epoxides. For example, Liu et al. used propylene oxide, 1,2-epoxybutane, 1,2-epoxyhexane, epichlorohydrin, epibromohydrin, styrene oxide and cyclohexene oxide to synthesize the corresponding cyclic carbonates [40,43]. The highest yields were obtained for epichlorohydrin (99%), epibromohydrin (97%), and propylene oxide (96%), while the lowest yield was obtained for cyclohexene oxide (19%) [40]. In contrast, styrene carbonate was obtained with a yield of 87%. Similarly, in the work of Lü et al., the yield of the corresponding cyclic carbonates was the highest for propylene oxide (98.6%) and the lowest for cyclohexene oxide (2.5%) [33]. Based on the research, it can be concluded that terminal linear aliphatic epoxides without steric hindrance and epoxides with electron-withdrawing substituents, such as epichlorohydrin, allow obtaining cyclic carbonates with high yields. In epichlorohydrin, the presence of a chlorine atom in its structure makes the exposed carbon atom in the epoxide ring more susceptible to nucleophile attack [40]. In contrast, the use of 1,2-epoxycyclohexane with a steric hindrance resulting from the presence of two rings in its molecule reduces its reactivity [24,40].

2.1.7. Recycling of the DES-Based Catalysts

The possibility of separating and reusing the catalyst is one of the key factors in developing new synthetic methods for industrial applications. Therefore, in most works on the use of DES as catalysts for the synthesis of cyclic carbonates, an attempt was made to recycle them. Extraction was the most commonly used method of catalyst separation. Vagnoni et al. added ethyl acetate with a small amount of water to the post-reaction mixture, resulting in two phases: organic, containing the product and unreacted substrate, and water with a dissolved catalyst [36]. After phase separation, the DES catalyst was recovered by water distillation and used directly in the next process. DES retained its activity in three consecutive cycles. The same method of catalyst separation was used, for example, by Yang et al. [23]. In other works, ethyl acetate [24] or MTBE alone [40] or a mixture of ethyl acetate and ether [49] were used for extraction. Wu et al., in turn, separated the mixture after the reaction, distilled off the cyclic carbonate, and the remaining catalyst was used directly for the next cycle [28]. In a similar manner, the catalyst was recovered by Lü et al. [33]. In the case of a supported DES (e.g., on molecular sieves [50] or lignin [51]), the catalyst can be separated by simple filtration.

2.2. DES-Based Catalytic Systems Used in the Synthesis of Cyclic Carbonates—An Overview

One of the first attempts to use deep eutectic solvents in the synthesis of cyclic carbonates from CO_2 and epoxides was reported by Zhu et al. [50]. They applied a mixture of ChCl:urea in a 1:2 molar ratio as a catalyst for the reaction of CO_2 and propylene oxide leading to propylene carbonate. The process was carried out in a 1.5 to 1.87 molar ratio of CO_2 to epoxide, at 110 °C and for 10 h. The new catalyst transpired to be not only biodegradable, but also active and selective in the studied reaction. The yield of propylene carbonate was higher (99%) than in the reaction with choline chloride alone (85%). However, when the DES was immobilized on molecular sieves (Si/Al 1:1, pores size 6.7 nm), the time was decreased to only 4 h with unchanged yield. The authors assume that the synergistic effect of the anion and the cation of ChCl, and the interaction of urea with Cl^- have a major impact on the activity and selectivity of DES. The developed catalytic system was used for the reaction of other epoxides (such as epichlorohydrin, styrene oxide, cyclohexene oxide and phenoxyoxirane) with CO_2, affording the corresponding cyclic carbonates in yields ranging from 80% to 99%. After the reaction, the catalyst constitutes a separate phase and can be easily regenerated and reused up to at least five times without a significant loss in activity.

In 2016, Liu et al. developed a catalyst consisting of urea and different zinc halides, for the synthesis of cyclic carbonates from epoxides and CO_2 [46]. The obtained eutectic-based ionic liquids exhibited lower melting points than the melting points of the individual components, which is a characteristic feature of DESs. In the case of the urea:ZnI_2 mixture, the lowest temperature was achieved for a molar ratio of 3:1 (Figure 3).

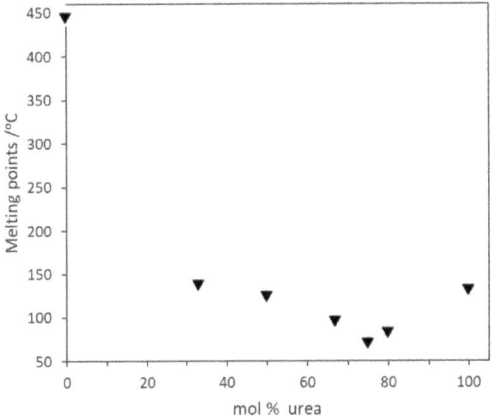

Figure 3. Dependence of DES ([urea-Zn]I_2) melting points on DES composition. Reprinted with permission from Ref. [46].

Combinations of zinc chloride, bromide and iodide with urea were tested. At a temperature of 120 °C, CO_2 pressure of 1.0 MPa and after 2 h, the selectivity to propylene carbonate always reached 98%, but only in the presence of the [urea-Zn]I_2 catalyst, the yield was also significant (84%). When the time was prolonged to 3 h, the yield was even higher (95%). It is worth noting that if ZnI_2 and urea were introduced separately into the reaction, a lower yield was obtained (72%). Other epoxides were also tested in this reaction and yields ranged from 81 to 97%. After the reaction, the catalyst could be recycled by separating it through centrifugation, washing and drying, and used for four consecutive runs, in which it retained its selectivity but the yield slowly diminished. The catalyst was efficient, easy to prepare, reusable, and in comparison to pure ZnI_2, [urea-Zn]I_2 DES was less sensitive to moisture. However, the main drawbacks are the separation of the product requiring the use of dichloromethane and the necessity of applying relatively high temperatures and pressure.

In 2016, Saptal and Bhanage introduced an efficient catalytic system consisting of an ionic liquid and a quaternary ammonium salt for the synthesis of cyclic carbonates [52]. The tested ionic liquids were composed of amino acids (such as proline [Pro], glycine [Gly], alanine [Ala], or histidine [His]) and choline chloride, while the ammonium salts were tetrabutylammonium bromide (TBAB) or tetrabutylammonium iodide (TBAI). The researchers chose ILs to be composed of inexpensive, natural, biodegradable, and non-toxic compounds, hence the use of amino acids and choline chloride. In addition, the presence of HBDs in ILs in combination with amine groups can form an active catalyst for the synthesis of cyclic carbonates. It was also assumed that the amine groups of amino acids could improve the absorption of CO_2 and thus the reaction parameters. The model reaction in this study was the cycloaddition of CO_2 to epichlorohydrin (EP) under ambient conditions—at room temperature and under atmospheric pressure. IL ([Ch][Pro]) alone turned out to be ineffective and no cyclic carbonate was formed. However, when it was mixed with TBAB or TBAI, the catalyst gave a good performance—the selectivity reached 99% and the conversion was around 60%. The authors observed the formation of low-viscosity DES from the ionic liquid [Ch][Pro] and quaternary ammonium salts. If the pressure was increased to 1 MPa, the conversion was 73%, while after an additional increase in temperature to 70 °C, the conversion reached 98% for the [Ch][His]/TBAI catalytic system. The influence of potassium iodide instead of quaternary ammonium salt was also investigated—it gave worse results than TBAI but similar results to TBAB. After the reaction, the catalyst was regenerated and reused in the next 5 runs under two different pressures. When the reactions were carried out under atmospheric pressure, the cyclic carbonate yield diminished with each consecutive run, while for the reactions at 1 MPa, the yield decreased only slightly. It was suggested that this may be the influence of the reaction time on the activity of the catalyst—at the lower pressure the reaction time was 30 h, and at the higher pressure only 2 h. The applied catalytic system gave satisfying results under ambient conditions, but required a very long reaction time. However, when the pressure and temperature were increased, the outcome was excellent.

Deep eutectic solvents based on amino acids as HBAs (such as alanine, glycine, L-proline) and dicarboxylic acids as HBDs (oxalic acid and propanedioic acid, succinic acid) in cycloaddition reactions of CO_2 to propylene oxide were investigated [33]. The DES based on L-proline and propanedioic acid was the most active; however, to obtain satisfactory results, it was necessary to use zinc bromide as a co-catalyst. The cyclic carbonates were obtained in good yields (56.6–98%) and selectivity (99%) at 150 °C and under a pressure of 1.2 MPa after 5 h. The authors emphasized that DES activity was influenced by the length of the carbon chain of the dicarboxylic acid. The chain length determines the strength of the hydrogen bond between DES and epoxide. Among the tested acids, the most preferred was propanedioic acid. Acids with shorter or longer chains formed too strong or too weak H-bonds. The DES and $ZnBr_2$ could be recovered after distilling off the cyclic carbonate, and could be used at least nine times without loss in its activity. The main drawback of this catalytic system is the need to use toxic $ZnBr_2$ as a co-catalyst, which has a negative impact on the environment. For this reason, the process cannot be considered fully "green" [28]. Catalyst separation after the reaction can also be problematic as the distillation process is energy-intensive, making the process not environmentally friendly and generating additional costs.

In 2017, García-Argüelles et al. conducted a reaction of carbon dioxide with epoxides using deep eutectic solvents based on superbases [53]. The superbases alone have already been applied in this process with good results [54]. In addition, superbase-based DESs show the ability to absorb CO_2 [55]. García-Argüelles et al. [53] used superbases as hydrogen bond acceptors (1,8-diazabicyclo[5.4.0]undec-7-ene (DBU) and 1,5,7-triazabicyclo[4.4.0]dec-5-ene (TBD)) and combined them with mono- and polyalcohols as hydrogen bond donors (ethylene glycol, methyldiethanolamine and benzyl alcohol). The tested reaction was the cycloaddition of CO_2 to epichlorohydrin. The highest yield (98%) and selectivity (98%) were achieved using DES consisting of TBD and benzyl alcohol in a molar ratio of 1:1. The reaction time was only 2 h and the process was carried out at a temperature of 100 °C and a pressure of 0.12 MPa. It is worth noting, however, that TBD and DBU alone used as catalysts also gave very good results (over 92% yield and over 97% selectivity), even at a lower temperature (50 °C), but with a much longer reaction time (20 h). As shown in Figure 4, according to the authors, DBU attacks the carbon atom of the epoxide ring, and the resulting intermediate I reacts with carbon dioxide.

One year later, Wu et al. demonstrated DESs containing choline chloride and poly (ethylene glycol) (PEG) of various chain lengths (PEG200–PEG1000) [28]. PEG is a thermally stable, non-toxic, cheap and widely used polymer. The researchers pointed out that PEGs in different catalytic systems had already been used to catalyze the reactions of CO_2 with epoxides with good results [56,57]. The addition of PEG diminishes the viscosity of the liquid phase and facilitates the diffusion between gaseous and liquid phases, which in turn can enhance the reaction rate due to better CO_2 adsorption. In the DESs used in this study, PEG plays the role of a hydrogen bond donor. The researchers evidenced that the activity of the studied DES was dependent on the length of the PEG chain. The longer the chain, the lower the yield of cyclic carbonates. It has been suggested that this may be due to mass transport limitations. Additionally, the CO_2 solubility was different in DESs consisting of different PEGs. The best results (99.1% yield of propylene carbonate) were obtained for ChCl:PEG200 DES in a molar ratio of 1:2, at 150 °C and 0.8 MPa for 5 h without any additional co-catalysts or solvents. Distillation was used to separate the post-reaction mixture. The recovered catalyst was used in subsequent cycles. After the fifth cycle, a drop in yield from 99.1 to 95.6% was observed with a negligible decrease in selectivity.

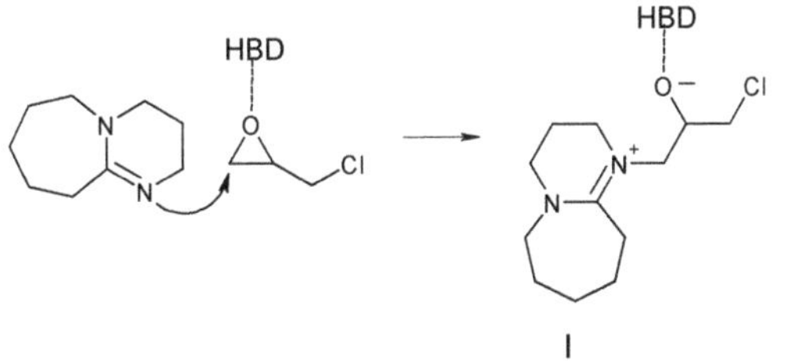

Figure 4. Activation of the epoxide by DES composed of DBU as a HBA proposed by García-Argüelles et al. [53].

In 2018, Tak et al. applied a DES-based catalytic system for the synthesis of spiro-cyclic carbonates by the cycloaddition of CO_2 to spiro-epoxy oxindoles [58]. There was no need for using any additional co-catalyst and the reaction conditions were fairly benign. The model reaction was the cycloaddition of carbon dioxide to n-benzyl spiro-epoxyoxindole to obtain the corresponding carbonate (Figure 5). All tested deep eutectic solvents were based on choline chloride with different HBDs, such as glycerol, urea, benzoic acid and ethylene glycol. The best result (98% yield) was achieved while using ChCl:urea DES (similarly as in the work by Zhu et al. [50]) with a reaction time equal to only 2 h, at a temperature of 70 °C and under atmospheric pressure. In addition to this, the used catalyst could be separated and then reused in at least the next four runs. However, the conducted separation process was more complicated and less environmentally friendly than the methods documented in previous works. 2-Methyl tetrahydrofuran and water were added to the mixture and stirred, and then the water phase was separated and washed three times with 2-methyl tetrahydrofuran. Finally, the water was evaporated and the DES could be recovered.

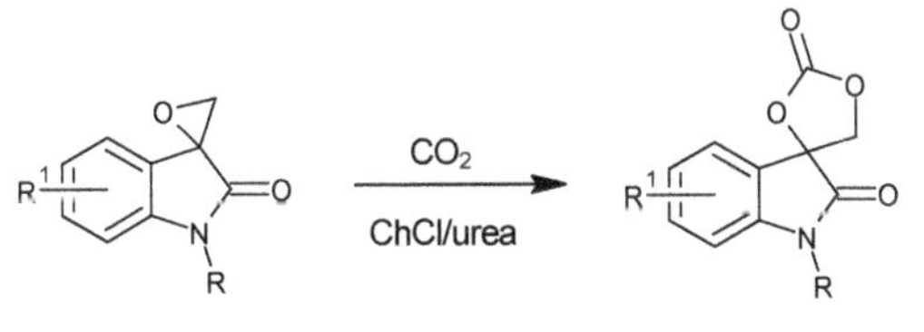

Figure 5. Structure of DES obtained from choline chloride and NBS in a molar ratio of 1:2. R^1 = CH_3, CH_3O, F; R = CH_3, C_4H_9, CH_2=$CHCH_2$.

In 2019, Liu et al. screened a series of DES based on N-hydroxysuccinimide (NHS) and choline halides (Figure 6) for the synthesis of cyclic carbonates from epoxides and CO_2 [43]. Such DESs proved to be highly effective catalysts. For example, DES obtained from choline iodide (ChI) and NHS in a molar ratio of 1:2 was the best catalyst for the reaction of propylene oxide and CO_2, affording PC in high yield (96%) and selectivity (99%) at 30 °C, and 1.0 MPa pressure, after 10 h. However, the reaction time can be reduced to 1 h if the temperature is increased to 60 °C, which allows a PC to be obtained in 92% yield with 99% selectivity. The catalyst can be easily recovered by extraction with ether and reused for at least in five consecutive runs without any significant loss in activity. In comparison to previous works on the use of deep eutectic solvents in the reaction of cycloaddition of CO_2

to epoxides, DESs based on NHS exhibit higher activity and made significant progress in terms of carrying out this synthesis under ambient conditions.

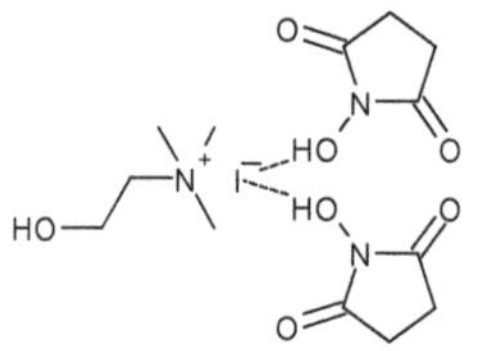

Figure 6. Structure of DES obtained from choline chloride and NBS in a molar ratio of 1:2.

Soon after, the same researchers tested a series of DESs comprising tetra-*n*-butylphosphonium bromide (TBPB) as HBA and different aromatic substrates (phenol, aminophenol (AP), aniline and benzoic acid) as HBDs in the synthesis of cyclic carbonates from CO_2 and epoxides [40]. They found that the TBPB:3-AP deep eutectic solvent was the most excellent catalyst, giving propylene carbonate in high yield (99%) and selectivity (96%) in fairly mild reaction conditions (80 °C, 1 MPa and only 1 h reaction time). A lower reaction temperature with a longer reaction time was also tested and gave results that were nearly as good. At a temperature of 30 °C after 24 h, the yield was slightly lower (95%) and the selectivity remained at the same level. Additionally, DESs with benzoic acid and 2- and 4-AP gave yields over 90% and selectivity over 99%. In addition, the catalyst could be recovered by solvent extraction and used for at least 5 consecutive cycles without significant loss in selectivity, although with a slight drop in yield. TBPB:3-AP as a catalyst was also applied in the synthesis of cyclic carbonates from different epoxides and gave favourable results as well (99–87%), except for the cyclohexene oxide (19% yield), where the steric hindrance probably impeded the epoxide ring opening and the CO_2 molecule incorporation. The researchers pointed out that phosphonium-based deep eutectic solvents could be such efficient catalysts for cyclic carbonate formation because of their Lewis basicity, nucleophilicity, and the ability to create hydrogen bonds.

Subsequently, Vagnoni et al. developed DESs based on choline chloride and iodide with different hydrogen bond donors such as polyols and carboxylic acids [36]. One of the goals of the work was to synthesize active DES from cheap and naturally derived HBDs. The model reaction was the synthesis of styrene carbonate (SC) from styrene oxide (SO) and CO_2. Experiments involving choline chloride-based DES were carried out at a pressure of 0.4 MPa at a temperature of 80 °C for 8 h. The highest yield (97%) was achieved for three different ChCl-based DESs: with maleic acid (selectivity 98%), malonic acid (selectivity 99%), and malic acid (selectivity also 99%). When the reaction was carried out under atmospheric pressure, SO conversion was not high (maximum 26%) but the selectivity reached 98%. The researchers compared the reactions conducted with DESs prepared before the process with those in which HBD and HBA were introduced separately into the reactor. Interestingly, some of the in situ generated DESs gave similar results to the preformed ones. For example, for DES based on ChCl and ethylene glycol, the PC yield was 82%, and for the components introduced separately, it was 80%, while in the other cases for previously prepared DES, the yield was higher by 12 (for ChCl:tartaric acid) or 21 percentage points (for ChCl:glycerol). DESs containing choline iodide were also tested. In this case, the DES components were added separately to the reaction mixture. The best outcome was reached for the mixture with glycerol in the reaction carried out 80 °C and under atmospheric pressure for 7 h. The SO conversion was 99% and the selectivity to SC was 96%. Such good results are due to the halide, which is responsible for opening the epoxide ring and is able to form hydrogen bonds that can stabilize intermediates. The catalyst can be recovered by extraction with water. After evaporation of the water, without further purification, the catalyst was ready for reuse for at least four more cycles, while maintaining its activity.

An interesting study was presented by Xiong et al. in 2020 [51]. As a catalyst for the cycloaddition reaction of CO_2 to epoxides, the group applied a deep eutectic solvent consisting of choline chloride and *p*-aminobenzoic acid (PABA) immobilized on lignin, together with TBAB as a co-catalyst. The model reaction was the cycloaddition of CO_2 to phenyl glycidyl ether. Under the most favourable conditions (110 °C, 1.0 MPa, 3 h), the product was obtained with a yield of 99%. Nevertheless, it was necessary to add TBAB in an amount of 10 mol% relative to the epoxide. The same catalytic system was also tested in the reactions with other epoxides, affording the corresponding products in high yields (90–99%) and selectivity (over 90%). It was suggested that hydroxyl and amino groups on the surface of the catalyst were responsible for such good performance of this catalytic system. The applied heterogeneous catalyst exhibited high activity and stability and could be simply recycled by filtration and then reused in five consecutive runs, during which the selectivity remained on the same level (99%), while the yield dropped from 99 to 84%. Scaling up of the reaction was also investigated. As the scale increased, the product yield decreased, but even at an 8-fold scale-up, the product yield was greater than 90%. However, the disadvantage of this method is the relatively high reaction temperature. Other drawbacks are the necessity of using the co-catalyst and the multistep time-consuming catalyst preparation procedure.

In 2020, Wang et al. developed an innovative type of deep eutectic solvent for the synthesis of cyclic carbonates [32]. The group reported the excellent performance of ternary DESs in the synthesis of these compounds from carbon dioxide and different epoxides. The researchers wanted to avoid the use of iodide and bromide anions due to their negative environmental impact. The best performance was achieved for the composition of 1-butyl-3-methylimidazolium chloride (bmimCl), boric acid (BA), and glutaric acid (GA) in a 7:1:1 molar ratio. The propylene carbonate yield reached 98.3% at the temperature of 70 °C, under a pressure of 0.8 MPa after 7 h. The catalytic performance of this DES was compared to the classic binary systems containing bmimCl:BA and bmimCl:GA. It was demonstrated that the ternary system was superior. The suggested explanation for this occurrence was the acidity of the applied catalysts, as the ternary system showed the highest acidity and bmimCl:BA the lowest. Strong acidity of DES promotes the formation of hydrogen bonds between DES and oxygen of the epoxy ring of PO, which activate the PO molecule (Figure 7). The ternary DES was also applied for the synthesis of other cyclic carbonates, affording very good results (yields: 87.9–99.0%), except for the cyclohexene oxide (39.5%). The catalyst could also be easily separated after the reaction and reused in the next run; however, the product yield decreased by about 20% after 5 consecutive runs.

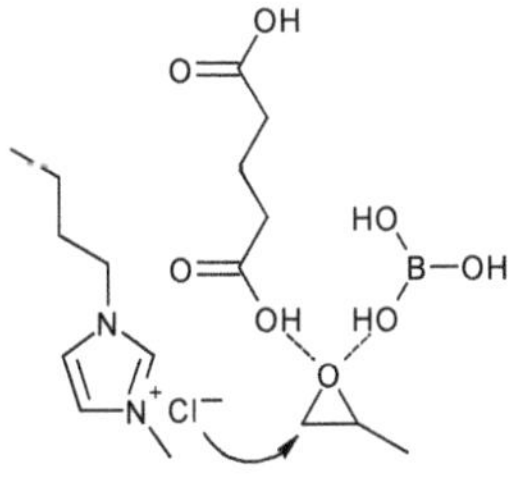

Figure 7. Activation of the epoxy ring by DES based on bmimCl, BA, and GA.

Even greater progress was demonstrated by Yang et al. in their 2021 paper [23]. Processes presented in their work seem to be almost flawless in terms of reaction conditions. They succeeded in applying the appropriate DES in the synthesis of cycloaddition of CO_2 to epoxides at room temperature and under atmospheric pressure, achieving product yields of up to 99%. However, the required synthesis time was 48 h. It is worth noting that no extra solvents or additives were used and the catalyst could be recycled and used in the next five consecutive runs without loss in activity. The researchers synthesized a deep eutectic solvent from protic ionic liquids and amines (Figure 8). It was assumed that such a catalytic system could give a great outcome, as amines can absorb carbon dioxide and act as an HBD. Additionally, protic ionic liquids have already been successfully applied as catalysts in the synthesis of cyclic carbonates [59,60]. DES consisting of these two components could potentially increase the concentration of CO_2 in the liquid phase and activate the carbon dioxide and epoxide molecules. It transpired that this assumption was right. After carrying out the reactions and investigating their mechanisms, it was confirmed that the high catalytic activity of the studied DESs was caused by synergistic catalysis of protic IL and amine.

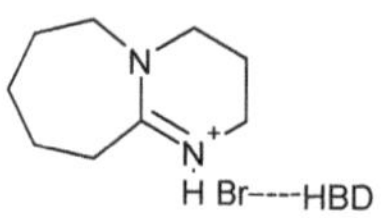

Figure 8. DES based on [DBUH][Br] and HBD.

The model reaction was the synthesis of styrene carbonate from styrene oxide and CO_2. The reaction was carried out under ambient conditions; however, the time was quite long (48 h). The best results were obtained for two different deep eutectic solvents: [DBUH][Br]/DEA (1,8-diazabicyclo[5.4.0]undec-7-ene bromide and diethanolamine) and [TMGH][Br]/DEA (1,1,3,3-tetramethylguanidine bromide and diethanolamine). In both cases, the yield reached 97% and the selectivity was more than 99%. Under the best conditions (25 °C, atmospheric pressure, 48 h), a series of cyclic carbonates was obtained with yields ranging from 94 to 99%. For [DBUH][Br]/DEA, the synthesis was also carried out at a temperature of 60 °C, which allowed a decrease in the reaction time to 5 h. The main drawback of this method is the significant amount of the catalyst that should be used in this process—20 mol%. Moreover, the DES components might be expensive. The influence of water on the reaction was also investigated, and it was shown that its addition negatively affects the reaction yield. This creates the necessity of drying the reagents before applying them to the reaction.

In 2021, Liu et al. pointed out that in many previous studies regarding deep eutectic solvents as catalysts in the cycloaddition reaction of CO_2 to epoxides, the DES incorporated the HBA with a nucleophilic site (usually halogen anion) and the HBD with Brønsted acidic site (active hydrogen) [49]. The group came up with the idea of creating a DES in which both the HBA and the HBD have Brønsted acidic sites, and thus the epoxide could be activated more efficiently. They chose to use imidazolium salts as HBAs, as it had already been proven that they could activate epoxides because of their Brønsted acidic site, and different aromatic compounds with active hydrogens as HBDs (e.g., benzoic acid, aniline, o-aminophenol, dihydroxybenzenes). Indeed, such a catalyst gave a great performance in the synthesis of styrene carbonate from styrene oxide and carbon dioxide. At room temperature, atmospheric pressure and a reaction time of 12 h, 97% yield and over 99% selectivity could be achieved using a DES consisting of 1-ethyl-3-methylimidazolium iodide (emimI) and *m*-dihydroxybenzene (*m*-DHB). The catalytic system was also tested for other epoxides, and each time (except for cyclohexene oxide) the yield reached 98 or 99%. The catalyst was separated from the reaction mixture by extraction with ethyl acetate and ether and then tested in five consecutive runs. The selectivity remained at the same level (>99%), while the yield decreased from 97% to 79%. It was suggested that this drop in yield may be due to the loss of the catalyst during its recovery process. The main advantage of this method is the mild synthesis conditions, while the drawbacks may still be a relatively long reaction time and the DES components—both are hazardous and toxic, which creates problems with their maintenance and disposal.

He et al. published a paper in which DESs based on compounds derived from biomass (biomass-derived deep eutectic solvents, BDESs) were used as catalysts for the synthesis of cyclic carbonates from epoxides and carbon dioxide [24]. It is an important issue, as the significance of biomass and its products is increasing in the chemical industry. Biomass is a virtually unlimited source of basic chemicals, which can be converted into valuable and highly desirable compounds. Biomass is an eco-friendly alternative for these kinds of substances, usually originating from fossil resources. The list of chemicals that can be obtained from biomass is very long and includes many compounds that can be used to synthesize deep eutectic solvents, such as malic acid, levulinic acid, glycerol or xylitol [61]. He et al. [24] used choline chloride, iodide and bromide as HBAs, which were mixed with different HBDs—citric acid, 3-methylglutaric acid, levulinic acid and 2-hydroxypropionic acid. The reaction of propylene oxide with CO_2 was carried out for 3 h, at a temperature of 70 °C, under a pressure of 1.0 MPa. The best results were achieved using DES obtained from choline iodide and citric acid; the reaction yield was 98% and the selectivity to propylene carbonate was over 99%. BDESs are highly stable substances, and despite being homogenous catalysts, they can be easily recovered and reused in the next five runs without relevant loss of activity. To recover the catalyst, ethyl acetate was added to the post-reaction mixture to extract the BDES, and after evaporation of the solvent, they were dried under vacuum (BDESs are highly soluble in polar solvents such as methanol, ethanol and water).

A similar study was carried out by Yang et al. [22]. They focused on bio-based deep eutectic solvents and their use in the reaction of CO_2 with epoxides leading to cyclic carbonates. This time, they investigated deep eutectic solvents based on choline bromide and also on acetylcholine bromide (AcChBr) as HBAs. Malonic acid, L-malic acid, succinic acid and glycerol were applied as HBDs. The model reaction of styrene oxide with CO_2 was carried out under atmospheric pressure at a temperature of 80 °C for 2 h. The authors also applied less of the catalyst than in their previous research—10 mol%. The best results were obtained using DES consisting of acetylcholine bromide and L-malic acid. The product yield was 98% and the selectivity reached over 99%. The catalyst could be easily separated and used in the next five runs without losing its activity. The studied DES also transpired to be very stable—comparing FTIR spectra of the freshly prepared catalyst and the one used in five cycles, there was no significant difference in their structures.

Very recently, TBAB-based DESs were used as catalysts for cyclic carbonate synthesis [62]. Wang et al. paid attention to the use of hydroxyl groups in the catalyst, as they can positively affect the reaction of hydrogen bonds. It was assumed that hydroxyl groups could potentially enhance the epoxide activation and have the ability to stabilize the intermediate. Therefore, as HBDs, glycerol, 1,3-propanediol (PG), malonic acid and citric acid (CA), and as HBAs, tetrabutylammonium salts with Cl, Br and F anions, were applied. In the model reaction between PO and CO_2, the most active was the DES composed of CA and TBAB in a molar ratio of 1:2.5. Reactions were carried out at a temperature of 80 °C, under a pressure of 0.8 MPa for 1 h. After the reaction, ethyl acetate and water (5:1 v/v) were added to the reaction mixture. Then, the separated aqueous layer was dried, and the recovered catalyst could be used for the next reaction. The recycled catalyst retained its selectivity in the next 5 cycles; however, during this time the yield diminished by about 5 percentage points.

Table 1 presents examples of results obtained under the best-established conditions for the synthesis of cyclic carbonates from CO_2 and epoxides using DESs as catalysts, while Figure 9 presents a summary of the structures of various DES used in this process. Turnover number (TON) and turnover frequency (TOF) values presented in Table 1 in most cases range from 4.6 to 49.1 and from 0.1 to 27.2, respectively. The highest TON values were obtained for the DES composed of TBAB and citric acid (102.3), ChCl/urea (95.0–99.0), TBD/BA (98.0) and [urea-Zn]I_2 (81.7). In all these cases, the reaction temperatures were above 80 °C. The highest TOF showed TBAB and citric acid (102.3) and TBD/BA (49.0) catalytic systems.

Table 1. Comparison of the results of cyclic carbonate synthesis with various DES-based catalysts.

DES HBA	DES HBD	HBA:HBD Molar Ratio	DES, mol%	Epoxide	Reaction Parameters Temperature, °C	Pressure, MPa	Time, h	Yield, %	Selectivity, %	TON[1]	TOF[2], h^{-1}	Ref.
ChCl	urea	1:2	1.0	PO	110	-[3] -[4]	10 4	99 99[5]	>99	95.0 99.0	9.5 24.8	[50]
ZnI$_2$	urea	1:3	1.2	PO	120	1.5	3	95	98	81.7	27.2	[46]
TBAI	[Chl][His]	1:1	20.0	EP[6]	70	0.1 1.0	30 2	92[7] 98[7]	99 99	4.6 4.9	0.2 2.5	[52]
L-Pro + ZnBr$_2$	propanedioic acid	1:2	2.0	PO	150	1.2	5	98	-	49.1	9.8	[33]
TBD	benzyl alcohol	1:1	1.0	EP	100	0.1	2	98	98	98.0	49.0	[53]
ChCl	PEG200	1:2	2.1	PO	150	0.8	5	99	>99	47.2	9.4	[28]
ChCl	urea	1:2	577.5	n-benzyl spiro-epoxyoxindole	70	0.1	2	98	-	0.2	0.1	[58]
ChI	NHS	1:2	6.0	PO	30 60	1.0	10 1	96 92	99 99	16.0 15.3	1.6 15.3	[43]
TBPB	3-AP	1:2	4.5	PO	30 80	1.0	24 1	95 96	>99 >99	21.1 21.3	0.9 21.3	[40]
ChI	glycerol	1:1	5.0	SO	80	0.1	7	90	-	18.0	2.6	[36]
ChCl	PABA[8]	1:1	10.0	phenyl glycidyl ether	110	1.0	3	99	>99	9.9	3.3	[51]
bmimCl	GA + BA[9]	7:1:1	7.0	PO	70	0.8	7	98	-	14.0	2.0	[32]
[DBUH][Br]	diethanoloamine	2:1	20.0	SO	r.t.[10] 60	0.1	48 5	97 >99	>99	4.9 5.0	0.1 1.0	[23]
emimI	m-DHB[11]	2:1	10.0	SO	r.t.	0.1	24	99	>99	9.9	0.4	[49]
ChI	citric acid	2:1	3.0	PO	70	0.5	3	98	>99	32.7	10.9	[24]
AcChBr	L-malic acid	2:1	10.0	SO	80	0.1	2	98	>99	9.8	4.9	[22]
TBAB	citric acid	2.5:1	0.9	PO	80	0.8	1	95	99	102.3	102.3	[62]

[1] moles of cyclic carbonate produced per mole of DES-based catalyst. [2] moles of cyclic carbonate produced per mole of DES-based catalyst per hour. [3] molar ratio of CO_2 to epoxide 2.45. [4] molar ratio of CO_2 to epoxide in the range of 1.5–1.87. [5] DES immobilized on molecular sieves. [6] epichlorohydrin. [7] conversion of EP. [8] p-aminobenzoic acid (PABA); DESs-modified lignin heterogeneous catalysts. [9] boric acid (BA), glutaric acid (GA). [10] room temperature. [11] m-dihydroxybenzene.

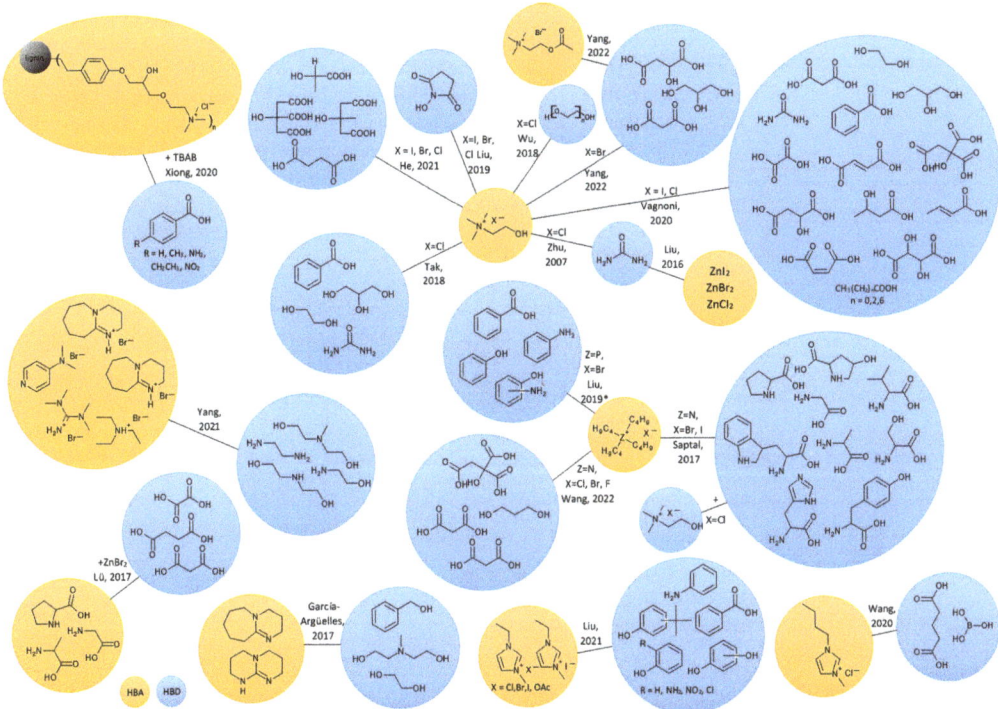

Figure 9. Various DESs used in the synthesis of cyclic carbonates from CO_2 and epoxides. Zhu, 2007 [50]; Liu, 2016 [46]; García-Argüelles, 2017 [53]; Lü, 2017 [33]; Saptal, 2017 [52]; Tak, 2018 [58]; Wu, 2018 [28]; Liu, 2019 [43]; Liu, 2019 * [40]; Wang, 2020 [32]; Vagnoni, 2020 [36]; Xiong, 2020 [51]; He, 2021 [24]; Liu, 2021 [49]; Yang, 2021 [23]; Wang, 2022 [62]; Yang, 2022 [22].

3. Conclusions

The amount of literature regarding the use of deep eutectic solvents in the synthesis of cyclic carbonates from CO_2 and epoxides increases every year. The main reasons to apply DESs in the cycloaddition of CO_2 to epoxides, among a variety of other different catalysts, are the simplicity of preparation, often simple and widely available starting materials, low price, stability and biodegradability. DESs also have the ability to capture carbon dioxide [63], which may positively affect the reaction, as CO_2 is one of the two main substrates.

The most commonly used deep eutectic solvents for the synthesis of cyclic carbonates, which give the best results, are the compositions based on choline halides as hydrogen bond acceptors. Other HBAs already applied are acetylcholine halides, quaternary ammonium halides, amino acids (such as L-proline), and protic ionic liquids. An even greater variety of compounds can be observed among the hydrogen bond donors. Common HBDs include carboxylic acids (e.g., oxalic acid, malonic acid, malic acid, levulinic acid), amines (such as diethanolamine), amides (e.g., urea), imides (e.g., *N*-hydroxysuccinimide), polyols (e.g., glycerol, ethylene glycol) and even polymers (the example of DESs with poly(ethylene glycol)).

Regarding the mechanism of the studied synthesis, it is currently believed that the formation of hydrogen bonds between the catalyst components and the epoxide, as well as the additional interaction of the epoxide with a nucleophile (usually halides), are of major significance. This way, the epoxy ring is opened and the CO_2 molecule can be incorporated. After this, the ring closes back and the catalyst is regenerated.

Currently, the main goals in the cyclic carbonate synthesis involving DESs are to reduce the reaction temperature and pressure to the ambient values, not to use any co-catalysts or additional agents (especially those containing metals), to reduce the amount of the catalyst and to use compounds that are benign and renewable, while achieving maximum activities and yields. Shortening the reaction time is also important, as well as the easiness of the catalyst and product separation after the reaction and purity of the product. Particular attention is paid to sustainable and eco-friendly deep eutectic solvents comprised of natural or bio-based substrates, especially derived from biomass. All these activities contribute to the development of energy-saving, material-saving, and environmentally safe methods for the synthesis of cyclic carbonates.

Author Contributions: Conceptualization, A.S. and D.M.; writing—original draft preparation, D.M. and A.S.; writing—review and editing, A.S. and D.M.; visualization, A.S.; supervision, A.S.; funding acquisition, A.S. All authors have read and agreed to the published version of the manuscript.

Funding: The APC was funded by Silesian University of Technology, Poland, Grant No. 04/050/BK_22/0139 in the framework of the BK program.

Institutional Review Board Statement: Not applicable.

Informed Consent Statement: Not applicable.

Data Availability Statement: Not applicable.

Conflicts of Interest: The authors declare no conflict of interest. The funders had no role in the design of the study; in the collection, analyses, or interpretation of data; in the writing of the manuscript; or in the decision to publish the results.

References

1. CAS SciFinder DataBase. Available online: https://scifinder-2n-1cas-1org-1q6ly29c00000.han.polsl.pl/search/reference/636e0 5ef0efde329e591e426/1 (accessed on 11 November 2022).
2. Dalpozzo, R.; Ca, N.D.; Gabriele, B.; Mancuso, R. Recent advances in the chemical fixation of carbon dioxide: A green route to carbonylated heterocycle synthesis. *Catalysts* **2019**, *9*, 115. [CrossRef]
3. Kamphuis, A.J.; Picchioni, F.; Pescarmona, P.P. CO_2-fixation into cyclic and polymeric carbonates: Principles and applications. *Green Chem.* **2019**, *21*, 406–448. [CrossRef]
4. Büttner, H.; Longwitz, L.; Steinbauer, J.; Wulf, C.; Werner, T. Recent developments in the synthesis of cyclic carbonates from epoxides and CO_2. *Top Curr. Chem.* **2017**, *375*, 50. [CrossRef] [PubMed]
5. Nasirov, F.; Nasirli, E.; Ibrahimova, M. Cyclic carbonates synthesis by cycloaddition reaction of CO_2 with epoxides in the presence of zinc-containing and ionic liquid catalysts. *J. Iran. Chem. Soc.* **2022**, *19*, 353–379. [CrossRef]
6. Reithofer, M.R.; Sum, Y.N.; Zhang, Y. Synthesis of cyclic carbonates with carbon dioxide and cesium carbonate. *Green Chem.* **2013**, *15*, 2086–2090. [CrossRef]
7. Rollin, P.; Soares, L.K.; Barcellos, A.M.; Araujo, D.R.; Lenardão, E.J.; Jacob, R.G.; Perin, G. Five-membered cyclic carbonates: Versatility for applications in organic synthesis, pharmaceutical, and materials sciences. *Appl. Sci.* **2021**, *11*, 5024. [CrossRef]
8. Wu, J.; Kozak, J.A.; Simeon, F.; Hatton, T.A.; Jamison, T.F. Mechanism-guided design of flow systems for multicomponent reactions: Conversion of CO_2 and olefins to cyclic carbonates. *Chem. Sci.* **2014**, *5*, 1227–1231. [CrossRef]
9. Jeffry, L.; Ong, M.Y.; Nomanbhay, S.; Mofijur, M.; Mubashir, M.; Show, P.L. Greenhouse gases utilization: A review. *Fuel* **2021**, *301*, 121017. [CrossRef]
10. Aresta, M.; Dibenedetto, A. Carbon dioxide fixation into organic compounds. In *Carbon Dioxide Recovery and Utilization*, 1st ed.; Aresta, M., Ed.; Springer: Dordrecht, The Netherlands, 2003; pp. 211–260.
11. Guo, L.; Lamb, K.J.; North, M. Recent developments in organocatalysed transformations of epoxides and carbon dioxide into cyclic carbonates. *Green Chem.* **2021**, *23*, 77–118. [CrossRef]
12. Caló, V.; Nacci, A.; Monopoli, A.; Fanizzi, A. Cyclic carbonate formation from carbon dioxide and oxiranes in tetrabutylammonium halides as solvents and catalysts. *Org. Lett.* **2002**, *4*, 2561–2563. [CrossRef]
13. Cokoja, M.; Wilhelm, M.E.; Anthofer, M.H.; Herrmann, W.A.; Kühn, F.E. Synthesis of cyclic carbonates from epoxides and carbon dioxide by using organocatalysts. *ChemSusChem* **2015**, *8*, 2436–2454. [CrossRef]
14. Siewniak, A.; Forajter, A.; Szymańska, K. Mesoporous silica-supported ionic liquids as catalysts for styrene carbonate synthesis from CO_2. *Catalysts* **2020**, *10*, 1363. [CrossRef]
15. Zhang, X.; Zhao, N.; Wei, W.; Sun, Y. Chemical fixation of carbon dioxide to propylene carbonate over amine-functionalized silica catalysts. *Catal. Today* **2006**, *115*, 102–106. [CrossRef]

16. Shiels, R.A.; Jones, C.W. Homogeneous and Heterogeneous 4-(N,N-Dialkylamino) pyridines as effective single component catalysts in the synthesis of propylene carbonate. *J. Mol. Catal. A Chem.* **2007**, *261*, 160–166. [CrossRef]
17. Wang, Y.-B.; Sun, D.-S.; Zhou, H.; Zhang, W.-Z.; Lu, X.-B. CO_2, COS and CS_2 adducts of n-heterocyclic olefins and their application as organocatalysts for carbon dioxide fixation. *Green Chem.* **2015**, *17*, 4009. [CrossRef]
18. Dibenedetto, A.; Angelini, A. Synthesis of organic carbonates. In *Advances in Inorganic Chemistry*, 1st ed.; Aresta, M., Eldik, R., Eds.; Academic Press Inc.: Waltham, MA, USA, 2014; Volume 66, pp. 25–81. [CrossRef]
19. Liu, M.; Liang, L.; Liang, T.; Lin, X.; Shi, L.; Wang, F.; Sun, J. Cycloaddition of CO_2 and epoxides catalyzed by dicationic ionic liquids mediated metal halide: Influence of the dication on catalytic activity. *J. Mol. Catal. A Chem.* **2015**, *408*, 242–249. [CrossRef]
20. Peng, J.; Deng, Y. Cycloaddition of carbon dioxide to propylene oxide catalyzed by ionic liquids. *New J. Chem.* **2001**, *25*, 639–641. [CrossRef]
21. Jasiak, K.; Siewniak, A.; Kopczyńska, K.; Chrobok, A.; Baj., S. Hydrogensulphate ionic liquids as an efficient catalyst for the synthesis of cyclic carbonates from carbon dioxide and epoxides. *J. Chem. Techol. Biotechnol.* **2016**, *91*, 2827–2833. [CrossRef]
22. Yang, X.; Liu, Z.; Chen, P.; Liu, F.; Zhao, T. Effective synthesis of cyclic carbonates from CO_2 and epoxides catalyzed by acetylcholine bromide-based deep eutectic solvents. *J. CO_2 Util.* **2022**, *58*, 101936. [CrossRef]
23. Yang, X.; Zou, Q.; Zhao, T.; Chen, P.; Liu, Z.; Liu, F.; Lin, Q. Deep eutectic solvents as efficient catalysts for fixation of CO_2 to cyclic carbonates at ambient temperature and pressure through synergetic catalysis. *ACS Sustain. Chem. Eng.* **2021**, *9*, 10437–10443. [CrossRef]
24. He, L.; Zhang, W.; Yang, Y.; Ma, J.; Liu, F.; Liu, M. Novel biomass-derived deep eutectic solvents promoted cycloaddition of CO_2 with epoxides under mild and additive-free conditions. *J. CO_2 Util.* **2021**, *54*, 101750. [CrossRef]
25. Yamaguchi, K.; Ebitani, K.; Yoshida, T.; Yoshida, H.; Kaneda, K. Mg-Al mixed oxides as highly active acid-base catalysts for cycloaddition of carbon dioxide to epoxides. *J. Am. Chem. Soc.* **1999**, *121*, 4526–4527. [CrossRef]
26. Tian, D.; Liu, B.; Gan, Q.; Li, H.; Darensbourg, D.J. Formation of cyclic carbonates from carbon dioxide and epoxides coupling reactions efficiently catalyzed by robust, recyclable one-component aluminum-salen complexes. *ACS Catal.* **2012**, *2*, 2029–2035. [CrossRef]
27. Shaikh, R.R.; Pornpraprom, S.; D'Elia, V. Catalytic strategies for the cycloaddition of pure, diluted, and waste CO_2 to epoxides under ambient conditions. *ACS Catal.* **2018**, *8*, 419–450. [CrossRef]
28. Wu, K.; Su, T.; Hao, D.; Liao, W.; Zhao, Y.; Ren, W.; Deng, C.; Lü, H. Choline chloride-based deep eutectic solvents for efficient cycloaddition of CO_2 with propylene oxide. *Chem. Commun.* **2018**, *54*, 9579–9582. [CrossRef]
29. Liu, J.; Li, X.; Row, K.H. Development of deep eutectic solvents for sustainable chemistry. *J. Mol. Liq.* **2022**, *362*, 119654. [CrossRef]
30. Mbous, Y.P.; Hayyan, M.; Hayyan, A.; Wong, W.F.; Hashim, M.A.; Looi, C.Y. Applications of deep eutectic solvents in biotechnology and bioengineering—Promises and challenges. *Biotechnol. Adv.* **2017**, *35*, 105–134. [CrossRef]
31. Hansen, B.B.; Spittle, S.; Chen, B.; Poe, D.; Zhang, Y.; Klein, J.M.; Horton, A.; Adhikari, L.; Zelovich, T.; Doherty, B.W.; et al. Deep eutectic solvents: A Review of fundamentals and applications. *Chem. Rev.* **2021**, *121*, 1232–1285. [CrossRef]
32. Wang, S.; Zhu, Z.; Hao, D.; Su, T.; Len, C.; Ren, W.; Lü, H. Synthesis cyclic carbonates with bmimcl-based ternary deep eutectic solvents system. *J. CO_2 Util.* **2020**, *40*, 101250. [CrossRef]
33. Lü, H.; Wu, K.; Zhao, Y.; Hao, L.; Liao, W.; Deng, C.; Ren, W. Synthesis of cyclic carbonates from CO_2 and Propylene Oxide (PO) with Deep Eutectic Solvents (DESs) based on Amino Acids (AAs) and dicarboxylic acids. *J. CO_2 Util.* **2017**, *22*, 400–406. [CrossRef]
34. Abbott, A.P.; Capper, G.; Davies, D.L.; Rasheed, R.K.; Tambyrajah, V. Novel solvent properties of choline chloride/urea mixtures. *Chem. Commun.* **2003**, *1*, 70–71. [CrossRef]
35. Hayyan, A.; Mjalli, F.S.; Alnashef, I.M.; Al-Wahaibi, T.; Al-Wahaibi, Y.M.; Hashim, M.A. Fruit sugar-based deep eutectic solvents and their physical properties. *Thermochim. Acta* **2012**, *541*, 70–75. [CrossRef]
36. Vagnoni, M.; Samorì, C.; Galletti, P. Choline-based eutectic mixtures as catalysts for effective synthesis of cyclic carbonates from epoxides and CO_2. *J. CO_2 Util.* **2020**, *42*, 101302. [CrossRef]
37. Claver, C.; Yeamin, M.B.; Reguero, M.; Masdeu-Bultó, A.M. Recent advances in the use of catalysts based on natural products for the conversion of CO_2 into cyclic carbonates. *Green Chem.* **2020**, *22*, 7665–7706. [CrossRef]
38. Ali, E.; Hadj-Kali, M.K.; Mulyono, S.; Alnashef, I. Analysis of operating conditions for CO_2 capturing process using deep eutectic solvents. *Int. J. Greenh. Gas Control* **2016**, *47*, 342–350. [CrossRef]
39. Dai, W.L.; Jin, B.; Luo, S.L.; Yin, S.F.; Luo, X.B.; Au, C.T. Cross-linked polymer grafted with functionalized ionic liquid as reusable and efficient catalyst for the cycloaddition of carbon dioxide to epoxides. *J. CO_2 Util.* **2013**, *3*, 7–13. [CrossRef]
40. Liu, F.; Gu, Y.; Xin, H.; Zhao, P.; Gao, J.; Liu, M. Multifunctional phosphonium-based deep eutectic ionic liquids: Insights into simultaneous activation of CO_2 and epoxide and their subsequent cycloaddition. *ACS Sustain. Chem. Eng.* **2019**, *7*, 16674–16681. [CrossRef]
41. Sultana, K.; Rahman, M.T.; Habib, K.; Das, L. Recent advances in deep eutectic solvents as shale swelling inhibitors: A comprehensive review. *ACS Omega* **2022**, *7*, 28723–28755. [CrossRef]
42. Abbott, A.P.; Barron, J.C.; Ryder, K.S.; Wilson, D. Eutectic-based ionic liquids with metal-containing anions and cations. *Chem. Eur. J.* **2007**, *13*, 6495–6501. [CrossRef]
43. Liu, F.; Gu, Y.; Zhao, P.; Xin, H.; Gao, J.; Liu, M. N-Hydroxysuccinimide based deep eutectic catalysts as a promising platform for conversion of CO_2 into cyclic carbonates at ambient temperature. *J. CO_2 Util.* **2019**, *33*, 419–426. [CrossRef]

44. Omar, K.A.; Sadeghi, R. Physicochemical properties of deep eutectic solvents: A review. *J. Mol. Liq.* **2022**, *360*, 119524. [CrossRef]
45. Yingcharoen, P.; Kongtes, C.; Arayachukiat, S.; Suvarnapunya, K.; Vummaleti, S.V.C.; Wannakao, S.; Cavallo, L.; Poater, A.; d'Elia, V. Assessing the pK_a-dependent activity of hydroxyl hydrogen bond donors in the organocatalyzed cycloaddition of carbon dioxide to epoxides: Experimental and theoretical study. *Adv. Synth. Catal.* **2019**, *361*, 366–373. [CrossRef]
46. Liu, M.; Li, X.; Lin, X.; Liang, L.; Gao, X.; Sun, J. Facile synthesis of [Urea-Zn]I$_2$ eutectic-based ionic liquid for efficient conversion of carbon dioxide to cyclic carbonates. *J. Mol. Catal. A Chem.* **2016**, *412*, 20–26. [CrossRef]
47. Adeleye, A.D.; Patel, D.; Niyogi, D.; Saha, B. Efficient and greener synthesis of propylene carbonate from carbon dioxide and propylene oxide. *Ind. Eng. Chem. Res.* **2014**, *53*, 18647–18657. [CrossRef]
48. Yang, Z.Z.; He, L.N.; Miao, C.X.; Chanfreau, S. Lewis basic ionic liquids-catalyzed conversion of carbon dioxide to cyclic carbonates. *Adv. Synth. Catal.* **2010**, *352*, 2233–2240. [CrossRef]
49. Liu, Y.; Cao, Z.; Zhou, Z.; Zhou, A. Imidazolium-based deep eutectic solvents as multifunctional catalysts for multisite synergistic activation of epoxides and ambient synthesis of cyclic carbonates. *J. CO$_2$ Util.* **2021**, *53*, 101717. [CrossRef]
50. Zhu, A.; Jiang, T.; Han, B.; Zhang, J.; Xie, Y.; Ma, X. Supported choline chloride/urea as a heterogeneous catalyst for chemical fixation of carbon dioxide to cyclic carbonates. *Green Chem.* **2007**, *9*, 169–172. [CrossRef]
51. Xiong, X.; Zhang, H.; Lai, S.L.; Gao, J.; Gao, L. Lignin modified by deep eutectic solvents as green, reusable, and bio-based catalysts for efficient chemical fixation of CO$_2$. *React. Funct. Polym.* **2020**, *149*, 104502. [CrossRef]
52. Saptal, V.B.; Bhanage, B.M. Bifunctional ionic liquids derived from biorenewable sources as sustainable catalysts for fixation of carbon dioxide. *ChemSusChem* **2017**, *10*, 1145–1151. [CrossRef]
53. García-Argüelles, S.; Ferrer, M.L.; Iglesias, M.; del Monte, F.; Gutiérrez, M.C. Study of superbase-based deep eutectic solvents as the catalyst in the chemical fixation of CO$_2$ into cyclic carbonates under mild conditions. *Materials* **2017**, *10*, 759. [CrossRef]
54. Yu, K.M.K.; Curcic, I.; Gabriel, J.; Morganstewart, H.; Tsang, S.C. Catalytic coupling of CO$_2$ with epoxide over supported and unsupported amines. *J. Phys. Chem. A* **2009**, *114*, 3863–3872. [CrossRef]
55. Sze, L.L.; Pandey, S.; Ravula, S.; Pandey, S.; Zhao, H.; Baker, G.A.; Baker, S.N. Ternary deep eutectic solvents tasked for carbon dioxide capture. *ACS Sustain. Chem. Eng.* **2014**, *2*, 2117–2123. [CrossRef]
56. Dou, X.Y.; Wang, J.Q.; Du, Y.; Wang, E.; He, L.N. Guanidinium salt functionalized peg: An effective and recyclable homogeneous catalyst for the synthesis of cyclic carbonates from CO$_2$ and epoxides under solvent-free conditions. *Synlett* **2007**, *2007*, 3058–3062. [CrossRef]
57. Du, Y.; Wang, J.-Q.; Chen, J.-Y.; Cai, F.; Tian, J.-S.; Kong, D.-L.; He, L.-N. A Poly(Ethylene Glycol)-supported quaternary ammonium salt for highly efficient and environmentally friendly chemical fixation of CO$_2$ with epoxides under supercritical conditions. *Tetrahedron Lett.* **2006**, *47*, 1271–1275. [CrossRef]
58. Tak, R.K.; Patel, P.; Subramanian, S.; Kureshy, R.I.; Khan, N.U.H. Cycloaddition reaction of spiro-epoxy oxindole with CO$_2$ at atmospheric pressure using deep eutectic solvent. *ACS Sustain. Chem. Eng.* **2018**, *6*, 11200–11205. [CrossRef]
59. Zhang, Z.; Fan, F.; Xing, H.; Yang, Q.; Bao, Z.; Ren, Q. Efficient synthesis of cyclic carbonates from atmospheric CO$_2$ using a positive charge delocalized ionic liquid catalyst. *ACS Sustain. Chem. Eng.* **2017**, *5*, 2841–2846. [CrossRef]
60. Xiao, L.; Su, D.; Yue, C.; Wu, W. Protic ionic liquids: A highly efficient catalyst for synthesis of cyclic carbonate from carbon dioxide and epoxides. *J. CO$_2$ Util.* **2014**, *6*, 1–6. [CrossRef]
61. Gérardy, R.; Debecker, D.P.; Estager, J.; Luis, P.; Monbaliu, J.C.M. Continuous flow upgrading of selected C2-C6 platform chemicals derived from biomass. *Chem. Rev.* **2020**, *120*, 7219–7347. [CrossRef]
62. Wang, Z.; Yan, T.; Guo, L.; Wang, Q.; Zhang, R.; Zhan, H.; Yi, L.; Chen, J.; Wu, X. Synthesis of TBAB-based deep eutectic solvents as the catalyst in the coupling reaction between CO$_2$ and epoxides under ambient temperature. *ChemistrySelect* **2022**, *7*, e202202091. [CrossRef]
63. Sarmad, S.; Mikkola, J.P.; Ji, X. Carbon dioxide capture with ionic liquids and deep eutectic solvents: A new generation of sorbents. *ChemSusChem* **2017**, *10*, 324–352. [CrossRef]

Article

Tunable Aryl Alkyl Ionic Liquid Supported Synthesis of Platinum Nanoparticles and Their Catalytic Activity in the Hydrogen Evolution Reaction and in Hydrosilylation

Dennis Woitassek [1,†], Till Strothmann [1,†], Harry Biller [2], Swantje Lerch [2], Henning Schmitz [1], Yefan Song [1], Stefan Roitsch [3], Thomas Strassner [2,*] and Christoph Janiak [1,*]

1. Institut für Anorganische Chemie und Strukturchemie, Heinrich-Heine-Universität Düsseldorf, 40204 Düsseldorf, Germany
2. Physikalische Organische Chemie, Technische Universität Dresden, 01062 Dresden, Germany
3. Institut für Physikalische Chemie, Universität zu Köln, 50939 Köln, Germany
* Correspondence: thomas.strassner@tu-dresden.de (T.S.); janiak@uni-duesseldorf.de (C.J.); Tel.: +49-2118112286 (C.J.)
† These authors contributed equally to this work.

Abstract: Tunable aryl alkyl ionic liquids (TAAILs) are ionic liquids (ILs) with a 1-aryl-3-alkylimidazolium cation having differently substituted aryl groups. Herein, nine TAAILs with the bis(trifluoromethylsulfonyl)imide anion are utilized in combination with and without ethylene glycol (EG) as reaction media for the rapid microwave synthesis of platinum nanoparticles (Pt-NPs). TAAILs allow the synthesis of small NPs and are efficient solvents for microwave absorption. Transmission electron microscopy (TEM) shows that small primary NPs with sizes of 2 nm to 5 nm are obtained in TAAILs and EG/TAAIL mixtures. The Pt-NPs feature excellent activity as electrocatalysts in the hydrogen evolution reaction (HER) under acidic conditions, with an overpotential at a current density of 10 mA cm^{-2} as low as 32 mV vs the reversible hydrogen electrode (RHE), which is significantly lower than the standard Pt/C 20% with 42 mV. Pt-NPs obtained in TAAILs also achieved quantitative conversion in the hydrosilylation reaction of phenylacetylene with triethylsilane after just 5 min at 200 °C.

Keywords: platinum; nanoparticles; hydrogen evolution reaction; ethylene glycol; microwave heating; ionic liquid; tunable aryl alkyl ionic liquid; hydrosilylation

1. Introduction

A consistently high interest in heterogeneous catalysis is dedicated to Pt nanostructures and their alloys, which are known for their high catalytic activity in oxidation, hydrogenation, hydrosilylation, and electrocatalysis reactions [1–9]. The large surface area in relation to their volume gives nanoparticles (NPs) a high mass-based activity. The catalytic activity and stability of nanoparticles are further determined by their size, composition, shape, surface structure, protection, and surface accessibility [1,10,11]. To prevent coalescence, agglomeration, or Ostwald ripening of NPs stabilizing capping ligands, surfactants or polymers are crucial for utilizing NPs [12–17]. One group of convenient nanoparticle stabilizers are ionic liquids (ILs), which can also function as reaction media for NP synthesis [12,18–21]. In addition, ILs are excellent solvents for microwave reactions due to their high absorptivity of microwave irradiation which allows for a combination of fast and homogenous microwave heating with the stabilizing effects of ILs, resulting in an effective way to produce small NPs [21–24]. Tunable aryl alkyl ionic liquids (TAAILs) are a newer class of ILs, which contain an *N*-aryl group as well as an *N*-alkyl chain on the imidazole ring (Figure 1). Both substituents can be tailored to influence the reaction environment and the resulting properties of NPs [18,21,24]. Another widespread method for the synthesis of metal NPs is the polyol process using ethylene glycol (EG) as a solvent and as a stabilizer [25–27]. EG is also suitable for microwave heating as it offers a high boiling point and a strong absorptivity of microwave irradiation [28]. Generally, size and shape control in EG-mediated synthesis is achieved by adding surfactants [25,29]. The addition of sodium hydroxide to EG allows a surfactant-free approach, which also enables reliable size control [30–32]. Although polyols are well-researched solvents and reducing agents for NP synthesis, mixtures of ILs and polyols as solvents are less common [33–38]. Such mixtures have been used as a solvent system for the synthesis of various mono- and multimetallic NPs and the sonochemical synthesis of different M-NPs [33–38], but not yet for Pt-NPs in a microwave setting. Moreover, polyol and IL mixtures are utilized for biomolecule extraction and as electrolytes for electrodeposited metal nanostructures [39–42].

In this article, we present the microwave-assisted synthesis of Pt-NPs in TAAILs with the 1-aryl-3-alkyl- imidazolium cation and novel TAAILs that contain an additional phenyl group on the imidazolium C2-position [43]. As reaction media, the TAAILs are used individually and in combination with EG.

The obtained Pt-NPs were tested for their activity toward the electrochemical hydrogen evolution reaction (HER) and toward the hydrosilylation of phenylacetylene with triethylsilane. It is known that Pt compounds can form catalytically active Pt-NP species in situ in ionic liquids [44–46]. Therefore, we also examined the use of potassium hexachloridoplatinate(IV) (K_2PtCl_6) dispersed in an EG/TAAIL phase for hydrosilylation without a preceding NP separation.

2. Results and Discussion

2.1. Tunable Aryl Alkyl Ionic Liquids (TAAILs)

All 1-aryl-3-alkylimidazolium bis(trifluoromethylsulfonyl)imide ([Ph_xImC_4][NTf_2]) and 1-aryl-2-aryl-3-alkylimidazolium bis(trifluoromethylsulfonyl)imide ([Ph_xImPhC_4][NTf_2]) TAAILs which were used for the Pt-NP synthesis are presented in Figure 1. With the exception of [$Ph_{4-Br}ImC_5$][NTf_2], which contains an n-pentyl substituent, only n-butyl (C_4) substituents were present as alkyl groups. The [NTf_2]$^-$ anion has been chosen because it induces a low melting temperature, high inertness, and hydrophobicity to the IL, with the latter being important to prevent water uptake that could cause the deactivation of the catalyst during hydrosilylation. It has also been shown that ILs containing the [NTf_2]$^-$ anion are most suitable for hydrosilylation reactions [45,47]. The synthesis and characterization of [$Ph_{4-Br}ImPhC_4$][NTf_2] and its precursors can be found in the Supplementary Materials while the synthesis of the other TAAILs has been described by Strassner et al. before [43,48,49]. NMR spectra of all TAAILs can be found in Sections S2.1 and S2.2 (Supplementary Materials).

TAAIL anion purity and temperature stability have been examined by ion chromatography (IC) and thermogravimetric analysis (TGA), respectively, with the results shown in Table 1 and in Section S2.3 and Figure S13, Section S2.4. Anion purities over 92% and IL purities of at least 97% were achieved, with traces of halogenides as residual anions remaining from the ion exchange. All ILs are stable up to at least 390 °C under a nitrogen atmosphere, similar to other ILs and TAAILs containing [NTf_2]$^-$ anions [24,48–51], and are, thus, suitable solvents for microwave reactions at 200 °C.

Table 1. Anion purity, IL purity, and decomposition temperature of TAAILs utilized in this work.

TAAIL	Color	Anion Purity [1] (wt%)	IL purity [2] (wt%)	T_{Dec} [3] (°C)
[$Ph_{4-Me}ImC_4$][NTf_2]	orange	98.4	92	403
[$Ph_{4-OMe}ImC_4$][NTf_2]	brown	>99	95	417
[$Ph_{4-Br}ImC_4$][NTf_2]	orange	97.6	98	410
[$Ph_{4-Br}ImC_5$][NTf_2]	brown	98.4	102	399
[$Ph_{2,4-F}ImC_4$][NTf_2]	brown	>99	97	394
[$PhImPhC_4$][NTf_2]	orange	>99	94	414
[$Ph_{2-Me}ImPhC_4$][NTf_2]	yellow	98.3	93	414

Table 1. Cont.

TAAIL	Color	Anion Purity [1] (wt%)	IL purity [2] (wt%)	T_{Dec} [3] (°C)
[Ph4-OMeImPhC4][NTf2]	brown	>99	102	410
[Ph4-BrImPhC4][NTf2]	black	97.6	94	413

[1] The anion purity as [NTf2]− wt% based on the sum of the determined anion ([NTf2]−, F−, Cl−, Br−) contributions (average from a double determination of two samples, that is four measurements). [2] The IL purity as wt% is assessed as the average (double determination of two samples) of the IC-determined triflimide concentration ([NTf2]−exp in mg/L) relative to the theoretical concentration ([NTf2]−theo in mg/L) based on the mass of IL (m_{exp}(IL)) in 100 mL of solution and assuming 100% purity. At very high IL purities, small measurement errors can result in values above 100%. For further information, see Section S2.3. [3] Decomposition temperature from thermogravimetric analysis at a heating rate of 5 K min^{-1} under a nitrogen atmosphere.

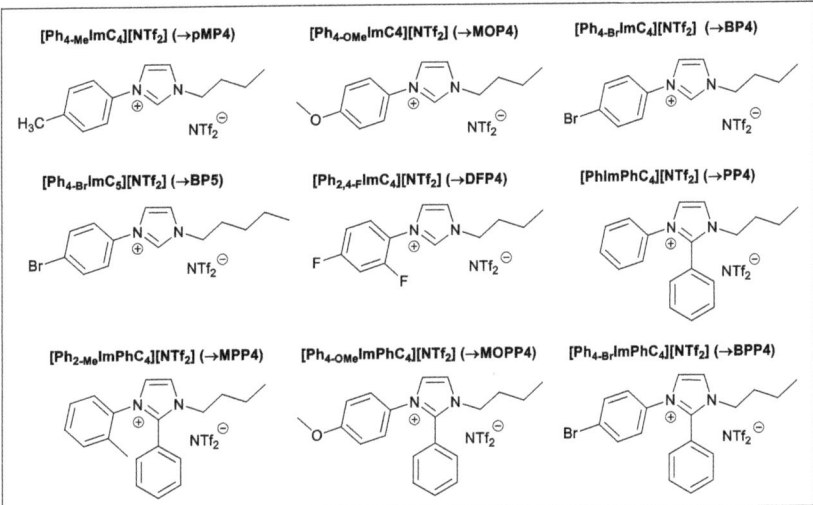

Figure 1. TAAILs utilized in this work with bis(trifluoromethylsulfonyl)imide, [NTf2]− as anion and the sample designation (in parentheses) for the (TAAIL)Pt-NPs.

2.2. Synthesis and Characterization of Pt-NPs in TAAILs

Scheme 1 presents the general synthetic approach of (TAAIL)Pt-NPs via microwave heating and follows the synthesis procedure of Pt-NPs in TAAILs previously reported [21,22,24]. Microwave conditions offer fast, uniform heating, resulting in homogenous Pt-NPs and ILs providing a stable, fast heating media and acting as stabilizers for the formed NPs. The Pt precursor (η^5-methylcyclopentadienyl)trimethylplatinum(IV) (MeCpPtMe3) can be decomposed at mild reaction conditions without additional reducing agents to Pt-NPs [16,22]. Together with the Pt-NPs, only volatile side products are obtained which are removed from the decomposition of MeCpPtMe3, resulting in a contaminant-free Pt-NP dispersion [52]. The Pt-content was set to one or two weight percent (wt%) in the TAAIL dispersion.

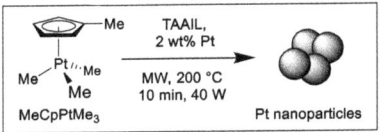

Scheme 1. Reaction conditions for the microwave-assisted synthesis of (TAAIL)Pt-NPs. The amount of precursor was set to achieve 1 or 2 wt% Pt-NP in IL.

The microwave reaction was carried out at 200 °C for 10 min. Afterward, the obtained black Pt-NP dispersion was washed several times with acetonitrile (ACN), separated by centrifugation, and dried in a vacuum, giving a nearly quantitative yield of Pt-NPs. Microwave-assisted heating of metal precursors in IL dispersions results in small M-NP sizes as was shown for Ir-NPs, Ru-NPs [24], and Pt-NPs [21,22]. Fast microwave heating and efficient energy absorption by the IL lead to rapid decomposition of the metal precursor and a high nucleation rate of metal NPs. These metal NPs themselves absorb microwave radiation very efficiently, leading to "hot spots" with a further locally increasing temperature [53,54]. In Figure 2, powder X-ray diffraction (PXRD) patterns of the Pt-NP samples show reflexes matching crystalline fcc-Pt. The Pt-NP sizes have been determined as crystallite sizes from the peak widths in the PXRD patterns with the Scherrer equation and as particle sizes from transmission electron microscopy (TEM) images. These values are listed in Table 2. The crystallite sizes from the Scherrer equation (see Section 3 Materials and Methods) range from 3 nm to 5 nm.

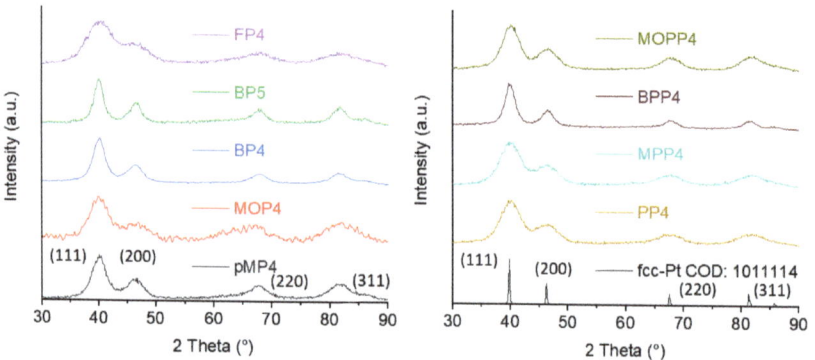

Figure 2. PXRD patterns of synthesized (TAAIL)Pt-NPs (cf. Figure 1 and Table 2). All obtained reflexes match the simulation for face-centered cubic, fcc-Pt with its indexed reflections (Crystallographic open database, COD fcc-Pt: 1011114).

Table 2. Summary of crystallite and particle sizes of (TAAIL)Pt-NPs.

Sample	TAAIL Used in Synthesis	Crystallite Size [1] (nm)	Particle Size [2] (nm)
pMP4	$[Ph_{4\text{-}Me}ImC_4][NTf_2]$	4	3.1 ± 0.6
MOP4	$[Ph_{4\text{-}OMe}ImC_4][NTf_2]$	3	1.8 ± 0.3
BP4	$[Ph_{4\text{-}Br}ImC_4][NTf_2]$	4	3.2 ± 0.5
BP5	$[Ph_{4\text{-}Br}ImC_5][NTf_2]$	5	3.3 ± 0.6
DFP4	$[Ph_{2,4\text{-}F}ImC_4][NTf_2]$	3	2.2 ± 0.6
PP4	$[PhImPhC_4][NTf_2]$	3	2.3 ± 0.4
MPP4	$[Ph_{2\text{-}Me}ImPhC_4][NTf_2]$	3	2.4 ± 0.4
MOPP4	$[Ph_{4\text{-}OMe}ImPhC_4][NTf_2]$	3 ± 1	2.9 ± 0.4
BPP4	$[Ph_{4\text{-}Br}ImPhC_4][NTf_2]$	4	5.0 ± 1.0

[1] From PXRD, average crystallite size and standard deviation are determined by applying the Scherrer equation to all reflexes observed in the respective sample. [2] From TEM, determined from over 200 evaluated particles.

The particle sizes from TEM (Table 2) are similar to the calculated crystallite sizes from PXRD, indicating the formation of isolated nanocrystals. Compared to Pt-NPs synthesized in TAAILs previously [21], the crystallite sizes obtained here for (TAAIL)Pt-NPs are very similar and seem independent of the TAAIL or of the targeted wt% Pt in IL (Table S3 and Figure S14 in Section S3). Pt-NPs which were synthesized from MeCpPtMe$_3$ in the TAAILs $[Ph_{2\text{-}Me}ImC_{4/5/9}][NTf_2]$, $[Ph_{2,4\text{-}Me}ImC_9][NTf_2]$ and $[Ph_{4\text{-}OMe}ImC_5][NTf_2]$ had a particle size of 2 to 5 nm [21]. When the Pt-NPs were deposited on reduced graphene oxide from the TAAILs, the particle size was found between 2 to 6 nm [21]. It is sometimes argued that metal NPs and imidazolium ILs form metal *N*-heterocyclic carbene (NHC) complexes on the NP surface [55,56]. For the NHC-metal complex formation, the C2 position between the imidazolium nitrogen atoms is deprotonated. In the 1-aryl-2-aryl-3-alkylimidazolium TAAILs [Ph$_x$ImPhC$_4$][NTf$_2$], the C2 position of the imidazolium core is blocked and the observed (TAAIL)Pt-NP particle size is not affected. This speaks against a carbene formation in the utilized TAAILs (although carbene formation in the C4 or C5 positions cannot be fully ruled out).

TEM images of Pt-NPs obtained in TAAILs are collected in Figures 3 and 4 for non-C2- and C2-substituted TAAILs, respectively. All [Ph$_x$ImC$_4$][NTf$_2$] and [Ph$_x$ImPhC$_4$][NTf$_2$] samples exhibit dense Pt-NP aggregates with the edges showing a thin layer from residual TAAILs that may hold the Pt-NPs together. This dense aggregation after microwave-assisted synthesis in ILs was not only seen for Pt-NPs before [21,22], but also for Ir-NPs and Ru-NPs [24]. The methoxy, bromo, or fluoro functionalization of the *N*-aryl group in the TAAILs of the (TAAIL)Pt-NPs MOP4, MOPP4, BP5, BPP4, and DFP4 does not affect the particle size or aggregation observed with TEM when compared to the alkyl-substituted aryl groups. This finding is similar to those observed for Pt-NPs in TAAILs before [21]. Only in the bromo-functionalized TAAILs of BP5 and BPP4 smaller aggregates with larger, more isolated NPs are seen, together with a larger amount of residual TAAIL. Noteworthy, all samples have been thoroughly washed with ACN until a clear centrifugate could be achieved. We conclude that the NP-adherent IL layer is difficult to remove and the bromo derivatives BP5 and BPP4 may feature an even lower solubility in ACN. TEM images of the other samples can be found in Section S3, Figures S16–S24.

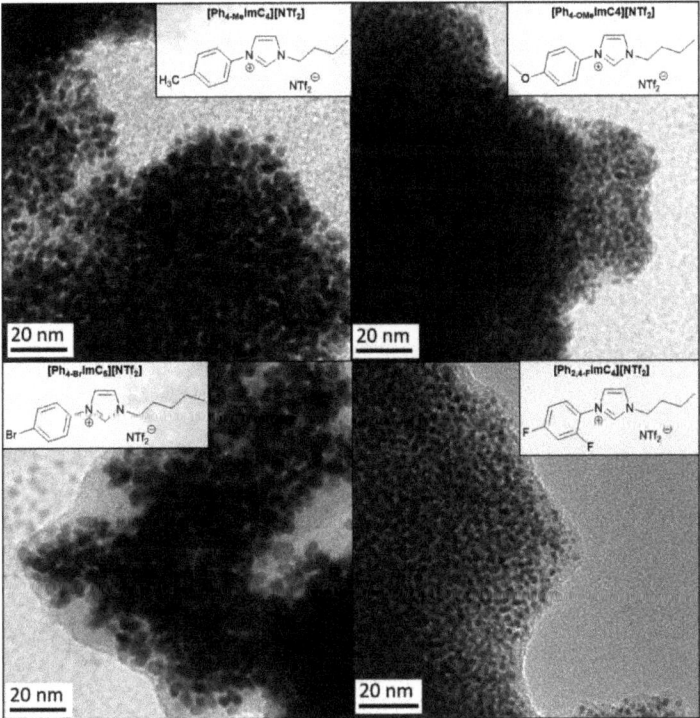

Figure 3. TEM images of (TAAIL)Pt-NPs obtained in the (non-C2-substituted) TAAILs [Ph$_{4\text{-Me}}$ImC$_4$][NTf$_2$] (MP4), [Ph$_{4\text{-OMe}}$ImC$_4$][NTf$_2$] (MOP4), [Ph$_{4\text{-Br}}$ImC$_5$][NTf$_2$] (BP5) and [Ph$_{2,4\text{-F}}$ImC$_4$][NTf$_2$] (DFP4).

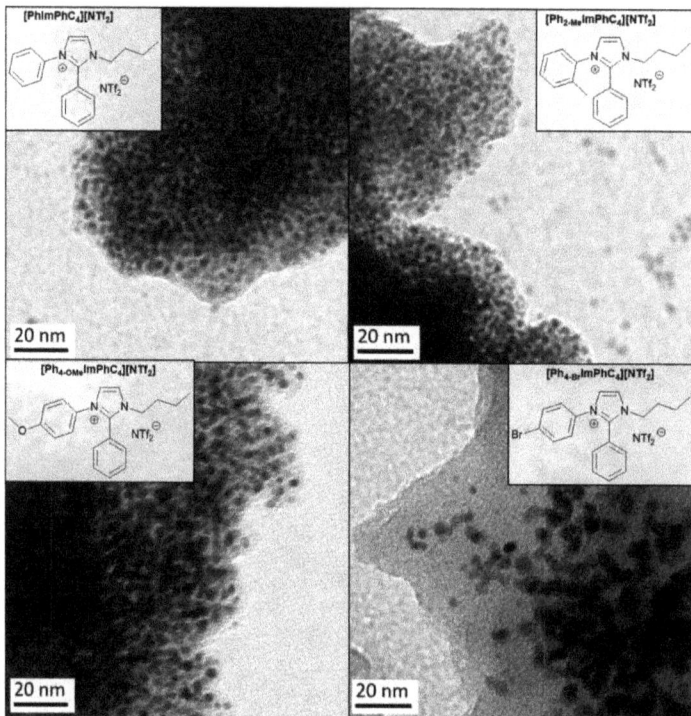

Figure 4. TEM images of (TAAIL)Pt-NPs obtained in the TAAILs [PhImPhC$_4$][NTf$_2$] (PP4), [Ph$_{2\text{-Me}}$ImPhC$_5$][NTf$_2$] (MPP4), [Ph$_{4\text{-OMe}}$ImPhC$_4$][NTf$_2$] (MOPP4) and [Ph$_{4\text{-Br}}$ImPhC$_4$][NTf$_2$] (BPP4).

2.3. Synthesis and Characterization of Pt-NPs in EG/TAAIL Mixtures

The synthesis of (EG/TAAIL)Pt-NPs is depicted in Scheme 2. This method is a modified version of the surfactant-free polyol process presented by Quinson et al. [30] with potassium hexachloridoplatinate(IV), K$_2$PtCl$_6$ as Pt precursor, and the addition of 10, 25, 50, or 75 wt% IL to EG, that is, using a 9/1, 3/1, 1/1 or 1/3 EG/TAAIL mass ratio, respectively. The Pt-content was set to 1 wt% Pt in EG/IL. The reaction was carried out in a glass vial under microwave irradiation with a reaction temperature of 170 °C. Reactions with MeCpPtMe$_3$ as a metal source were unsuccessful; even at 195 °C, no Pt-NP formation was observed. K$_2$PtCl$_6$ was chosen as Pt precursor instead because it is a common Pt source for the Pt-NP synthesis [57,58] and can be effectively reduced by EG at the temperature of 170 °C.

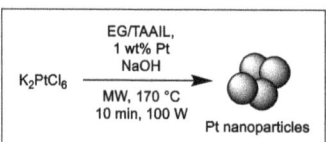

Scheme 2. Reaction conditions for synthesizing (EG/TAAIL)Pt-NPs via microwave heating in EG/TAAIL mixtures. The amount of precursor used has been chosen to achieve 1 wt% Pt-NP in EG/IL.

The addition of NaOH is an established procedure to limit the growth of NPs in the polyol process. The effect depends on the ratio between precursor and NaOH and is suggested to derive from the coordination of hydroxide ions onto the NP surface [30].

The synthesis of (EG/TAAIL)Pt-NPs without NaOH produced significantly larger particles as was already observed before for M-NP formation (M = Pt, Ir) in EG [30,59]. Quinson et al. have shown that a NaOH/H$_2$PtCl$_6$ molar ratio of ~12/1 produces Pt-NPs with a size of ~2 nm in neat EG [30], which is why this ratio was also used in this work.

After microwave heating, the resulting black dispersions were washed multiple times with ACN until a clear solution after centrifugation could be separated. The remaining sodium salts were removed afterward by washing twice with methanol. The (EG/TAAIL)Pt-NPs were dried in a vacuum and obtained quantitative yields, as the (TAAIL)Pt-NPs above. PXRD patterns in Figure 5 confirm the nanocrystallinity of the platinum particles. The crystallite sizes, given in Table 3, were determined from the peak widths in the PXRD patterns via the Scherrer equation. The PXRD patterns exhibit no reflexes that could be attributed to sodium or potassium chloride residues. Higher amounts of IL in EG (25, 50, and 75 wt%) led to larger crystallite sizes (Table S4 and Figure S15). At very low EG/IL-ratios (1/9, 90 wt% IL), no Pt-NP formation was observed anymore from K$_2$PtCl$_6$, presumably due to the low concentration of the EG-reducing agent.

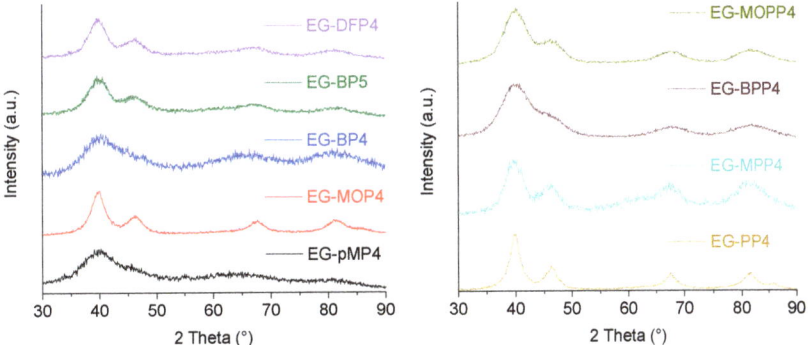

Figure 5. PXRD patterns of (EG/TAAIL)Pt-NPs (cf. Figure 1 and Table 3). All obtained reflexes match the simulation for fcc-Pt and its indexed reflections (cf. Figure 2).

Table 3. Summary of crystallite and particle sizes of (EG/TAAIL)Pt-NPs.

Sample Name	TAAIL Used in Synthesis	Crystallite Size [1] (nm)	Particle Size [2] (nm)
EG-pMP4	[Ph$_{4-Me}$ImC$_4$][NTf$_2$]	2	/
EG-MOP4	[Ph$_{4-OMe}$ImC$_4$][NTf$_2$]	4	1.8 ± 0.3
EG-BP4	[Ph$_{4-Br}$ImC$_4$][NTf$_2$]	2	/
EG-BP5	[Ph$_{4-Br}$ImC$_5$][NTf$_2$]	3	3.3 ± 0.6
EG-DFP4	[Ph$_{2,4-F}$ImC$_4$][NTf$_2$]	3	2.2 ± 0.6
EG-PP4	[PhImPhC$_4$][NTf$_2$]	5	2.3 ± 0.4
EG-MPP4	[Ph$_{2-Me}$ImPhC$_4$][NTf$_2$]	3	2.4 ± 0.4
EG-MOPP4	[Ph$_{4-OMe}$ImPhC$_4$][NTf$_2$]	2	2.9 ± 0.4
EG-BPP4	[Ph$_{4-Br}$ImPhC$_4$][NTf$_2$]	2	5.0 ± 1.0

[1] From PXRD, average crystallite size and standard deviation are determined by applying the Scherrer equation to all reflexes observed in the respective sample. [2] From TEM, determined from over 200 counted particles.

The crystallite sizes calculated from PXRD patterns, and the particle sizes observed from the TEM images are given in Table 3. Comparable to the particles in pure TAAIL, the (EG/TAAIL)Pt-NPs all show similar crystallite sizes between 2 nm and 5 nm. TEM images of two (EG/TAAIL)Pt-NP samples are given in Figure 6. In general, the particles form large and dense agglomerates of several 100 nm in size. Different from the Pt-NPs in neat TAAILs which were depicted in Figure 4, the individual Pt-NPs in EG/TAAIL can hardly be differentiated anymore. This indicates a lower particle-separating effect of EG in comparison to ILs. With an excess of EG over TAAIL, the outer solvent layer adherent to the aggregated NPs is smaller and less regular compared to neat TAAIL.

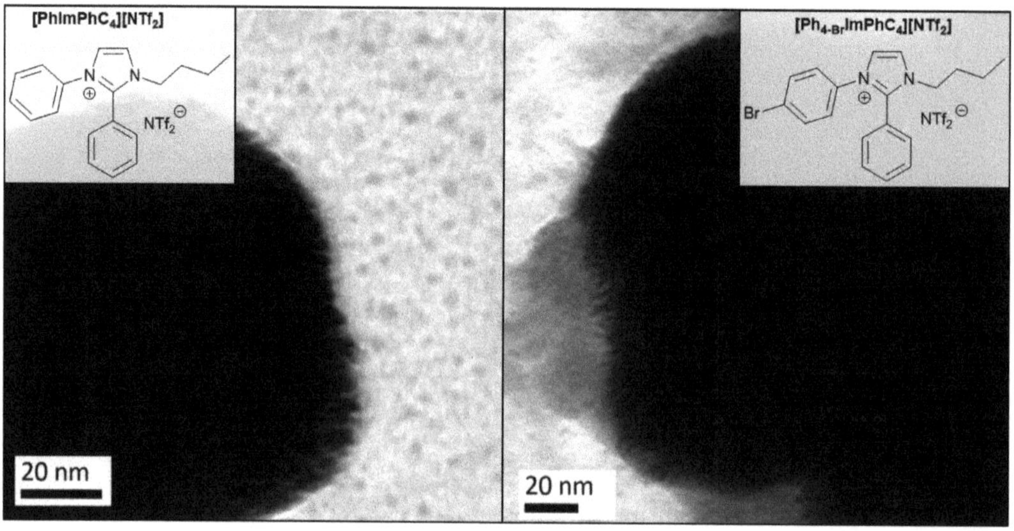

Figure 6. TEM images of (EG/TAAIL)Pt-NPs obtained in [PhImPhC$_4$][NTf$_2$] (PP4) and [Ph$_{4\text{-Br}}$ImPhC$_4$][NTf$_2$] (BPP4), respectively. TEM images of the other samples can be found in Figures S25–S33.

2.4. Hydrogen Evolution Reaction (HER)

HER is one of the half-reactions for water splitting to generate molecular hydrogen for the storage of renewable wind or solar electricity [60,61]. Platinum is known as a highly active electrocatalyst for this reaction in acid media, yet its scarcity and high cost hinder its deployment in large-scale industrial applications [62,63]. The electrocatalytic activity towards HER of (TAAIL)Pt-NPs and (EG/TAAIL)Pt-NPs in 0.5 mol L^{-1} sulfuric acid was investigated. Activation of the samples was achieved by cyclovoltammetry (see Section 3.3 Materials and Methods). The samples that showed an overpotential of less than 60 mV at 10 mA cm^{-2} after activation were also subjected to a stability test. As reference material, commercially available Pt on carbon (Pt/C 20 wt%) was used and its electrochemical data agreed with literature reports [64,65].

Figure 7a displays the polarization curves of the (EG/TAAIL)Pt-NP samples and the reference material after activation. The electrochemical parameters are summarized in Table 4. EG-MPP4 reached the lowest overpotential of 32 mV, outperforming Pt/C 20 wt% with an overpotential of 42 mV. The overpotential of EG-MPP4 was similar to those of single-atom Pt-Catalysts (Pt$_1$/OLC and ALD50Pt/NGNs) and Pt-Ni nanowires (Pt$_3$Ni$_2$ NWs-S/C) with overpotentials of 38, 50 and 27 mV, respectively [66–68]. Additionally, EG-BPP4, with an overpotential of 39 mV, still performed slightly better than the reference material Pt/C 20 wt%. EG-BP4 and EG-BP5 both have somewhat higher overpotentials of 54 mV and 58 mV, respectively. The remaining samples showed fairly high overpotentials or did not reach the necessary current. (EG/TAAIL)Pt-NP probes with TAAILs substituted at the C2 position display lower overpotentials and also bromo functionalization the N-aryl group produces NPs with lower overpotentials. (TAAIL)Pt-NPs have also been analyzed electrochemically but are mostly inactive, with most samples not reaching a current density of 10 mA cm^{-2} under the measurement conditions (polarization curves are displayed in Section S4). In general, (EG/TAAIL)Pt-NP samples performed better than those in TAAIL alone.

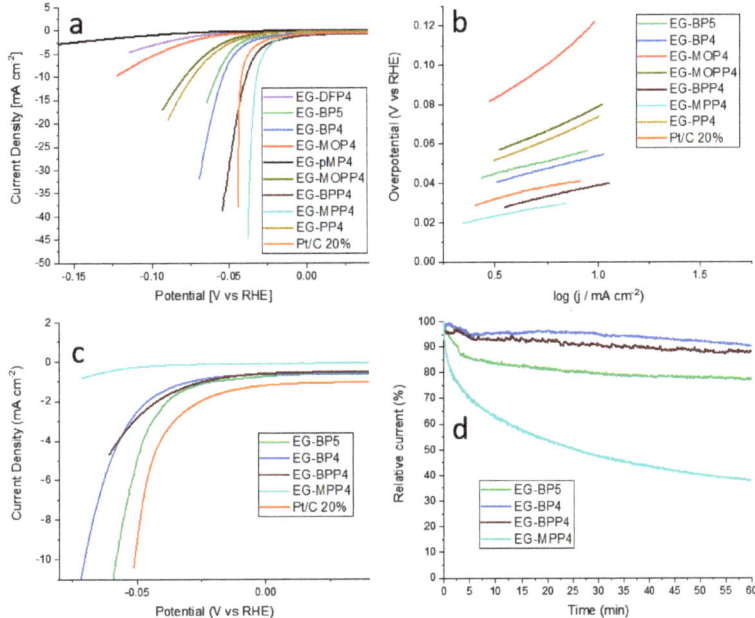

Figure 7. HER polarization curves of (EG/TAAIL)Pt-NPs (**a**), Tafel plots of the samples that reached an overpotential of 10 mA cm^{-2} (**b**), polarization curves after a 1000 CV stability test (**c**), and the chronoamperometric current loss of selected samples (**d**).

Table 4. Summary of electrochemical parameters for (TAAIL)Pt-NPs and (EG/TAAIL)Pt-NPs.

Sample Name	$\eta_{10\,mA\,cm^{-2}}$ after Activation (mV)	Tafel Slope (mV dec^{-1}) [3]	$\eta_{10\,mA\,cm^{-2}}$ after Stability Test (mV)
(TAAIL)Pt-NPs			
pMP4	64	24	/
BP4	n.a. [1]	/	/
DFP4	n.a. [1]	/	/
PP4	n.a. [1]	/	/
MPP4	n.a. [1]	/	/
MOPP4	83	54	/
BPP4	108	81	/
(EG/TAAIL)Pt-NPs			
EG-pMP4	n.a. [1]	/	/
EG-MOP4	123	78	/
EG-BP4	54	27	70
EG-BP5	58	26	/
EG-DFP4	n.a. [1]	/	/
EG-PP4	74	44	/
EG-MPP4	32	20	n.a. [1]
EG-MOPP4	79	46	/
EG-BPP4	39	24	n.a. [1]
Reference materials			
Pt/C 20 wt%	42	25	51
Pt$_1$/OLC [66]	38	35	38 [4]
Pt$_3$Ni$_2$ NWs-S/C [67]	27	/	/
ALD50Pt/NGNs [68]	50 [2]	29	50

[1] The sample did not reach a current density of 10 mA cm^{-2}. [2] Overpotential reported at 16 mA cm^{-2}. [3] Based on the kinetically controlled area at low overpotentials. [4] The stability test consisted of 6000 CV cycles instead of 5000.

Figure 7b displays the Tafel plots based on the kinetically controlled areas at low overpotentials for those samples that reached a current density of 10 mA cm^{-2}. The Tafel slope describes the increase of the overpotential required for a ten-fold increase of the current density [69]. A low Tafel slope is a good indicator of an effective electrocatalyst [65]. The (EG/TAAIL)Pt-NP sample with the lowest overpotential, EG-MPP4, also has the lowest Tafel slope of 20 mV dec^{-1}. All samples with brominated TAAILs, that is EG-BPP4, EG-BP4, and EG-BP5, give similar Tafel slopes of 24 mV dec^{-1}, 27 mV dec^{-1} and 26 mV dec^{-1}, respectively, similar to Pt/C 20% with 25 mV dec^{-1}. Similar Tafel slopes have also been reported for single-atom Pt catalysts (Pt$_1$/OLC and ALD50Pt/NGNs) [66,68]. Much higher Tafel slopes are seen for EG-PP4 with 44 mV dec^{-1} and EG-MOPP4 with 46 mV dec^{-1}, with EG-MOP4 having the highest value of 78 mV dec^{-1}. The long-term stability of the catalysts was verified via a cyclic voltammetry (CV) durability test comprising 1000 CV cycles. The polarization curves after the stability tests are plotted in Figure 7c. EG-BP4 revealed a significant decrease in activity and reached an overpotential of 70 mV. All other samples ended with larger activity losses and did not reach 10 mA cm^{-2} anymore under the measurement conditions, including EG-MPP4, which was the most active (EG/TAAIL)Pt-NP prior to the stability test. Chronoamperometry has been performed as an alternative stability test for the samples that also underwent CV stability tests. The relative current density losses over time are displayed in Figure 7d. Similar to CV stability tests, EG-BP5, EG-BP4, and EG-BPP4 all show a moderate activity loss. The activity of EG-BP5, the sample with the lowest activity loss during CV stability tests, lost almost 15% of its activity within 3 min but stays more stable afterward with a total current loss of 22% after 60 min. In contrast to CV stability tests, both EG-BP4 and EG-BPP4 show slightly less activity loss than EG-BP5, with 5% reduced activity after 7 min and 10% and 12% after 60 min, respectively. All three samples exhibit similar long-term behavior, after the initial activity changes within the first minutes. EG-MPP4 degenerates much more rapidly, losing over 60% activity within 60 min, and is in agreement with the CV stability test.

2.5. Hydrosilylation Reaction

The hydrosilylation of phenylacetylene with triethylsilane, has been chosen as a proof of principle to determine the catalytic activity of Pt-NPs in conjunction with ionic liquids. Hydrosilylation is of high importance for modern silicone chemistry for the addition of Si-H bonds to C-C multiple bonds [70–72]. For industrial applications of the hydrosilylation reaction, almost exclusively noble-metal catalysts containing Ir, Ru, Pd, or the Speier and Karstedt Pt catalysts, are employed [73,74]. Until now, the Chalk-Harrod mechanism is the most accepted mechanism for heterogeneous hydrosilylation [70–72,75], which we assume to apply to our IL/NP system as well. The terminal acetylene can be hydrosilylated at both carbon atom positions, resulting in a proximal and distal product. On the lab scale, microwave conditions offer a fast, energy-efficient alternative compared to conventional oil bath heating. We chose three different microwave-assisted methods to investigate the catalytic activity of the catalyst systems as sketched in Scheme 3. Method 1 describes a reaction at 110 °C for 15 min consisting of an EG/TAAIL liquid, K$_2$PtCl$_6$, and the substrate phase. Pt-salts in IL have already been shown as promising catalyst systems for the hydrosilylation reaction [44–46]. For method 2 and method 3, the (EG/TAAIL)Pt-NP and (TAAIL)Pt-NP samples have been heated with the substrate phase at 110 °C for 15 min and 200 °C for 5 min, respectively. The conversions of the substrates and the ratios between the distal and proximal products were determined by ^{1}H NMR spectroscopy and gas chromatography coupled with mass spectrometry (GC-MS) (see Sections S5.2 and S5.4, respectively).

In Tables 5 and 6, the catalytic conversions with selected catalyst samples and reference catalysts are summarized, respectively (see Table S5 for the full list). The IL-containing probes together with microwave heating led to a significant reduction in reaction time to achieve high conversions compared to literature reports with conventional thermal heating [44,45,75–80]. The catalyst derived from EG/IL K$_2$PtCl$_2$ with method 1 generally achieved quantitative conversion after 15 min, with some exceptions for non-C2-substituted TAAILs (Section S5.1). Distal/proximal product ratios were between two to three. EG/TAAIL mixtures without Pt catalyst and (TAAIL)K$_2$PtCl$_6$ showed no product formation (see Section S5). The sample (EG)K$_2$PtCl$_6$ without TAAIL gave a conversion of only 50%. It is assumed in the literature that the catalytically active Pt species from Pt-salts in IL are in situ formed Pt-NPs [44–46]. Yet, the (EG/TAAIL)Pt-NP samples used with method 2 (110 °C for 15 min) yielded significantly less conversion (at most 38 to 75%) than the reaction with K$_2$PtCl$_6$ after method 1. The distal/proximal ratios range from 3.1 to 3.6. (TAAIL)Pt-NPs according to method 2 as well as reference Pt-NPs obtained in the IL [BMIm][NTf$_2$] [21] did not show any conversion at all.

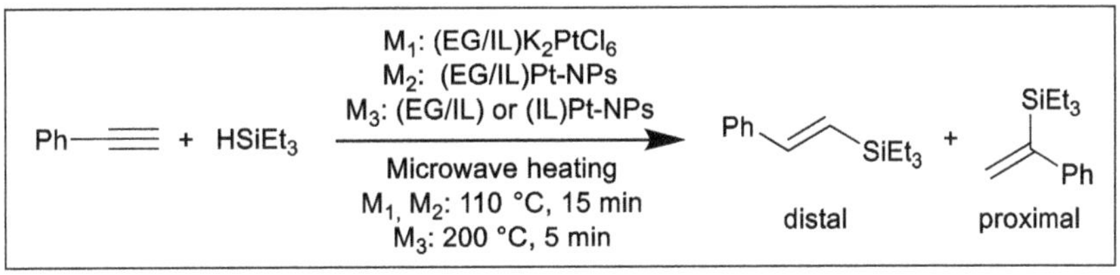

Scheme 3. Hydrosilylation of phenylacetylene with triethylsilane in the presence of Pt-catalysts with three methods (labeled M1 to M3). The proximal triethyl(1-phenylvinyl)silane and distal triethyl(styryl)silane [triethyl(2-phenylvinyl)silane] products are obtained.

Table 5. Catalytic hydrosilylation of phenylacetylene with triethylsilane [1].

Sample Name	Time (h)	Temp. (°C)	Molar Substrate/Pt Ratio [2]	Conversion (%) [3]	d/p Ratio [4]
Method 1, (EG/TAAIL)K$_2$PtCl$_6$					
(EG/[Ph$_{4-Br}$ImC$_4$][NTf$_2$])K$_2$PtCl$_6$	0.25	110	9620	96	3.1
(EG/[Ph$_{4-Br}$ImC$_5$][NTf$_2$])K$_2$PtCl$_6$	0.25	110	9620	96	2.5
(EG/[PhImPhC$_4$][NTf$_2$])K$_2$PtCl$_6$	0.25	110	9940	99	3.0
(EG/[Ph$_{2-Me}$ImPhC$_4$][NTf$_2$])K$_2$PtCl$_6$	0.25	110	10,100	>99	3.1
(EG/[Ph$_{4-Br}$ImPhC$_4$][NTf$_2$])K$_2$PtCl$_6$	0.25	110	8790	>99	2.2
Method 2, (EG/TAAIL)Pt-NPs					
EG-BP4	0.25	110	1050	38	3.6
EG-PP4	0.25	110	980	57	3.1
EG-BPP4	0.25	110	980	75	3.4
Method 3					
(EG/TAAIL)Pt-NPs					
EG-BP4	0.083	200	880	>99	1.5
EG-BP5	0.083	200	980	67	2.2
EG-MPP4	0.083	200	950	>99	2.0
EG-DPP4	0.083	200	990	99	1.8
(TAAIL)Pt-NPs					
BP4	0.083	200	1030	82	2.6
BP5	0.083	200	1140	95	2.2
PP4	0.083	200	880	99	1.5
MPP4	0.083	200	990	>99	2.0

[1] Additional data for all reactions including the distal/proximal ratio determined with GC can be found in Table S5, Section S5. [2] Molar ratios of phenylacetylene substrate to Pt content. A molar ratio of triethylsilane to phenylacetylene of 1.0 was chosen for all reactions. [3] Conversion determined from the reaction mixture by ^{1}H NMR spectroscopy. The highest TOF values achieved for each method in h^{-1} (turnover frequency (mol$_{product}$ mol$_{Pt}^{-1}$ h^{-1})) were 43,300 ((EG/[Ph$_{2-Me}$ImPhC$_4$][NTf$_2$])K$_2$PtCl$_6$), 2900 (EG-BPP4) and 12,900 (BP5) for method 1 to 3, respectively. [4] Molar ratios of distal (d) to proximal (p) product, determined from ^{1}H NMR spectra of the reaction mixture. For GC-determined ratios see Table S5 and Section S5.

Table 6. Reference reactions for the catalytic hydrosilylation of phenylacetylene with triethylsilane [1].

Sample Name	Time (h)	Temp. (°C)	Molar Substrate/Pt Ratio [2]	Conversion (%) [3]	d/p Ratio [4]
[P$_{44414}$][NTf$_2$]/Karstedt [44] [1]	1	110	10,000	>99	-
[S$_{222}$][NTf$_2$]/K$_2$PtCl$_6$ [45] [1]	1	110	10,000	~88	-
Pt$_1$/NaY [75]	24	110	2440	82	0.3
7.0 nm Pt/SBA-15 [76]	6	70	390	6.8	1.8
Pt/C [77]	4.5	70	4880	91	3.3
Pt-NP [78]	1.3	60	200	82	4.9
Pt-NP [78]	24	60	200	94	9.0
Pt-NP [79]	10	rt	1000	98	6.7
C-Pt/ImIP-2BrB [80]	4	80	2000	79	0.8

[1] Differing from the other reactions, [P$_{44414}$][NTf$_2$] and [S$_{222}$][NTf$_2$] were carried out as hydrosilylation reactions of 1-octene and 1,1,1,3,5,5,5-heptamethyltrisiloxane with the Karstedt catalyst and K$_2$PtCl$_6$ as catalyst, respectively. Further experiments and information can be found in refs. [44,45]. [2] Molar ratios of phenylacetylene (1-octene) substrate to Pt content. For both substrates, a molar ratio of 1.0 was chosen with the exception of refs. [75,78], where a ratio of 1.2 of phenylacetylene to triethylsilane was utilized. [3] Conversion determined from the reaction mixture by ^{1}H NMR spectroscopy. [4] Molar ratios of distal (d) to proximal (p) product.

We have shown before that an increased temperature can lead to quantitative conversions after just 5 min [81]. We prepared a similar approach for our samples with method 3. In general, the conversions were significantly higher while often reaching quantitative yields for (EG/TAAIL)Pt-NPs and (TAAIL)Pt-NPs. A Pt-free reaction resulted in no conversion while K$_2$PtCl$_6$ and Pt-NPs synthesized in [BMIm][NTf$_2$] also gave high yields (Table S5). In general, the distal/proximal ratios detected are lower than for the other two methods, with a minimum of 1.5 and an average of two. GC-MS generally resulted in slightly reduced distal/proximal ratios compared to the ratios determined by ^{1}H NMR (see Table S5 for the full list).

In comparison to literature results for Pt-NPs collected in Table 6 [44,45,75–80], the catalysis following method 1 and 3 resulted in quantitative yields in remarkably shorter reaction times. However, both methods only achieved the preferred formation of the distal product with a distal/proximal ratio of ~2–3 while reference reactions can achieve stronger preferences for one specific product with distal/proximal ratios as low as 0.3 [75] or as high as 9.0 [78]. Many Pt-NP catalysts in the literature yielded d/p ratios between 3.3 and 9.0. Only Pt$_1$/NaY and C-Pt/ImIP-2BrB yielded distal/proximal ratios below one (Table 6).

Samples used in method 1 achieved similar conversions in notably shorter reaction time using the same substrate/Pt ratio and temperature as systems with Pt catalysts dispersed in IL reacting 1,1,1,3,5,5,5-heptamethyltrisiloxane with 1-octene [44,45]. The short reaction time and highly diluted dispersion of our (EG/TAAIL) and (TAAIL)Pt catalysts resulted in high turnover frequencies (TOF, highest values under footnote 3 in Table 5), especially for method 1, showing a maximum value of 43,300 h^{-1} for (EG/[Ph$_{2\text{-Me}}$ImPhC$_4$][NTf$_2$]) K$_2$PtCl$_6$. Samples containing the other TAAILs still achieved high TOF values of at least 25,500 h^{-1}. Samples used for method 2 and method 3 yielded maximum TOF values of up to 2900 h^{-1} and 12,900 h^{-1}, respectively.

Heavy-metal impurities are a challenge for the application of silicones in pharmaceutical or medical products [82,83] and received, for example, high attention for the still contested "breast implant illness" [84]. Contrary to our expectations, graphite furnace atomic absorption spectrometry (GF-AAS) of the majority of the (EG/TAAIL)Pt samples revealed a high Pt leaching into the product solution, up to over 20% of the amount of Pt used for the catalysis (see Table S6). Only (TAAIL)Pt-NP samples from method 3 gave leaching below 1%, which is usually interpreted as no leaching [75,78].

To determine the catalytic stability, the EG/IL phase was recovered for method 1 while for methods 2 and 3, the catalyst was regained after separation from the product via centrifugation, and the recovered catalysts were reused for two additional hydrosilylation reactions. Unexpectedly, all reactions resulted in less conversion compared to the first reaction. The post-mortem TEM images after the third catalysis run of (EG/[Ph$_{2\text{-Me}}$ImPhC$_4$][NTf$_2$])-

K$_2$PtCl$_6$ and (EG/[Ph$_{4\text{-Br}}$ImPhC$_4$][NTf$_2$])-K$_2$PtCl$_6$ for method 1 and EG-BPP4 for method 2 (Figures S39–S41) show the presence of Pt-NPs. These particles demonstrate a similar degree of aggregation as the (TAAIL)Pt-NPs from which we conclude that, also from EG/TAAIL-K$_2$PtCl$_6$, platinum nanoparticles form under the catalysis conditions. Zielinski et al. reported hydrosilylation reactions with Pt catalysts dispersed in different ILs and observed a drastic loss of catalytic stability when C=C double bonds were present in the IL [45]. Competitive side reactions between silanes and double bonds were suspected. Catalyst leaching reduces the remaining catalytic activity as well.

In summary, all three presented methods allow the successful hydrosilylation of phenylacetylene with triethylsilane. Method 1 and method 3 can achieve quantitative conversion and high TOF values after the respective reaction time. Reactions carried out after method 2 are not quantitative. Method 1 can be carried out at a temperature of 110 °C, which is more suitable for industrial applications and commonly used in literature experiments [44,45,75]. In the literature, hydrosilylation catalysis is typically performed at temperatures below 110 °C, applying reaction times from 1.3 to 24 h. The reaction times in the literature are longer, but in many cases, only ppm amounts of Pt precursors were used. The reaction times appear to be set to reach the high conversion. The lower reaction temperature makes method 1 superior to method 3. We conclude that the in situ preparation of Pt-NP species in method 1 is, thus, more advantageous in comparison to the independent preparation of Pt-NPs before the catalysis run.

3. Materials and Methods

3.1. Chemicals and Equipment

All starting materials and solvents were obtained from commercial sources and used as delivered unless mentioned otherwise (Table S1).

Tunable aryl alkyl ionic liquids (TAAIL) were provided by the group of Prof. Dr. Thomas Strassner, Technische Universität Dresden. The synthetic procedure for the TAAIL [Ph$_{4\text{-Br}}$ImPhC$_4$][NTf$_2$] is described in Section S2.1, while the syntheses of the other TAAILs have been documented in the works of T. Strassner [43,48,49]. (η^5-Methylcyclopentadienyl)trimethylplatinum(IV), MeCpPtMe$_3$ was synthesized and characterized using a method described by Xue et al. [22,85].

Transmission electron microscopy (TEM) measurements were carried out with a Zeiss LEO912 (Zeiss, Oberkochen, Germany) at 120 kV accelerating voltage. The microscope features a theoretical spatial resolution of 0.1 nm. The samples were prepared using 200 μm carbon-coated copper grids. 0.05 mL of the NP/IL dispersion was diluted in 0.5 mL acetonitrile (ACN) and one drop of the diluted dispersion was placed on the grid. After 30 min, the grid was washed with 3 mL of ACN and dried in ambient air. The images were analyzed by the program Gatan Microscopy Suite (Version: 3.3, Gatan Inc., Pleasanton, CA, USA) and the particle size distribution was determined from at least 200 individual particles at different positions on the TEM grid within the same magnification.

Powder X-ray diffractograms (PXRDs) were measured at ambient temperature on a Bruker D2-Phaser (Bruker, Billerica, MA, USA) using a flat sample holder and Cu-Kα radiation (λ = 1.54182 Å, 35 kV). The program Diffrac.Eva V4.2 was used to evaluate the PXRD data. Particle sizes were calculated with the Scherrer equation $L = K \times \lambda/(\Delta(2\theta) \times \cos\theta)$ where L is the average crystallite size (in nm), K the dimensionless shape factor (here 1), λ the wavelength (in nm), $\Delta(2\theta)$ the full width at half maximum (FWHM) in radians and θ the Bragg angle (in °).

A CEM-Discover SP microwave reactor (CEM GmbH, Kamp-Lintfort, Germany), with a power range of 0–300 W (±30 W) was used for all microwave reactions.

Thermogravimetric analysis (TGA) was carried out with a Netzsch TG 209 F3 Tarsus (Netzsch, Selb, Germany) in Al crucibles applying a heating rate of 5 K min^{-1} under a nitrogen atmosphere. Determined decomposition temperatures can deviate up to 2 K.

NMR spectra were recorded on a Bruker Avance III-300 (Bruker, Karlsruhe, Germany) and a Bruker Avance III-600 (Bruker, Karlsruhe, Germany) spectrometer (NMR spectra in Sections S2.1 and S5.2). CDCl$_3$ was used as a solvent. Chemical shifts were referenced on the residual solvent peak versus TMS (^{1}H NMR δ = 7.26 ppm for CHCl$_3$, ^{13}C NMR δ = 77.16 ppm for CHCl$_3$).

Ion chromatography (IC) measurements were performed with a Dionex ICS 1100 instrument (Dionex, Idstein, Germany) with suppressed conductivity detection (chromatograms in Section S2.3). The suppressor (AERS 500, Dionex) was regenerated with an external water module. The system was equipped with the analytical column IonPac AS 22 from Dionex (4 mm × 250 mm) and the corresponding guard column AG 22 (4 mm × 50 mm). The instrument was controlled by Chromeleon® software (Version: 7.1.0.898, Thermo Fisher Scientific GmbH, Dreieich, Germany). The injection volume was 25 µL. The standard eluent used was a 4.5 mmol L^{-1} Na$_2$CO$_3$ + 1.0 mmol L^{-1} NaHCO$_3$ mixture with an addition of 30 vol% ACN. NTf$_2$-anion purity could be determined within an error range of up to 0.5% while the IL purity could be determined within an error range of up to 10%.

For the analysis of Pt leaching or Pt residues after catalysis, graphite furnace atomic absorption spectrometry (GF-AAS) was made using a Perkin Elmer PinAAcle 900T (Perkin Elmer LAS GmbH, Rodgau-Jügesheim, Germany) spectrometer. Solutions of 0.050 mg$_{Pt}$ L^{-1}, 0.100 mg$_{Pt}$ L^{-1}, 0.200 mg$_{Pt}$ L^{-1}, and 0.400 mg$_{Pt}$ L^{-1} were prepared from an AAS Pt standard (Fluka, 1000 ± 4 mg L^{-1}, 5% HCl) for calibration. The samples contained 0.2 mL of the product solution and were further diluted with ethanol to achieve values within the calibration range of 0.050 to 0.400 mg$_{Pt}$ L^{-1}. The obtained values can deviate within a range of ±10%.

Gas chromatography (GC) was performed with a Thermo Finnigan Trace GC Ultra, Column BPX5 (column length: 15 m), combined with the mass spectrometer (MS) Thermo Finnigan Trace DSQ (Thermo Fischer Scientific GmbH, Dreieich, Germany), using the EI ionization method with 70 eV and a source temperature of 200 °C.

3.2. Synthesis of Pt-NPs in IL and EG/IL mixtures

(TAAIL)Pt-NPs: Pt-NPs in TAAILs were synthesized as described previously [21,22]. In general, MeCpPtMe$_3$ and the corresponding IL were placed in a 10 mL microwave vessel. The mass of the Pt precursor was set to achieve 2 wt% of Pt-NPs in IL when assuming quantitative conversion in a batch of about 500 mg IL (~0.4 mL). The dispersion was stirred for at least 6 h and afterward heated in the microwave reactor (200 °C, 40 W, 10 min holding time). To remove the IL several washing steps (with ultrasonication and centrifugation) were performed with 3 mL of ACN per washing step until a clear colorless centrifugate was obtained. The (TAAIL)Pt-NP residue was dried in a high vacuum (5 × 10^{-3} mbar) for 2 h. The yield of Pt-NPs was quantitative.

(EG/TAAIL)Pt-NP: Pt-NPs in mixtures of EG and TAAILs were synthesized using a modified version of the surfactant-free polyol process by Quinson et al. [30]. In general, K$_2$PtCl$_6$, NaOH, EG, and the TAAIL (with 10, 25, 50, and 75 wt% IL in EG/IL) were placed in a 10 mL microwave glass vessel. Then 12 equivalents of NaOH to Pt were added. The amount of Pt precursor was set to achieve 1 wt% of Pt-NPs in EG/TAAIL at quantitative conversion, with batch sizes of about 600 mg EG/TAAIL. The dispersion was stirred for at least 6 h and heated afterward in the microwave reactor (170 °C, 100 W, 10 min holding time). To remove the EG and IL the black dispersion was washed (ultrasonicated and centrifugated) several times with 3 mL of ACN each until a clear colorless centrifugate was obtained. The black solid was then washed twice (with ultrasonication and centrifugation) with MeOH to remove NaOH residues. The remaining black product was dried in a high vacuum (5 × 10^{-3} mbar) for 2 h to give a quantitative yield of Pt-NPs.

3.3. Electrochemical Measurements

For all measurements, a conventional three-electrode cell with a glassy carbon rotating disk working electrode (5 mm diameter), a Pt sheet as a counter electrode (1.5 × 1.5 cm^2), and a silver/silver chloride reference electrode (Ag/AgCl in 3 mol L^{-1} NaCl solution) was used with 0.5 mol L^{-1} H$_2$SO$_4$ electrolyte solution and an Interface 1010 potentiate by Gamry Instruments.

As electrochemically active material fresh NP inks were prepared similarly to Beermann et al. [86] where 0.2 mg of the NP component was first mixed with 0.8 mg Vulcan XC-72R. Further, 1 mg of this solid was dispersed in 1 mL of a 1/5 (*v*/*v*) isopropanol/water mixture containing 5 µL Nafion™ 1100W 5 wt% and sonicated for at least 30 min. Next, 20 µL of the ink was deposited onto the working electrode and dried at room temperature with a rotation speed of 120 rpm to form a thin film. The resulting platinum loading on the electrode was 20 µg$_{Pt}$ cm^{-2}.

All following measurements were completed under a protective gas atmosphere at a rotation speed of 3600 rpm. Before electrochemical measurements were started, the electrolyte solution was purged with N$_2$ for 10 min. The catalyst was activated via potential cycling between −0.10 and 0.30 V$_{Ag/AgCl}$ for 30 cycles with a scan rate of 100 mV s^{-1}. To determine the activities of the catalysts, linear sweep voltammograms (LSV) were recorded in a potential range between 0.1 and −0.35 V$_{Ag/AgCl}$ with a scan rate of 10 mV s^{-1}. The overpotential was determined at a current density of 10 mA cm^{-2}. Polarization curves vs Ag/AgCl were corrected by iR compensation and converted to a reversible hydrogen electrode (RHE), according to *E (RHE) = E(Ag/AgCl) + E° + 0.059 V·pH* with *E°* = 0.211 V. Stability tests were conducted via potential scanning between 0.1 and −0.3 V$_{Ag/AgCl}$ for 1000 cycles at 100 mV s^{-1}. Chronoamperometry has been performed as an alternative stability test at a controlled voltage of 63 mV for 1 h at room temperature. Due to the parameters of the measurement, given voltages can deviate by up to 1 mV for all measurements.

3.4. Hydrosilylation Reactions

Method 1: The catalytic reactions were performed as a two-phase system in a microwave reactor using quartz glass vials of 10 mL. A mixture of K$_2$PtCl$_6$ (~1.3 µmol Pt, see Table S5 for the molar ratio of substrate/Pt), EG, and IL (~0.2 wt% Pt in a 9/1 ratio of EG/IL) was placed in the glass vial and degassed under vacuum. Afterward, 1.37 mL (12.5 mmol) of phenylacetylene and 2.00 mL (12.5 mmol) of triethylsilane were added to the glass vial under an N$_2$ atmosphere, followed by a reaction at 110 °C for 15 min under 30 W of microwave irradiation. The upper product phase was syringed off after centrifugation and analyzed by ^{1}H NMR, ^{13}C NMR, and GC for the different product species and the conversion. The statistical error of the distal/proximal product ratio and substrate conversion determined by signal intensities in ^{1}H NMR is about 5% for both determinations. Conversions above 99% result in significantly larger deviations due to the low intensity of the remaining starting material and are only mentioned as >99%. The statistical error of the distal/proximal product ratio determined by signal intensities in GC is roughly up to 10%. To test the stability of the catalyst, the same number of starting materials was added again to the remaining EG/IL phase and the procedure was repeated.

Method 2: The catalytic reactions were performed as a one-phase system in the same glass vials as in method 1. (EG/TAAIL)Pt-NP probes (~5.0 µmol Pt, see Table S5 for the molar ratio of substrate/Pt) were placed in the glass vial followed by the addition of 0.55 mL (5.0 mmol) of phenylacetylene and 0.80 mL (5.0 mmol) of triethylsilane. The reaction was carried out as described in method 1 at 110 °C for 15 min but under 200 W microwave irradiation for 5 min. The product solution was syringed off from the solid catalyst after centrifugation and analyzed as described above. The solid catalyst was reused, and the procedures were repeated to test the catalyst's stability.

Method 3: The catalytic reactions, washing, and characterization were carried out in the same manner as described in method 2 but with (EG/TAAIL)Pt-NP and (TAAIL)Pt-NP probes (see Table S5 for the molar ratio of substrate/Pt) at 200 °C for 5 min under 200 W of microwave irradiation.

4. Conclusions

Nine tunable aryl alkyl ionic liquids (TAAIL), including TAAILs with an additional phenyl substitution at the imidazole C2 position, were utilized as reaction media and stabilizer for the microwave-assisted synthesis of Pt-nanoparticles (Pt-NPs) from MeCpPtMe$_3$. In an ethylene glycol (EG)/TAAILs mixture the precursor K$_2$PtCl$_6$ was used. Small Pt-NPs were obtained whose calculated crystallite sizes from PXRD with the Scherrer equation of 3 nm to 5 nm correspond to particle sizes observed by TEM. TEM further illustrated that all samples formed large aggregates of the primary NPs.

The (TAAIL)Pt-NPs and (EG/TAAIL)Pt-NPs showed competitive activities in the electrocatalytic hydrogen evolution reaction. In particular, the Pt-NP sample with EG/[Ph$_{2\text{-Me}}$ImPhC$_4$][NTf$_2$] (EG-MPP4) exhibited a very low overpotential of 32 mV at 10 mA cm^{-2}, outperforming the reference material Pt/C 20 wt% with 42 mV. The sample EG-MPP4 also had a low Tafel slope of 19 mV dec^{-1}.

The (TAAIL)Pt-NP and (EG/TAAIL)Pt-NP probes could function as catalysts for the hydrosilylation of phenylacetylene with triethylsilane with quantitative conversion in a short time of 5 min. In addition, a two-phase system with an EG/TAAIL phase containing the salt K$_2$PtCl$_6$ also achieved quantitative conversion in 15 min. In all cases, the short reaction time for quantitative conversion resulted from microwave heating while in literature references work significantly higher reaction times of 1 h for two-phase reactions [44,45] and over 2 h for Pt-NP catalysts were needed [69,75–79]. The samples achieved very high TOF values of up to 43,300 h^{-1}. The distal hydrosilylation product was preferentially obtained over the proximal one in all reactions, with a ratio of up to 3.5. However, the recycling and reuse of the catalysts could still not be successfully implemented, in part due to an unexpectedly high degree of Pt leaching into the product solution. Finding the right reaction conditions for IL/Pt-NP catalysts to prevent leaching and deactivation is a challenge for future work. Only then can the full design potential of ionic liquids as reaction media in hydrosilylation catalysis be utilized.

Supplementary Materials: The following supporting information can be downloaded at https://www.mdpi.com/article/10.3390/molecules28010405/s1. Section S1: Sources of chemicals; S2: Synthesis and characterization of TAAILs; S3: Synthesis parameters and analyses of platinum-nanoparticles (Pt-NPs) in TAAILs and ethylene glycol (EG); S4: Additional electrochemical measurements; S5: Hydrosilylation conversion and product analysis. Reference [87] is cited in the supplementary materials.

Author Contributions: Conceptualization, C.J. and D.W.; methodology, D.W. and T.S. (Till Strothmann); validation, D.W. and T.S. (Till Strothmann); formal analysis, D.W. and T.S. (Till Strothmann); investigation, D.W., T.S. (Till Strothmann), H.B., S.L., H.S., Y.S. and S.R.; resources, C.J. and T.S. (Thomas Strassner); data curation, D.W. and T.S. (Till Strothmann); writing—original draft preparation, D.W. and T.S. (Till Strothmann); writing—review and editing, C.J. and T.S. (Thomas Strassner); visualization, D.W. and T.S. (Till Strothmann); supervision, C.J.; project administration, C.J.; funding acquisition, C.J. and T.S. (Thomas Strassner). All authors have read and agreed to the published version of the manuscript.

Funding: This research was supported by a joint National Natural Science Foundation of China–Deutsche Forschungsgemeinschaft (NSFC-DFG) project (DFG JA466/39-1). T.S. is grateful for funding by the Deutsche Forschungsgemeinschaft (SPP 1708: STR 526/20-2).

Institutional Review Board Statement: Not applicable.

Informed Consent Statement: Not applicable.

Data Availability Statement: The data presented in this study are available on request from the corresponding author.

Acknowledgments: We would like to thank Yangyang Sun for the TGA measurements and Marius Otten for his help with TEM measurements. The authors also thank the Center for Molecular and Structural Analytics at Heinrich Heine University (CeMSA@HHU) for recording the mass spectrometric and NMR-spectrometric data.

Conflicts of Interest: The authors declare that they have no known competing financial interest or personal relationship that could have appeared to influence the work reported in this paper.

References

1. Xie, C.; Niu, Z.; Kim, D.; Li, M.; Yang, P. Surface and Interface Control in Nanoparticle Catalysis. *Chem. Rev.* **2020**, *120*, 1184–1249. [CrossRef] [PubMed]
2. Hartley, F.R. *Chemistry of the Platinum Group Metals. Recent Developments*; Elsevier: Amsterdam, The Netherlands, 2010; ISBN 9780080933955.
3. Liu, L.; Corma, A. Metal Catalysts for Heterogeneous Catalysis: From Single Atoms to Nanoclusters and Nanoparticles. *Chem. Rev.* **2018**, *118*, 4981–5079. [CrossRef]
4. de Almeida, L.D.; Wang, H.; Junge, K.; Cui, X.; Beller, M. Recent Advances in Catalytic Hydrosilylations: Developments beyond Traditional Platinum Catalysts. *Angew. Chem. Int. Ed.* **2021**, *60*, 550–565. [CrossRef] [PubMed]
5. Mahata, A.; Nair, A.S.; Pathak, B. Recent advancements in Pt-nanostructure-based electrocatalysts for the oxygen reduction reaction. *Catal. Sci. Technol.* **2019**, *9*, 4835–4863. [CrossRef]
6. Li, M.; Zhao, Z.; Cheng, T.; Fortunelli, A.; Chen, C.-Y.; Yu, R.; Zhang, Q.; Gu, L.; Merinov, B.V.; Lin, Z.; et al. Ultrafine jagged platinum nanowires enable ultrahigh mass activity for the oxygen reduction reaction. *Science* **2016**, *354*, 1414–1419. [CrossRef] [PubMed]
7. Shao, M.; Chang, Q.; Dodelet, J.-P.; Chenitz, R. Recent Advances in Electrocatalysts for Oxygen Reduction Reaction. *Chem. Rev.* **2016**, *116*, 3594–3657. [CrossRef]
8. Chen, H.; Wang, G.; Gao, T.; Chen, Y.; Liao, H.; Guo, X.; Li, H.; Liu, R.; Dou, M.; Nan, S.; et al. Effect of Atomic Ordering Transformation of PtNi Nanoparticles on Alkaline Hydrogen Evolution: Unexpected Superior Activity of the Disordered Phase. *J. Phys. Chem. C* **2020**, *124*, 5036–5045. [CrossRef]
9. Bao, J.; Wang, J.; Zhou, Y.; Hu, Y.; Zhang, Z.; Li, T.; Xue, Y.; Guo, C.; Zhang, Y. Anchoring ultrafine PtNi nanoparticles on N-doped graphene for highly efficient hydrogen evolution reaction. *Catal. Sci. Technol.* **2019**, *9*, 4961–4969. [CrossRef]
10. Vollath, D.; Fischer, F.D.; Holec, D. Surface energy of nanoparticles—Influence of particle size and structure. *Beilstein J. Nanotechnol.* **2018**, *9*, 2265–2276. [CrossRef]
11. Garlyyev, B.; Kratzl, K.; Rück, M.; Michalička, J.; Fichtner, J.; Macak, J.M.; Kratky, T.; Günther, S.; Cokoja, M.; Bandarenka, A.S.; et al. Optimizing the Size of Platinum Nanoparticles for Enhanced Mass Activity in the Electrochemical Oxygen Reduction Reaction. *Angew. Chem. Int. Ed.* **2019**, *58*, 9596–9600. [CrossRef]
12. Xu, D.; Lv, H.; Liu, B. Encapsulation of Metal Nanoparticle Catalysts Within Mesoporous Zeolites and Their Enhanced Catalytic Performances: A Review. *Front. Chem.* **2018**, *6*, 550. [CrossRef]
13. Bera, S.; Mondal, D. A role for ultrasound in the fabrication of carbohydrate-supported nanomaterials. *J. Ultrasound* **2019**, *22*, 131–156. [CrossRef]
14. Heinz, H.; Pramanik, C.; Heinz, O.; Ding, Y.; Mishra, R.K.; Marchon, D.; Flatt, R.J.; Estrela-Lopis, I.; Llop, J.; Moya, S.; et al. Nanoparticle decoration with surfactants: Molecular interactions, assembly, and applications. *Surf. Sci. Rep.* **2017**, *72*, 1–58. [CrossRef]
15. Ong, S.Y.; Zhang, C.; Dong, X.; Yao, S.Q. Recent Advances in Polymeric Nanoparticles for Enhanced Fluorescence and Photoacoustic Imaging. *Angew. Chem. Int. Ed.* **2021**, *60*, 17797–17809. [CrossRef]
16. Marquardt, D.; Beckert, F.; Pennetreau, F.; Tölle, F.; Mülhaupt, R.; Riant, O.; Hermans, S.; Barthel, J.; Janiak, C. Hybrid materials of platinum nanoparticles and thiol-functionalized graphene derivatives. *Carbon* **2014**, *66*, 285–294. [CrossRef]
17. Wegner, S.; Janiak, C. Metal Nanoparticles in Ionic Liquids. *Top. Curr. Chem.* **2017**, *375*, 65. [CrossRef]
18. Seidl, V.; Romero, A.H.; Heinemann, F.W.; Scheurer, A.; Vogel, C.S.; Unruh, T.; Wasserscheid, P.; Meyer, K. A New Class of Task-Specific Imidazolium Salts and Ionic Liquids and Their Corresponding Transition-Metal Complexes for Immobilization on Electrochemically Active Surfaces. *Chemistry* **2022**, *28*, e202200100. [CrossRef]
19. Migowski, P.; Machado, G.; Texeira, S.R.; Alves, M.C.M.; Morais, J.; Traverse, A.; Dupont, J. Synthesis and characterization of nickel nanoparticles dispersed in imidazolium ionic liquids. *Phys. Chem. Chem. Phys.* **2007**, *9*, 4814–4821. [CrossRef] [PubMed]
20. Jang, H.; Lee, J.R.; Kim, S.J.; Jeong, H.; Jung, S.; Lee, J.-H.; Park, J.-C.; Kim, T.-W. Concerns and breakthroughs of combining ionic liquids with microwave irradiation for the synthesis of Ru nanoparticles via decarbonylation. *J. Colloid Interface Sci.* **2021**, *599*, 828–836. [CrossRef] [PubMed]
21. Woitassek, D.; Lerch, S.; Jiang, W.; Shviro, M.; Roitsch, S.; Strassner, T.; Janiak, C. The Facile Deposition of Pt Nanoparticles on Reduced Graphite Oxide in Tunable Aryl Alkyl Ionic Liquids for ORR Catalysts. *Molecules* **2022**, *27*, 1018. [CrossRef] [PubMed]
22. Marquardt, D.; Barthel, J.; Braun, M.; Ganter, C.; Janiak, C. Weakly-coordinated stable platinum nanocrystals. *CrystEngComm* **2012**, *14*, 7607. [CrossRef]
23. Scheeren, C.W.; Machado, G.; Teixeira, S.R.; Morais, J.; Domingos, J.B.; Dupont, J. Synthesis and characterization of Pt0 nanoparticles in imidazolium ionic liquids. *J. Phys. Chem. B* **2006**, *110*, 13011–13020. [CrossRef] [PubMed]
24. Schmolke, L.; Lerch, S.; Bülow, M.; Siebels, M.; Schmitz, A.; Thomas, J.; Dehm, G.; Held, C.; Strassner, T.; Janiak, C. Aggregation control of Ru and Ir nanoparticles by tunable aryl alkyl imidazolium ionic liquids. *Nanoscale* **2019**, *11*, 4073–4082. [CrossRef] [PubMed]
25. Rao, B.G.; Mukherjee, D.; Reddy, B.M. Novel approaches for preparation of nanoparticles. In *Nanostructures for Novel Therapy*; Elsevier: Amsterdam, The Netherlands, 2017; ISBN 9780323461429.

26. Liu, Z.; Lee, J.Y.; Chen, W.; Han, M.; Gan, L.M. Physical and electrochemical characterizations of microwave-assisted polyol preparation of carbon-supported PtRu nanoparticles. *Langmuir* **2004**, *20*, 181–187. [CrossRef]
27. Fang, B.; Chaudhari, N.K.; Kim, M.-S.; Kim, J.H.; Yu, J.-S. Homogeneous deposition of platinum nanoparticles on carbon black for proton exchange membrane fuel cell. *J. Am. Chem. Soc.* **2009**, *131*, 15330–15338. [CrossRef]
28. Zhu, Y.-J.; Chen, F. Microwave-assisted preparation of inorganic nanostructures in liquid phase. *Chem. Rev.* **2014**, *114*, 6462–6555. [CrossRef] [PubMed]
29. Şen, F.; Gökağaç, G. Improving Catalytic Efficiency in the Methanol Oxidation Reaction by Inserting Ru in Face-Centered Cubic Pt Nanoparticles Prepared by a New Surfactant, tert-Octanethiol. *Energy Fuels* **2008**, *22*, 1858–1864. [CrossRef]
30. Quinson, J.; Inaba, M.; Neumann, S.; Swane, A.A.; Bucher, J.; Simonsen, S.B.; Theil Kuhn, L.; Kirkensgaard, J.J.K.; Jensen, K.M.O.; Oezaslan, M.; et al. Investigating Particle Size Effects in Catalysis by Applying a Size-Controlled and Surfactant-Free Synthesis of Colloidal Nanoparticles in Alkaline Ethylene Glycol: Case Study of the Oxygen Reduction Reaction on Pt. *ACS Catal.* **2018**, *8*, 6627–6635. [CrossRef]
31. Wang, Y.; Ren, J.; Deng, K.; Gui, L.; Tang, Y. Preparation of Tractable Platinum, Rhodium, and Ruthenium Nanoclusters with Small Particle Size in Organic Media. *Chem. Mater.* **2000**, *12*, 1622–1627. [CrossRef]
32. Neumann, S.; Grotheer, S.; Tielke, J.; Schrader, I.; Quinson, J.; Zana, A.; Oezaslan, M.; Arenz, M.; Kunz, S. Nanoparticles in a box: A concept to isolate, store and re-use colloidal surfactant-free precious metal nanoparticles. *J. Mater. Chem. A* **2017**, *5*, 6140–6145. [CrossRef]
33. Dewan, M.; Kumar, A.; Saxena, A.; De, A.; Mozumdar, S. Using hydrophilic ionic liquid, bmimBF$_4$-ethylene glycol system as a novel media for the rapid synthesis of copper nanoparticles. *PLoS ONE* **2012**, *7*, e29131. [CrossRef] [PubMed]
34. Kim, T.Y.; Kim, W.J.; Hong, S.H.; Kim, J.E.; Suh, K.S. Ionic-liquid-assisted formation of silver nanowires. *Angew. Chem. Int. Ed.* **2009**, *48*, 3806–3809. [CrossRef]
35. Freire, M.G.; Louros, C.L.S.; Rebelo, L.P.N.; Coutinho, J.A.P. Aqueous biphasic systems composed of a water-stable ionic liquid + carbohydrates and their applications. *Green Chem.* **2011**, *13*, 1536–1545. [CrossRef]
36. Rodríguez, H.; Rogers, R.D. Liquid mixtures of ionic liquids and polymers as solvent systems. *Fluid Phase Equilib.* **2010**, *294*, 7–14. [CrossRef]
37. Xiao, S.; Lu, Y.; Li, X.; Xiao, B.-Y.; Wu, L.; Song, J.-P.; Xiao, Y.-X.; Wu, S.-M.; Hu, J.; Wang, Y.; et al. Hierarchically Dual-Mesoporous TiO$_2$ Microspheres for Enhanced Photocatalytic Properties and Lithium Storage. *Chem. Eur. J.* **2018**, *24*, 13246–13252. [CrossRef] [PubMed]
38. Lee, H.S.; Kim, J.F.; Kim, T.; Suh, K.S. Ionic liquid-assisted synthesis of highly branched Ag:AgCl hybrids and their photocatalytic activity. *J. Alloys Compd.* **2015**, *621*, 378–382. [CrossRef]
39. Li, M.; Bo, X.; Zhang, Y.; Han, C.; Guo, L. One-pot ionic liquid-assisted synthesis of highly dispersed PtPd nanoparticles/reduced graphene oxide composites for nonenzymatic glucose detection. *Biosens. Bioelectron.* **2014**, *56*, 223–230. [CrossRef]
40. Okoli, C.U.; Kuttiyiel, K.A.; Cole, J.; McCutchen, J.; Tawfik, H.; Adzic, R.R.; Mahajan, D. Solvent effect in sonochemical synthesis of metal-alloy nanoparticles for use as electrocatalysts. *Ultrason. Sonochem.* **2018**, *41*, 427–434. [CrossRef]
41. He, X.; Sun, Z.; Zou, Q.; Wu, L.; Jiang, J. Electrochemical Behavior of Co(II) Reduction for Preparing Nanocrystalline Co Catalyst for Hydrogen Evolution Reaction from 1-ethyl-3-methylimidazolium Bisulfate and Ethylene Glycol System. *J. Electrochem. Soc.* **2019**, *166*, D57–D64. [CrossRef]
42. He, X.; Sun, Z.; Zou, Q.; Yang, J.; Wu, L. Codeposition of Nanocrystalline Co-Ni Catalyst Based on 1-ethyl-3-methylimidazolium Bisulfate and Ethylene Glycol System for Hydrogen Evolution Reaction. *J. Electrochem. Soc.* **2019**, *166*, D908–D915. [CrossRef]
43. Biller, H.; Strassner, T. Synthesis and Physical Properties of Tunable Aryl Alkyl Ionic Liquids (TAAILs) Comprising Imidazolium Cations Blocked with Methyl-, Propyl- and Phenyl-Groups at the C2 Position. *Chem. Eur. J.* **2022**, *29*, e202202795. [CrossRef]
44. Kukawka, R.; Pawlowska-Zygarowicz, A.; Dutkiewicz, M.; Maciejewski, H.; Smiglak, M. New approach to hydrosilylation reaction in ionic liquids as solvent in microreactor system. *RSC Adv.* **2016**, *6*, 61860–61868. [CrossRef]
45. Zielinski, W.; Kukawka, R.; Maciejewski, H.; Smiglak, M. Ionic Liquids as Solvents for Rhodium and Platinum Catalysts Used in Hydrosilylation Reaction. *Molecules* **2016**, *21*, 1115. [CrossRef]
46. Geldbach, T.J.; Zhao, D.; Castillo, N.C.; Laurenczy, G.; Weyershausen, B.; Dyson, P.J. Biphasic hydrosilylation in ionic liquids: A process set for industrial implementation. *J. Am. Chem. Soc.* **2006**, *128*, 9773–9780. [CrossRef] [PubMed]
47. Hofmann, N.; Bauer, A.; Frey, T.; Auer, M.; Stanjek, V.; Schulz, P.S.; Taccardi, N.; Wasserscheid, P. Liquid-Liquid Biphasic, Platinum-Catalyzed Hydrosilylation of Allyl Chloride with Trichlorosilane using an Ionic Liquid Catalyst Phase in a Continuous Loop Reactor. *Adv. Synth. Catal.* **2008**, *350*, 2599–2609. [CrossRef]
48. Lerch, S.; Strassner, T. Expanding the Electrochemical Window: New Tunable Aryl Alkyl Ionic Liquids (TAAILs) with Dicyanamide Anions. *Chem. Eur. J.* **2019**, *25*, 16251–16256. [CrossRef] [PubMed]
49. Lerch, S.; Strassner, T. Synthesis and Physical Properties of Tunable Aryl Alkyl Ionic Liquids (TAAILs). *Chem. Eur. J.* **2021**, *27*, 15554–15557. [CrossRef] [PubMed]
50. Tokuda, H.; Hayamizu, K.; Ishii, K.; Susan, M.A.B.H.; Watanabe, M. Physicochemical properties and structures of room temperature ionic liquids. 2. Variation of alkyl chain length in imidazolium cation. *J. Phys. Chem. B* **2005**, *109*, 6103–6110. [CrossRef]
51. Maton, C.; de Vos, N.; Stevens, C.V. Ionic liquid thermal stabilities: Decomposition mechanisms and analysis tools. *Chem. Soc. Rev.* **2013**, *42*, 5963–5977. [CrossRef]

52. Lubers, A.M.; Muhich, C.L.; Anderson, K.M.; Weimer, A.W. Mechanistic studies for depositing highly dispersed Pt nanoparticles on carbon by use of trimethyl(methylcyclopentadienyl)platinum(IV) reactions with O_2 and H_2. *J. Nanopart. Res.* **2015**, *17*, 85. [CrossRef]
53. Hill, J.M.; Marchant, T.R. Modelling microwave heating. *Appl. Math. Model.* **1996**, *20*, 3–15. [CrossRef]
54. Tierney, J.P.; Lidström, P. *Microwave Assisted Organic Synthesis*; Blackwell Publishing: Oxford, UK, 2005. [CrossRef]
55. Prechtl, M.H.G.; Scholten, J.D.; Dupont, J. Carbon-carbon cross coupling reactions in ionic liquids catalysed by palladium metal nanoparticles. *Molecules* **2010**, *15*, 3441–3461. [CrossRef] [PubMed]
56. Vollmer, C.; Janiak, C. Naked metal nanoparticles from metal carbonyls in ionic liquids: Easy synthesis and stabilization. *Coord. Chem. Rev.* **2011**, *255*, 2039–2057. [CrossRef]
57. Yang, R.; Qiu, X.; Zhang, H.; Li, J.; Zhu, W.; Wang, Z.; Huang, X.; Chen, L. Monodispersed hard carbon spherules as a catalyst support for the electrooxidation of methanol. *Carbon* **2005**, *43*, 11–16. [CrossRef]
58. Meng, H.; Zhan, Y.; Zeng, D.; Zhang, X.; Zhang, G.; Jaouen, F. Factors Influencing the Growth of Pt Nanowires via Chemical Self-Assembly and their Fuel Cell Performance. *Small* **2015**, *11*, 3377–3386. [CrossRef] [PubMed]
59. Bizzotto, F.; Quinson, J.; Zana, A.; Kirkensgaard, J.J.K.; Dworzak, A.; Oezaslan, M.; Arenz, M. Ir nanoparticles with ultrahigh dispersion as oxygen evolution reaction (OER) catalysts: Synthesis and activity benchmarking. *Catal. Sci. Technol.* **2019**, *9*, 6345–6356. [CrossRef]
60. Edwards, P.P.; Kuznetsov, V.L.; David, W.I.F. Hydrogen energy. *Philos. Trans. Royal Soc. A* **2007**, *365*, 1043–1056. [CrossRef]
61. Chen, W.-F.; Sasaki, K.; Ma, C.; Frenkel, A.I.; Marinkovic, N.; Muckerman, J.T.; Zhu, Y.; Adzic, R.R. Hydrogen-evolution catalysts based on non-noble metal nickel-molybdenum nitride nanosheets. *Angew. Chem. Int. Ed.* **2012**, *51*, 6131–6135. [CrossRef]
62. Ravula, S.; Zhang, C.; Essner, J.B.; Robertson, J.D.; Lin, J.; Baker, G.A. Ionic Liquid-Assisted Synthesis of Nanoscale $(MoS_2)_x(SnO_2)_{1-x}$ on Reduced Graphene Oxide for the Electrocatalytic Hydrogen Evolution Reaction. *ACS Appl. Mater. Interfaces* **2017**, *9*, 8065–8074. [CrossRef]
63. Rademacher, L.; Beglau, T.H.Y.; Karakas, Ö.; Spieß, A.; Woschko, D.; Heinen, T.; Barthel, J.; Janiak, C. Synthesis of tin nanoparticles on Ketjen Black in ionic liquid and water for the hydrogen evolution reaction. *Electrochem. Commun.* **2022**, *136*, 107243. [CrossRef]
64. Qiao, S.; Zhang, B.; Li, Q.; Li, Z.; Wang, W.; Zhao, J.; Zhang, X.; Hu, Y. Pore Surface Engineering of Covalent Triazine Frameworks@MoS2 Electrocatalyst for the Hydrogen Evolution Reaction. *ChemSusChem* **2019**, *12*, 5032–5040. [CrossRef] [PubMed]
65. Rademacher, L.; Beglau, T.H.Y.; Heinen, T.; Barthel, J.; Janiak, C. Microwave-assisted synthesis of iridium oxide and palladium nanoparticles supported on a nitrogen-rich covalent triazine framework as superior electrocatalysts for the hydrogen evolution and oxygen reduction reaction. *Front. Chem.* **2022**, *10*, 945261. [CrossRef] [PubMed]
66. Liu, D.; Li, X.; Chen, S.; Yan, H.; Wang, C.; Wu, C.; Haleem, Y.A.; Duan, S.; Lu, J.; Ge, B.; et al. Atomically dispersed platinum supported on curved carbon supports for efficient electrocatalytic hydrogen evolution. *Nat. Energy* **2019**, *4*, 512–518. [CrossRef]
67. Cheng, N.; Stambula, S.; Wang, D.; Banis, M.N.; Liu, J.; Riese, A.; Xiao, B.; Li, R.; Sham, T.-K.; Liu, L.-M.; et al. Platinum single-atom and cluster catalysis of the hydrogen evolution reaction. *Nat. Commun* **2016**, *7*, 13638. [CrossRef]
68. Wang, P.; Zhang, X.; Zhang, J.; Wan, S.; Guo, S.; Lu, G.; Yao, J.; Huang, X. Precise tuning in platinum-nickel/nickel sulfide interface nanowires for synergistic hydrogen evolution catalysis. *Nat. Commun* **2017**, *8*, 14580. [CrossRef]
69. Zeng, M.; Li, Y. Recent advances in heterogeneous electrocatalysts for the hydrogen evolution reaction. *J. Mater. Chem. A* **2015**, *3*, 14942–14962. [CrossRef]
70. Chalk, A.J.; Harrod, J.F. Homogeneous Catalysis. II. The Mechanism of the Hydrosilation of Olefins Catalyzed by Group VIII Metal Complexes 1. *J. Am. Chem. Soc.* **1965**, *87*, 16–21. [CrossRef]
71. Naganawa, Y.; Inomata, K.; Sato, K.; Nakajima, Y. Hydrosilylation reactions of functionalized alkenes. *Tetrahedron Lett.* **2020**, *61*, 151513. [CrossRef]
72. Troegel, D.; Stohrer, J. Recent advances and actual challenges in late transition metal catalyzed hydrosilylation of olefins from an industrial point of view. *Coord. Chem. Rev.* **2011**, *255*, 1440–1459. [CrossRef]
73. Nakajima, Y.; Shimada, S. Hydrosilylation reaction of olefins: Recent advances and perspectives. *RSC Adv.* **2015**, *5*, 20603–20616. [CrossRef]
74. Komiyama, T.; Minami, Y.; Hiyama, T. Recent Advances in Transition-Metal-Catalyzed Synthetic Transformations of Organosilicon Reagents. *ACS Catal.* **2017**, *7*, 631–651. [CrossRef]
75. Rivero-Crespo, M.; Oliver-Meseguer, J.; Kapłońska, K.; Kuśtrowski, P.; Pardo, E.; Cerón-Carrasco, J.P.; Leyva-Pérez, A. Cyclic metal(oid) clusters control platinum-catalysed hydrosilylation reactions: From soluble to zeolite and MOF catalysts. *Chem. Sci.* **2020**, *11*, 8113–8124. [CrossRef] [PubMed]
76. Dobó, D.G.; Sipos, D.; Sipos, D.; Sápi, A.; London, G.; Juhász, K.; Kukovecz, Á.; Kónya, Z. Tuning the Activity and Selectivity of Phenylacetylene Hydrosilylation with Triethylsilane in the Liquid Phase over Size Controlled Pt Nanoparticles. *Catalysts* **2018**, *8*, 22. [CrossRef]
77. Chauhan, M.; Hauck, B.J.; Keller, L.P.; Boudjouk, P. Hydrosilylation of alkynes catalyzed by platinum on carbon. *J. Organomet. Chem.* **2002**, *645*, 1–13. [CrossRef]
78. Fernández, G.; Pleixats, R. Soluble Pt Nanoparticles Stabilized by a Tris-imidazolium Tetrafluoroborate as Efficient and Recyclable Catalyst for the Stereoselective Hydrosilylation of Alkynes. *ChemistrySelect* **2018**, *3*, 11486–11493. [CrossRef]
79. Chauhan, B.P.S.; Sarkar, A. Functionalized vinylsilanes via highly efficient and recyclable Pt-nanoparticle catalysed hydrosilylation of alkynes. *Dalton Trans.* **2017**, *46*, 8709–8715. [CrossRef]
80. Fang, H.; Chen, J.; Xiao, Y.; Zhang, J. Platinum nanoparticles confined in imidazolium-based ionic polymer for assembling a microfluidic reactor with enhanced catalytic activity. *Appl. Catal. A Gen.* **2019**, *585*, 117186. [CrossRef]

81. Woitassek, D.; Moya-Cancino, J.G.; Sun, Y.; Song, Y.; Woschko, D.; Roitsch, S.; Janiak, C. Sweet, Sugar-Coated Hierarchical Platinum Nanostructures for Easy Support, Heterogenization and Separation. *Chemistry* **2022**, *4*, 1147–1160. [CrossRef]
82. Abramov, A.; Diaz Diaz, D. Katalysatoren immobilisieren. *Nachr. Chem.* **2022**, *70*, 75–78. [CrossRef]
83. Lambert, J.M. The nature of platinum in silicones for biomedical and healthcare use. *J. Biomed. Mater. Res. B Appl. Biomater.* **2006**, *78*, 167–180. [CrossRef]
84. Wixtrom, R.; Glicksman, C.; Kadin, M.; Lawrence, M.; Haws, M.; Ferenz, S.; Sung, J.; McGuire, P. Heavy Metals in Breast Implant Capsules and Breast Tissue: Findings from the Systemic Symptoms in Women-Biospecimen Analysis Study: Part 2. *Aesthet. Surg. J.* **2022**, *42*, 1067–1076. [CrossRef] [PubMed]
85. Xue, Z.; Strouse, M.J.; Shuh, D.K.; Knobler, C.B.; Kaesz, H.D.; Hicks, R.F.; Williams, R.S. Characterization of (methylcyclopentadienyl)trimethylplatinum and low-temperature organometallic chemical vapor deposition of platinum metal. *J. Am. Chem. Soc.* **1989**, *111*, 8779–8784. [CrossRef]
86. Beermann, V.; Gocyla, M.; Kühl, S.; Padgett, E.; Schmies, H.; Goerlin, M.; Erini, N.; Shviro, M.; Heggen, M.; Dunin-Borkowski, R.E.; et al. Tuning the Electrocatalytic Oxygen Reduction Reaction Activity and Stability of Shape-Controlled Pt-Ni Nanoparticles by Thermal Annealing—Elucidating the Surface Atomic Structural and Compositional Changes. *J. Am. Chem. Soc.* **2017**, *139*, 16536–16547. [CrossRef] [PubMed]
87. Yong, L.; Kirleis, K.; Butenschön, H. Stereodivergent Formation of Alkenylsilanes:syn oranti Hydrosilylation of Alkynes Catalyzed by a Cyclopentadienylcobalt(I) Chelate Bearing a Pendant Phosphane Tether. *Adv. Synth. Catal.* **2006**, *348*, 833–836. [CrossRef]

Disclaimer/Publisher's Note: The statements, opinions and data contained in all publications are solely those of the individual author(s) and contributor(s) and not of MDPI and/or the editor(s). MDPI and/or the editor(s) disclaim responsibility for any injury to people or property resulting from any ideas, methods, instructions or products referred to in the content.

Article

Studies on the Prediction and Extraction of Methanol and Dimethyl Carbonate by Hydroxyl Ammonium Ionic Liquids

Xiaokang Wang [1,2], Yuanyuan Cui [2], Yingying Song [2], Yifan Liu [2,3], Junping Zhang [2], Songsong Chen [2,*], Li Dong [2,3,*] and Xiangping Zhang [2,3]

1. Henan Institute of Advanced Technology, Zhengzhou University, Zhengzhou 450003, China
2. Beijing Key Laboratory of Ionic Liquids Clean Process, CAS Key Laboratory of Green Process and Engineering, State Key Laboratory of Multiphase Complex Systems, Institute of Process Engineering, Chinese Academy of Sciences, Beijing 100190, China
3. Advanced Energy Science and Technology Guangdong Laboratory, Huizhou 516003, China
* Correspondence: sschen@ipe.ac.cn (S.C.); ldong@ipe.ac.cn (L.D.)

Abstract: The separation of dimethyl carbonate (DMC) and methanol is of great significance in industry. In this study, ionic liquids (ILs) were employed as extractants for the efficient separation of methanol from DMC. Using the COSMO-RS model, the extraction performance of ILs consisting of 22 anions and 15 cations was calculated, and the results showed that the extraction performance of ILs with hydroxylamine as the cation was much better. The extraction mechanism of these functionalized ILs was analyzed by molecular interaction and the σ-profile method. The results showed that the hydrogen bonding energy dominated the interaction force between the IL and methanol, and the molecular interaction between the IL and DMC was mainly Van der Waals force. The molecular interaction changes with the type of anion and cation, which in turn affects the extraction performance of ILs. Five hydroxyl ammonium ILs were screened and synthesized for extraction experiments to verify the reliability of the COSMO-RS model. The results showed that the order of selectivity of ILs predicted by the COSMO-RS model was consistent with the experimental results, and ethanolamine acetate ([MEA][Ac]) had the best extraction performance. After four regeneration and reuse cycles, the extraction performance of [MEA][Ac] was not notably reduced, and it is expected to have industrial applications in the separation of methanol and DMC.

Keywords: COSMO-RS; ionic liquids (ILs); dimethyl carbonate (DMC); methanol; extraction separation

1. Introduction

Dimethyl carbonate (DMC) is an important high value-added chemical [1]. Due to its high dielectric constant, DMC can be used as electrolyte for lithium electronic batteries [2]. DMC is considered a potential fuel additive due to its high oxygen content (53%), high octane number, high gasoline/water partition coefficient, low toxicity, and fast biodegradation [3]. Engine soot particle emissions may be decreased with the right quantity of DMC added to diesel, which will lower environmental pollution [4]. Since DMC contains the methoxyl, methyl and carbonyl groups, it is also a promising replacement for toxic dimethyl sulfate and phosgene in the synthesis of isocyanates, polyurethanes and polycarbonates [5].

Currently, phosgene [6], methanol oxidation carbonylation [7], transesterification [8], urea alcoholysis [9], and direct carbon dioxide synthesis [10] are the main methods to synthesize DMC. In these routes, the separation of DMC and methanol is a difficult problem due to the formation of azeotrope [11]. Conventional separation techniques, such as pressure-swing distillation [11], azeotropic distillation [12], extractive distillation [13], membrane separation [14] and adsorption [15], have been used to separate methanol-DMC azeotropic mixtures. However, these separation techniques have high energy consumption, which is typical of thermal-based separation techniques. Liquid-liquid extraction is a form of separation technique that is popular in the chemical industry, with excellent

Citation: Wang, X.; Cui, Y.; Song, Y.; Liu, Y.; Zhang, J.; Chen, S.; Dong, L.; Zhang, X. Studies on the Prediction and Extraction of Methanol and Dimethyl Carbonate by Hydroxyl Ammonium Ionic Liquids. *Molecules* **2023**, *28*, 2312. https://doi.org/10.3390/molecules28052312

Academic Editor: Mara G. Freire

Received: 13 January 2023
Revised: 5 February 2023
Accepted: 11 February 2023
Published: 2 March 2023

Copyright: © 2023 by the authors. Licensee MDPI, Basel, Switzerland. This article is an open access article distributed under the terms and conditions of the Creative Commons Attribution (CC BY) license (https:// creativecommons.org/licenses/by/ 4.0/).

environmental friendliness and low energy performance. To date, some research on the extraction and separation of alcohols and esters has been reported. Yang et al. [16] evaluated the solubility and liquid-liquid equilibrium data of the ternary system of dimethyl adipate + 1,6-hexanediol + water or ethylene glycol. The results showed that water is an appropriate solvent for 1,6-hexanediol extraction. Liu et al. [17] used deep eutectic solvents consisting of choline chloride and ethylene glycol to separate methanol and DMC. The experimental results showed that the extractant can realize the separation of DMC and methanol, and the maximum partition coefficient of methanol in the two phases is 0.1516. In general, using organic solvents as extractants is not environmentally friendly and it is difficult to regenerate and recycle.

Ionic liquid (IL), characterized by low saturated vapor pressure, good stability and designability, has excellent solubility for many systems and is an ideal solvent, frequently substituting for organic solvents in numerous industrial fields, particularly the extraction process [18]. However, the high viscosity and high cost of ILs have limited their industrial development to some extent. Cai et al. [19] reported the liquid-liquid equilibrium data of the ternary system of methanol + DMC + 1-alkyl-3-methylimidazolium dialkylphosphate. However, the selectivity and partition coefficient of the IL are low. Wen et al. [20] measured the liquid-liquid equilibrium (LLE) data of the ternary system of methanol + DMC + 1-methylimidazole hydrogen sulfate ([MIM][HSO_4]) at 298.15 K and 318.15 K. The results showed that [MIM][HSO_4] is a potential solvent for the separation of methanol and DMC. Most of the existing studies have focused on the determination of phase equilibrium data of ternary systems, and it is urgent to develop new extractants.

By introducing ILs into the methanol-DMC binary azeotropic system, the interaction between methanol and DMC can be regulated by the design of IL anions and cations [21]. The composition of ILs is diverse, and it is unrealistic to experiment one by one. Based on quantum chemistry theory and statistical thermodynamics theory, the COSMO-RS model is suitable for predicting the thermodynamic properties of LLE systems containing ILs, which can greatly reduce the blindness of IL screening and the workload of experiments [22]. To date, there has been much work on COSMO-RS screening of ILs for extraction. Jiang et al. [23] used COSMO-RS to screen suitable ILs from imidazole and pyridine ILs to extract 1,5-pentanediamine from aqueous solution. Zhao et al. [24] successfully obtained an IL with high efficiency for extracting lithium from aqueous solution by COSMO-RS simulation and experimental validation. To obtain the best IL for the drying of chloromethane, Wang et al. [25] screened 210 ILs using the COSMO-RS model and selected [EMIM][BF_4] as the water-removing agent. There has been no systematic study on extractant screening and extraction separation for the liquid-liquid extraction separation of a methanol-DMC azeotropic system.

In this work, 330 ILs composed of 15 cations and 22 anions were screened by using the COSMO-RS model. The selectivity (S), solvent loss (SL), and solvent solubility property (SP) of ILs for extracting methanol from DMC were evaluated. The effects of anion and cation species on the extraction performance of ILs were investigated by analyzing the molecular interaction and σ-profile. Subsequently, five hydroxyl ammonium ILs were selected and synthesized, and then extraction experiments were carried out.

2. Results and Discussion

2.1. COSMO-RS Prediction Results

Infinite dilution activity coefficients for methanol and DMC in different ILs were predicted by COSMO-RS. S, SP and SL were calculated and are illustrated in Figures 1–3. The specific data are listed in Tables S1–S3. The anions and cations involved in the prediction are listed in Tables 1 and 2, whose changes have a considerable impact on the selectivity. When choosing ILs, the value of SL should be considered, which must be extremely small. The smaller the value of SL is, the lower the loss of DMC in the extraction is. In addition, we prefer to choose ILs with a high value of SP. The higher the value of SP is, the stronger the extraction capacity of IL for methanol is. The results show that the ILs of alcohol amine cations and carboxylic acid anions have good performance and can be used to extract and separate methanol and DMC systems. Therefore, we selected these ILs for further study.

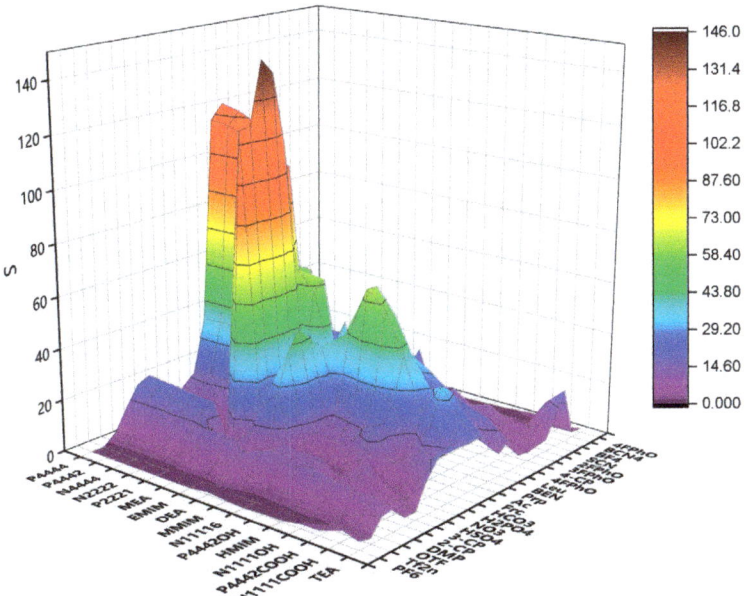

Figure 1. S of ILs predicted by COSMO-RS.

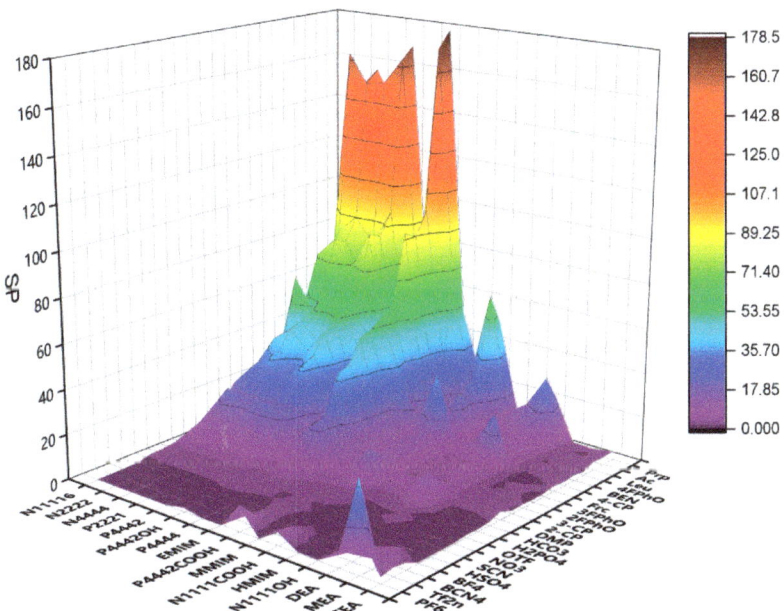

Figure 2. SP of ILs predicted by COSMO-RS.

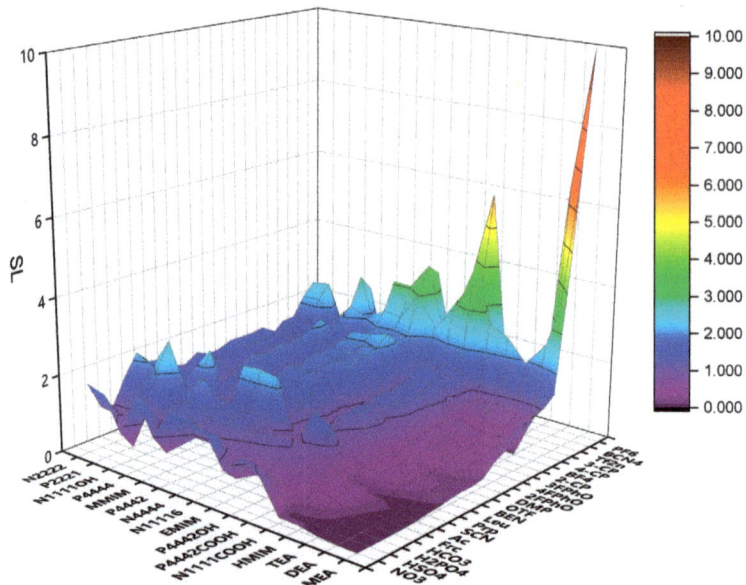

Figure 3. SL of ILs predicted by COSMO-RS.

Table 1. Liquid-liquid experiment results for ILs with different types of anions at 293.15 K.

Extractant	Initial Concentration of Methanol	DMC Phase (%)			IL Phase (%)			D	S	E
		x_{MeOH}	x_{DMC}	x_{IL}	x_{MeOH}	x_{DMC}	x_{IL}			
[MEA][Frc]	10%	2.54	97.44	0.02	9.46	4.25	86.29	3.72	85.27	79.20%
	20%	5.50	94.46	0.03	17.72	4.59	77.69	3.22	66.23	79.25%
	30%	8.65	91.30	0.05	24.50	4.70	70.80	2.83	54.97	79.29%
[MEA][Ac]	10%	2.13	97.85	0.02	9.98	14.69	75.33	4.70	31.29	83.31%
	20%	4.10	95.86	0.04	19.08	15.32	65.60	4.65	29.10	85.84%
	30%	6.86	93.09	0.05	27.60	16.34	56.06	4.02	22.92	86.81%
[MEA[Prp]	10%	2.23	97.52	0.25	10.54	15.40	74.07	4.73	29.93	83.24%
	20%	4.73	94.99	0.27	18.95	16.90	64.15	4.00	22.50	83.76%
	30%	7.24	92.42	0.35	28.28	17.99	53.74	3.91	20.07	86.93%

Table 2. Liquid-liquid experiment results for ILs with different types of cations at 293.15 K.

Extractant	Initial Concentration of Methanol	DMC Phase (%)			IL Phase (%)			D	S	E
		x_{MeOH}	x_{DMC}	x_{IL}	x_{MeOH}	x_{DMC}	x_{IL}			
[MEA][Ac]	10%	2.13	97.85	0.02	9.98	14.69	75.33	4.70	31.29	83.31%
	20%	4.10	95.86	0.04	19.08	15.32	65.60	4.65	29.10	85.84%
	30%	6.86	93.09	0.05	27.60	16.34	56.06	4.02	22.92	86.81%
[DEA][Ac]	10%	2.66	97.05	0.29	12.44	15.75	71.81	4.68	28.84	80.77%
	20%	4.70	94.95	0.35	19.70	20.58	59.72	4.20	19.35	82.97%
	30%	7.99	91.62	0.39	31.14	21.12	47.74	3.90	16.91	86.13%
[TEA][Ac]	10%	2.50	97.04	0.46	13.25	22.53	64.22	5.29	22.80	80.04%
	20%	5.06	94.46	0.48	23.61	23.09	53.30	4.67	19.10	81.78%
	30%	8.31	91.16	0.53	33.25	23.70	43.04	4.00	15.39	83.99%

2.2. Molecular Interaction Analysis

One of the most significant molecule-specific characteristics in COSMO-RS theory is the σ-profile, which is often divided into three regions: the nonpolar region (-0.0082 e/Å^2 $< \sigma < 0.0082$ e/Å^2), the hydrogen bond acceptor (HBA) region (s > 0.0082 e/Å^2), and the hydrogen bond donor (HBD) region (s < -0.0082 e/Å^2) [23]. The impacts of the cation and anion species on the effectiveness of IL extraction were investigated by σ-profiles and molecular interactions.

The σ-profiles of methanol and DMC are shown in Figure 4. It can be seen from the figure that DMC has a peak at 0.0125 e/Å^2, representing the carbonyl group and methoxy group in the molecule, which has strong hydrogen bond receiving capabilities. Methanol has peaks at -0.016 e/Å^2 and 0.017 e/Å^2, respectively, because the hydroxyl group in the molecular structure has strong hydrogen bond receiving (oxygen atom) and hydrogen bond supplying (hydrogen atom) capabilities at the same time, which results in strong intermolecular interactions between methanol and DMC, making it difficult to separate them. However, the possibility of separation by extraction is increased by the variation in the σ-profile distribution between DMC and methanol in the HB donor region.

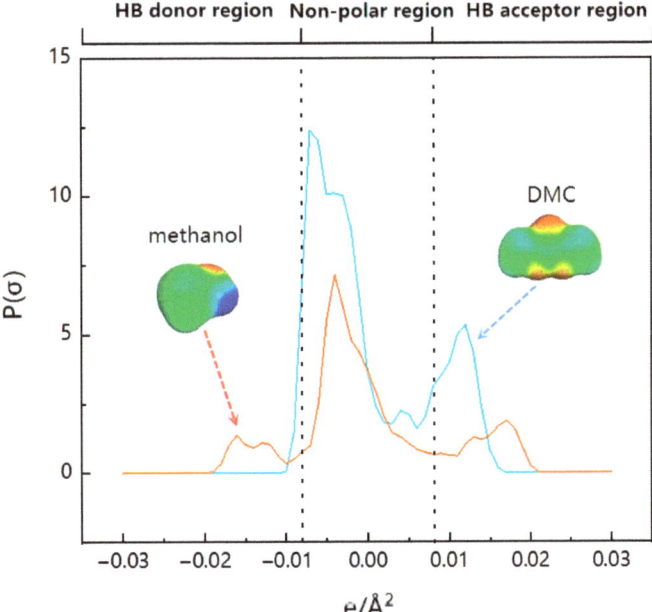

Figure 4. The σ-profiles of methanol and DMC.

2.2.1. Effect of Anions on Molecular Interaction

To further understand the influence of molecular interactions on the extraction performance of ILs, COSMOthermX was used to calculate the magnitude of three kinds of interaction forces of ILs on methanol and DMC, which are misfit, van der Waals (VDW), and hydrogen-bond (HB) interactions according to COSMO-RS theory [26]. Positive energy means repulsion, while negative energy means attraction [27].

The molecular interactions of ILs composed of [MEA]$^+$ and six anions on methanol and DMC were investigated, and are shown in Figures 5 and 6. From Figure 5 we can see that the total interaction between ILs and methanol was much greater than that between ILs and DMC, which is necessary for the effective separation of methanol from DMC. Figure 6a shows that the molecular interaction between methanol and ILs is mainly hydrogen bonding. With the enhancement of acid radical ion acidity, the hydrogen bonding force gradually increases, resulting in a slightly increasing *SP* value. Figure 6b shows that the dominant molecular interaction between DMC and ILs is the VDW interaction. With the enhancement of acid radical ion acidity, the VDW interaction gradually increases, which leads to a significant increase in the *SL* value, resulting in a decrease in the *S* value. The *S* value follows the order: [MEA][NO$_3$] > [MEA][HSO$_4$] > [MEA][HCO$_3$] > [MEA][Frc] > [MEA][Ac] > [MEA][Prp].

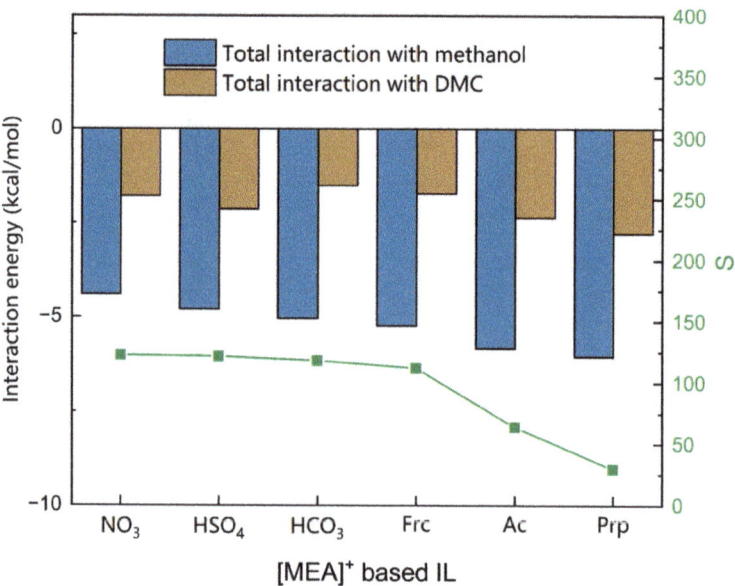

Figure 5. Total interaction between solutes and [MEA]$^+$ based ILs calculated by COSMO-RS.

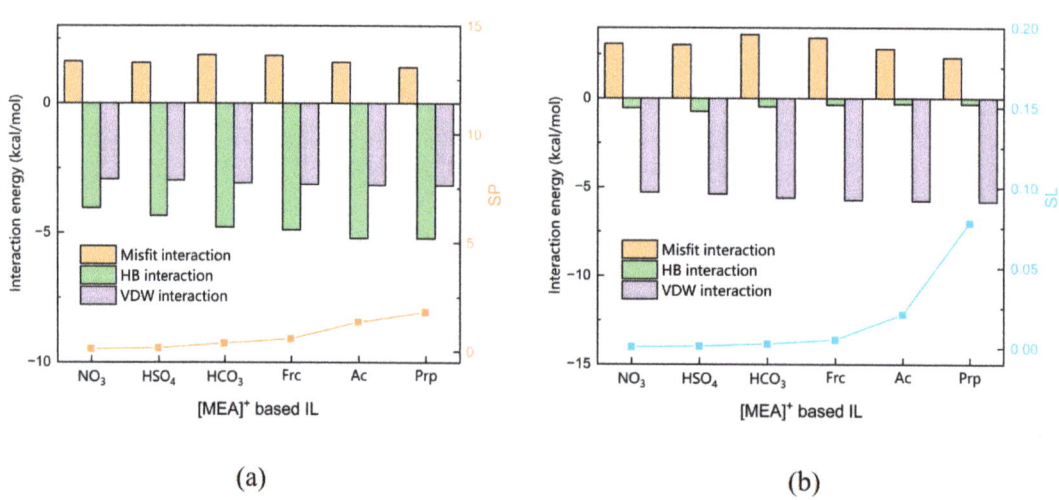

Figure 6. Molecular interaction calculated by COSMO-RS. (**a**) Molecular interaction between methanol and [MEA]$^+$ based ILs. (**b**) Molecular interaction between DMC and [MEA]$^+$ based ILs.

In addition, the σ-profiles of the six anions were also analyzed and are shown in Figure 7. The six anions have peaks mainly in the HBA region, and the peak height in the HBA region is very high, which indicates that these anions are more capable of accepting hydrogen bonds. As mentioned above, methanol has a high capacity for supplying hydrogen bonds, so methanol can be extracted by forming a strong hydrogen bond with the IL. Meanwhile, the entrainment of DMC during extraction can be effectively reduced due to the weak ability of DMC to provide hydrogen bonds.

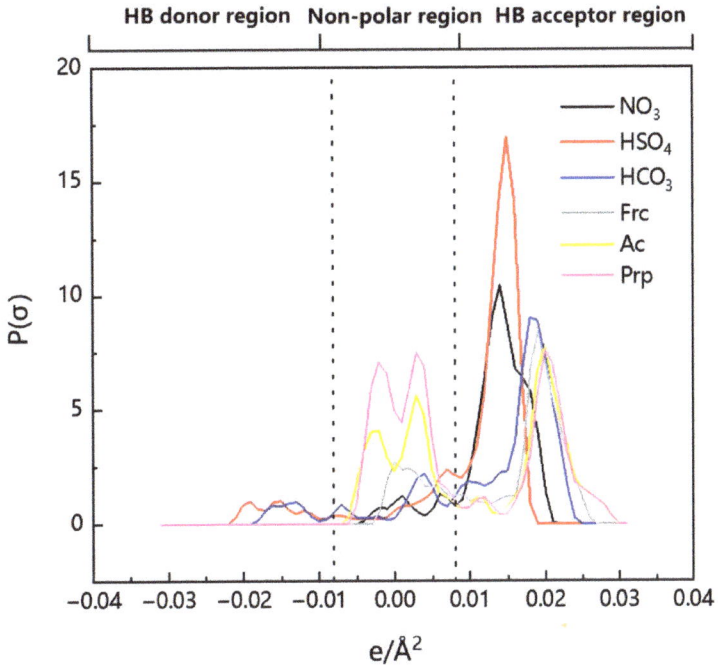

Figure 7. The σ-profiles of six different anions.

2.2.2. Effect of Cations on Molecular Interaction

The effect of cations on the extraction ability of ILs is less significant than that of anions. Fixing the anion to [Ac]$^-$, the effects of the number of hydroxyethyl groups on the cation on the interactions between ILs and methanol and between ILs and DMC were investigated.

It can be seen from Figure 8 that the total interaction between ILs and methanol is much stronger than that between ILs and DMC. Figure 9a shows that there is a weak increase in the magnitude of the hydrogen bonding force between the IL and methanol, as the number of hydroxyethyl groups rises, which increases the value of SP. Figure 9b shows that the increase in the number of hydroxyethyl groups also caused an increase in the magnitude of van der Waals and hydrogen bonding forces between DMC and IL, which caused a significant increase in the value of SL. Since SL rises faster than SP, the S value keeps decreasing. The S value follows the order: [MEA][Ac] > [DEA][Ac] > [TEA][Ac].

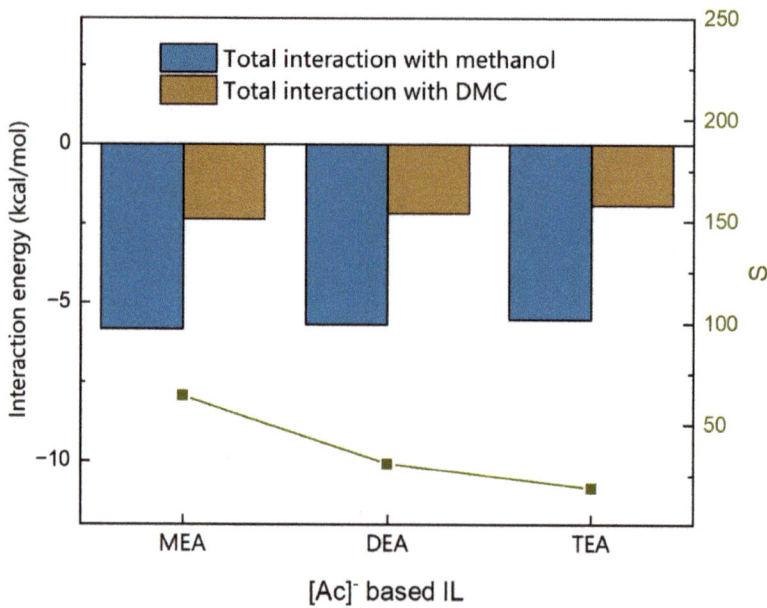

Figure 8. Total interaction between solutes and [Ac]⁻ based ILs calculated by COSMO-RS.

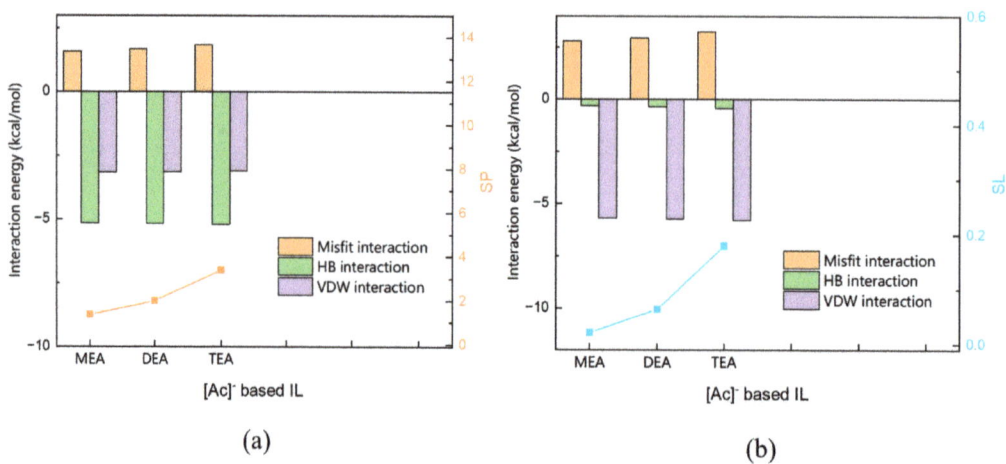

Figure 9. Molecular interaction calculated by COSMO-RS. (**a**) Molecular interaction between methanol and [Ac]⁻ based ILs. (**b**) Molecular interaction between DMC and [Ac]⁻ based ILs.

Similarly, the σ-profiles of the three cations were analyzed and are shown in Figure 10. The result shows that as the number of hydroxyethyl groups rises, the peak of the IL extends further in the HB donor region and higher in the HB acceptor region. This implies that the IL has a greater capacity for donating and accepting hydrogen bonds, which significantly improves its affinity for both DMC and methanol. Inevitably, more DMC is entrained despite the increased IL extraction efficiency for methanol.

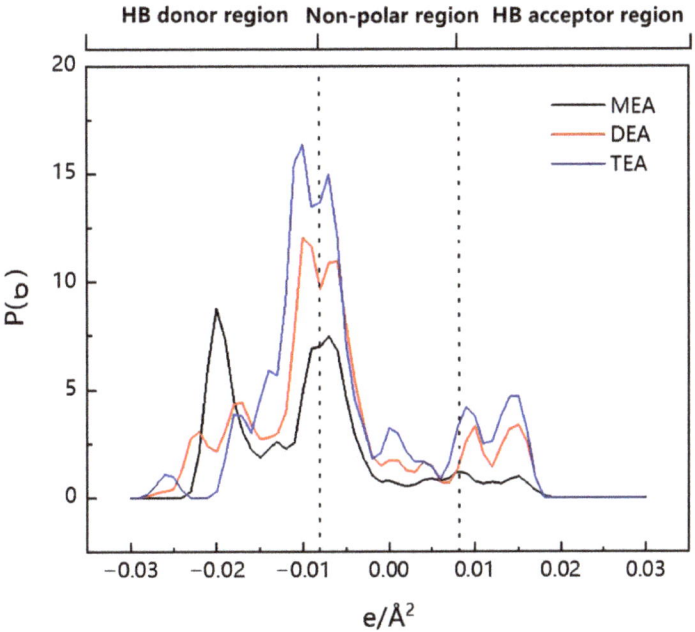

Figure 10. The σ-profiles of three different cations.

2.3. Extraction Experiment

Due to the instability of [MEA][HCO$_3$] and the difficulty of synthesizing [MEA][NO$_3$] and [MEA][HSO$_4$], five ILs: [MEA][Frc], [MEA][Ac], [MEA][Prp], [DEA][Ac] and [TEA][Ac] were synthesized. All ILs are characterized by ^{1}H NMR spectra, which are listed in Figures S1–S5. This confirmed the structures of the ILs and that there was no impurity in the ILs. The decomposing temperatures of the synthesized ILs were more than 140 °C, which is relatively stable under experimental conditions [28].

Five ILs were used to carry out extraction experiments. Through the experiments, the separation performance of typical ILs was examined to confirm the validity of COSMO-RS. At 293.15 K, the two-phase composition after the experiment was measured, and the experimental results are listed in Tables 1 and 2.

Among the ILs, [MEA][Frc], [MEA][Ac] and [MEA][Prp] were employed to investigate the impact of various anions on the extraction effect. The results showed that the three ILs had high extraction efficiency and selectivity, and the order of S was [MEA][Frc] > [MEA][Ac] > [MEA][Prp]. As the anion changes from [Frc]$^-$ to [Ac]$^-$ and then to [Prp]$^-$, S is significantly reduced. This is because as the anion's alkyl chain lengthens, the volume of the entire IL expands and its polarity decreases [29]. As a result, the van der Waals force between DMC and the IL develops, which lowers S.

Similarly, to investigate the effect of cation types on the extraction performance of ILs, we selected three ILs [MEA][Ac], [DEA][Ac] and [TEA][Ac] for extraction experiments, and the results are listed in Table 2. It can be seen from the table that the order of S is [MEA][Ac] > [DEA][Ac] > [TEA][Ac]. Compared with anions, the variety of cation types does not bring a noteworthy change to the selectivity of the IL. The reasons may include the following aspects. On the one hand, the increase in hydroxyl groups greatly strengthens the hydrogen bonding between the IL and methanol. On the other hand, when the number of hydroxyethyl groups attached to the N atom rises, the IL polarity declines, increasing the van der Waals force between the IL and DMC. Given these two factors, it is not unexpected that the value of S only marginally decreases.

In addition, it can be found in experiments that with the increase in the initial concentration of methanol, the S value of ILs decreases, but E slightly increases. Most ILs show a high extraction efficiency of more than 80%.

2.4. Recyclability

In the extraction experiment, the recycling of the extractant is also an important aspect. It is necessary to regenerate and recycle ILs from the mixture of methanol, DMC and ILs. After extraction, IL was regenerated from the IL phase by vacuum evaporation in 343.15 K for 5 h. Taking [MEA][Ac] as an example, using recycled [MEA][Ac] each time, we carried out extraction experiments under the same conditions four times. As shown in Figure 11, after four cycles, the extraction efficiency of [MEA][Ac] decreased slightly from 86.1% to 84.2%, proving the stability and efficiency of [MEA][Ac]. The fresh and recycled [MEA][Ac] were then analyzed by ^{1}H NMR, and the results are shown in Figure S6. The results showed that the ^{1}H NMR spectra of fresh and recycled ILs did not change, indicating that the structure of recycled ILs did not change.

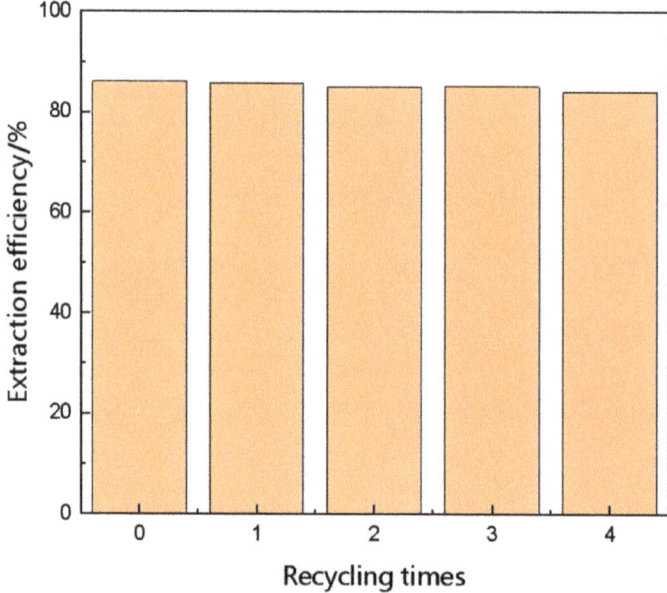

Figure 11. Extraction performance of recycled IL.

3. Materials and Methods

3.1. Computational Methods

The COSMOthermX program was used to perform all COSMO-RS calculations at the BVP86/TZVP quantum chemical level and related parameterization (BP TZVP C30 1401). The .cosmo files of methanol, DMC, and some anions and cations are obtained directly from the database. For ILs that do not exist in the database, Turbomole was used to generate their corresponding .cosmo files. The molecular model was computed using Turbomole software at the RI-DFT level of theory and the def-TZVP basis set. Because the parameterization divergence between TZVP and TZVPD-FINE is very similar, the TZVP operation is much faster than that of TZVPD-FINE [30]. In this work, 330 ILs composed of 15 cations and 22 anions were screened. The anions and cations involved in the prediction are listed in Tables 3 and 4.

Table 3. The names of the cations.

Number	Names	Abbreviations
1	2-Hydroxyethylammonium	[MEA]$^+$
2	Bis(2-hydroxyethyl)ammonium	[DEA]$^+$
3	Tris(2-hydroxyethyl)ammonium	[TEA]$^+$
4	1-Ethyl-3-methylimidazolium	[EMIM]$^+$
5	3-Methylimidazolium	[HMIM]$^+$
6	1-Methyl-3-methylimidazolium	[MMIM]$^+$
7	Hexadecyltrimethylammonium	[N$_{11116}$]$^+$
8	Hydroxyltrimethylammonium	[N$_{1111OH}$]$^+$
9	Carboxyltrimethylammonium	[N$_{1111COOH}$]$^+$
10	Tetraethylammonium	[N$_{2222}$]$^+$
11	Tetrabutylammonium	[N$_{4444}$]$^+$
12	Methyltriethylphosphorus	[P$_{2221}$]$^+$
13	Ethyltributylphosphorus	[P$_{4442}$]$^+$
14	Hydroxyethyltributylphosphorus	[P$_{4442OH}$]$^+$
15	Tetrabutylphosphorus	[P$_{4444}$]$^+$

Table 4. The names of the anions.

Number	Names	Abbreviations
1	Hexafluorophosphate	[PF$_6$]$^-$
2	Bis(trifluoromethylsulfonyl)imide	[Tf$_2$N]$^-$
3	Tetracyanoboric acid	[BCN$_4$]$^-$
4	Tetrafluoroborate	[BF$_4$]$^-$
5	Hydrogen sulfate	[HSO$_4$]$^-$
6	Thiocyanate thiocyanide	[SCN]$^-$
7	Nitrate	[NO$_3$]$^-$
8	Trifluoromethanesulfonate	[OTf]$^-$
9	Dihydrogen phosphate	[H$_2$PO$_4$]$^-$
10	Hydrocarbonate	[HCO$_3$]$^-$
11	Dimethylphosphate	[DMP]$^-$
12	2-Chlorophenol	[2-CP]$^-$
13	3-Chlorophenol	[3-CP]$^-$
14	Difluorophosphate	[2FPhO]$^-$
15	Trifluorophosphate	[3FPhO]$^-$
16	Formate	[Frc]$^-$
17	4-Chlorophenol	[4-CP]$^-$
18	Benzoate	[BEN]$^-$
19	Tetrafluorophosphate	[4FPhO]$^-$
20	Leucinate	[Leu]$^-$
21	Acetate	[Ac]$^-$
22	Propionate	[Prp]$^-$

3.2. Extraction Performance Index of ILs Using the COSMO-RS Model

The performance of methanol extraction from DMC by IL was predicted by the COSMO-RS model. Three important parameters, S, SP, and SL, were determined by

$$S = \frac{\gamma_{DMC}^\infty}{\gamma_{methanol}^\infty} \tag{1}$$

$$SP = \frac{1}{\gamma_{methanol}^\infty} \tag{2}$$

$$SL = \frac{1}{\gamma_{DMC}^\infty} \tag{3}$$

where γ_{DMC}^∞ and $\gamma_{Methanol}^\infty$ represent the infinite dilution activity coefficients of DMC and methanol in ILs, respectively. S stands for selectivity, which reflects the ability of ILs to extract methanol from DMC. The higher the value of S, the better the extraction separation effect. When $S = 1$, which means $\gamma_{DMC}^{\infty} = \gamma_{Methanol}^{\infty}$, IL cannot be used for extraction. SP stands for the solvent solubility property, and the higher the value is, the better the solubility of the IL to methanol is. SL stands for solvent loss, which reflects the loss of DMC after extraction. The higher the value is, the greater the loss.

3.3. Materials and Reagents

The specific compounds utilized for the IL synthesis and extraction studies are shown in Table 5. All chemicals were purchased from Aladdin, and they were not purified further prior to use.

Table 5. Materials.

Names	CAS	Purity	Supplier
Dimethyl carbonate	616-38-6	≥99%	Aladdin
Methanol	67-56-1	≥99%	Aladdin
Ethanolamine	141-43-5	≥99%	Aladdin
Diethanolamine	111-42-2	≥99%	Aladdin
Triethanolamine	102-71-6	≥99%	Aladdin
Formic acid	64-18-6	≥99%	Aladdin
Acetic acid	64-19-7	≥99%	Aladdin
Propanoic acid	79-09-4	≥99%	Aladdin

3.4. Synthesis of ILs

The synthesis of the ILs used in this study followed the acid-base neutralization reaction and was prepared by a one-step synthesis method [31]. Taking [MEA][AC] as an example, 100 mL of ethanol was used to dissolve 0.5 mol of 2-aminoethanol, which was then added into a 500 mL flask as a liquid mixture. As the reaction was intense and exothermic, the flask was placed in an ice water bath at 278.15 K under vigorous stirring with a magnetic stirrer. Next, 100 mL of mixed ethanol and 0.5 mol of acetic acid were slowly added to the flask within 90 min. The reaction persisted for 20 h. Following the reaction, the solvent was eliminated by vacuum evaporation, and the unreacted reactants were eliminated using repeated acetone washing. After being dried under a vacuum for 48 h at 333.15 K, a clear and colorless product was produced. It is worth mentioning that in the synthesis of these ILs, the synthesis process requires adequate cooling and strict control of the addition rate, as the reaction between formic acid and ethanolamine is the most exothermic. Otherwise, under high heat, dehydration of the salt may occur, resulting in the corresponding formamide, as in the case of nylon salt (salt of diamine and dicarboxylate) [32]. ^{1}H NMR spectra were used to confirm the structure of the ILs, and there was no impurity.

3.5. Extraction Experiments

The specific experimental methods are described below. Prepare DMC solution with 10%, 20%, and 30% mole fractions of methanol. Add 20 g solution to a 50 mL jacketed extraction bottle, and then add the same mass of IL. Seal the jacketed glass bottle to prevent the solution from volatilizing, and circulate water outside to keep the temperature of the system stable at 293.15 K. The solution was magnetically stirred at 500 rpm for 2 h and then remained still for 10 h to guarantee balance. Finally, the lower sampling port was unscrewed to take the lower sample, and the upper sample was removed from the upper layer with a straw, and the samples were then analyzed.

3.6. Sample Analysis

Gas chromatography (Agilent 8890A) outfitted with a DB-624 capillary column and a flame ionization detector (FID), along with the internal standard method, was used to measure the concentrations of methanol and DMC in two phases. ILs are nonvolatile, so the composition of IL in the sample was determined by the weight loss method. The volatile organic compounds were removed using a vacuum drying oven, and the weight of each sample was precisely determined both before and after evaporation. The evaporated samples' GC analysis showed no residues of organic materials [20].

3.7. Evaluation Index of Extraction Performance in Experiments

The distribution coefficient (D), selectivity (S), and extraction efficiency (E) are used to evaluate the extraction performance of ILs, and they are calculated as follows:

$$D = \frac{x^E_{methanol}}{x^R_{methanol}} \quad (4)$$

$$S = \frac{x^E_{methanol}}{x^R_{methanol}} \times \frac{x^R_{DMC}}{x^E_{DMC}} \quad (5)$$

$$E = \frac{n^E_{methanol}}{n^E_{methanol} + n^R_{methanol}} \quad (6)$$

where $x_{methanol}$ and x_{DMC} represent the mole fractions of methanol and DMC, $n_{methanol}$ represents the molar amount of methanol, and superscripts R and E represent the raffinate phase and extraction phase, respectively.

4. Conclusions

In this work, 330 ILs composed of 15 cations and 22 anions were screened by the COSMO-RS model to find ILs with good extraction performance and realize the effective separation of DMC and methanol. It can be seen from the results that the type of anions have a greater influence on the extraction performance of ILs than that of cations, and carboxylic acid anions deliver the most optimal performance when paired with cations such as [MEA]⁺, [DEA]⁺ or [TEA]⁺. Through the analysis of molecular interactions, it can be concluded that the HB interaction is dominant between methanol and ILs, while VDW is the main force between DMC and ILs. Five ILs are synthesized for extraction experiments to verify the reliability of the COSMO-RS model. The order of IL selectivity obtained from the experiment is consistent with that predicted by the COSMO-RS model. All selected ILs have high selectivity and distribution coefficients. Among these ILs, [MEA][Ac] has the highest extraction efficiency, which can reach 86.81%. In addition, [MEA][Ac] has the advantages of easy regeneration and recyclability, and is expected to be applied in the separation of methanol and DMC in industrial production.

Supplementary Materials: The following supporting information can be downloaded at: https://www.mdpi.com/article/10.3390/molecules28052312/s1, Figure S1: 1H NMR spectra of [MEA][Frc]; Figure S2: 1H NMR spectra of [MEA][Ac]; Figure S3: 1H NMR spectra of [MEA][Prp]; Figure S4: 1H NMR spectra of [DEA][Ac]; Figure S5: 1H NMR spectra of [TEA][Ac]; Figure S6: 1H NMR spectra of fresh and recycled [MEA][Ac]; Table S1: S of ILs predicted by COSMO-RS; Table S2: SP of ILs predicted by COSMO-RS; Table S3: SL of ILs predicted by COSMO-RS.

Author Contributions: Conceptualization, X.W. and Y.C.; methodology, X.W. and Y.C.; software, Y.S.; validation, S.C. and L.D.; formal analysis, X.W. and Y.C.; investigation, X.W.; resources, X.W., S.C. and L.D.; data curation, Y.S., Y.L. and J.Z.; writing—original draft preparation, X.W.; writing—review and editing, Y.L., J.Z., S.C., X.Z. and L.D.; visualization, X.Z. and L.D.; supervision, L.D. and X.Z. All authors have read and agreed to the published version of the manuscript.

Funding: This work was supported by the National Natural Science Foundation of China (No. 22178356, No. 21890763 and No. 22078329), "Clean Coal Combustion and Low Carbon Utilization", Strategic Priority Research Program of the Chinese Academy of Sciences, Grant No. XDA 29030202, and the Key-Area Research and Development Program of Guangdong Province (No.2020B0101370002).

Institutional Review Board Statement: Not applicable.

Informed Consent Statement: Not applicable.

Data Availability Statement: Not applicable.

Acknowledgments: We sincerely appreciate Suojiang Zhang (IPE, CAS) for his careful academic guidance and great support.

Conflicts of Interest: The authors declare that they have no conflict of interest.

Sample Availability: Samples of the compounds are available from the authors.

References

1. Zhang, M.; Xu, Y.; Williams, B.L.; Xiao, M.; Wang, S.; Han, D.; Sun, L.; Meng, Y. Catalytic materials for direct synthesis of dimethyl carbonate (DMC) from CO_2. *J. Clean. Prod.* **2021**, *279*, 123344. [CrossRef]
2. Wang, X.; Hu, H.; Chen, B.; Dang, L.; Gao, G. Efficient synthesis of dimethyl carbonate via transesterification of ethylene carbonate catalyzed by swelling poly(ionic liquid)s. *Green Chem. Eng.* **2021**, *2*, 423–430. [CrossRef]
3. Pacheco, M.A.; Marshall, C.L. Review of dimethyl carbonate (DMC) manufacture and its characteristics as a fuel additive. *Energy Fuels* **1997**, *11*, 2–29. [CrossRef]
4. Zi, K.; Tu, X.K.; Huang, Y.Q.; Shen, R.Y.; Pan, K.Y. Effects of diesel oil-DMC blend fuel on performances of diesel engine. *Neiranji Gongcheng/Chin. Intern. Combust. Engine Eng.* **2007**, *28*, 63–66+70.
5. Huang, S.; Yan, B.; Wang, S.; Ma, X. Recent advances in dialkyl carbonates synthesis and applications. *Chem. Soc. Rev.* **2015**, *44*, 3079–3116. [CrossRef]
6. Aresta, M.; Galatola, M. Life cycle analysis applied to the assessment of the environmental impact of alternative synthetic processes. The dimethylcarbonate case: Part 1. *J. Clean. Prod.* **1999**, *7*, 181–193. [CrossRef]
7. Figueiredo, M.C.; Trieu, V.; Eiden, S.; Koper, M.T.M. Spectro-Electrochemical Examination of the Formation of Dimethyl Carbonate from CO and Methanol at Different Electrode Materials. *J. Am. Chem. Soc.* **2017**, *139*, 14693–14698. [CrossRef]
8. Wang, J.Q.; Sun, J.; Cheng, W.G.; Shi, C.Y.; Dong, K.; Zhang, X.P.; Zhang, S.J. Synthesis of dimethyl carbonate catalyzed by carboxylic functionalized imidazolium salt via transesterification reaction. *Catal. Sci. Technol.* **2012**, *2*, 600–605. [CrossRef]
9. Shi, L.; Wang, S.J.; Wong, D.S.H.; Huang, K. Novel Process Design of Synthesizing Propylene Carbonate for Dimethyl Carbonate Production by Indirect Alcoholysis of Urea. *Ind. Eng. Chem. Res.* **2017**, *56*, 11531–11544. [CrossRef]
10. Bian, J.; Wei, X.W.; Jin, Y.R.; Wang, L.; Luan, D.C.; Guan, Z.P. Direct synthesis of dimethyl carbonate over activated carbon supported Cu-based catalysts. *Chem. Eng. J.* **2010**, *165*, 686–692. [CrossRef]
11. Wei, H.-M.; Wang, F.; Zhang, J.-L.; Liao, B.; Zhao, N.; Xiao, F.-k.; Wei, W.; Sun, Y.-H. Design and Control of Dimethyl Carbonate–Methanol Separation via Pressure-Swing Distillation. *Ind. Eng. Chem. Res.* **2013**, *52*, 11463–11478. [CrossRef]
12. Matsuda, H.; Inaba, K.; Nishihara, K.; Sumida, H.; Kurihara, K.; Tochigi, K.; Ochi, K. Separation Effects of Renewable Solvent Ethyl Lactate on the Vapor–Liquid Equilibria of the Methanol + Dimethyl Carbonate Azeotropic System. *J. Chem. Eng. Data* **2017**, *62*, 2944–2952. [CrossRef]
13. Shen, Y.; Su, Z.; Zhao, Q.; Shan, R.; Zhu, Z.; Cui, P.; Wang, Y. Molecular simulation and optimization of extractive distillation for separation of dimethyl carbonate and methanol. *Process Saf. Environ. Prot.* **2022**, *158*, 181–188. [CrossRef]
14. Vopička, O.; Pilnáček, K.; Friess, K. Separation of methanol-dimethyl carbonate vapour mixtures with PDMS and PTMSP membranes. *Sep. Purif. Technol.* **2017**, *174*, 1–11. [CrossRef]
15. Číhal, P.; Dendisová, M.; Švecová, M.; Hrdlička, Z.; Durďáková, T.-M.; Budd, P.M.; Harrison, W.; Friess, K.; Vopička, O. Sorption, swelling and plasticization of PIM-1 in methanol-dimethyl carbonate vapour mixtures. *Polymer* **2021**, *218*, 123509. [CrossRef]
16. Yang, X.; Li, H.; Cao, C.; Xu, L.; Liu, G. Experimental and correlated liquid-liquid equilibrium data for dimethyl adipate + 1,6-hexanediol + water or ethylene glycol. *J. Mol. Liq.* **2019**, *284*, 39–44. [CrossRef]
17. Liu, X.; Xu, D.; Diao, B.; Gao, J.; Zhang, L.; Ma, Y.; Wang, Y. Separation of Dimethyl Carbonate and Methanol by Deep Eutectic Solvents: Liquid–Liquid Equilibrium Measurements and Thermodynamic Modeling. *J. Chem. Eng. Data* **2018**, *63*, 1234–1239. [CrossRef]
18. Brennecke, J.F.; Maginn, E.J. Ionic liquids: Innovative fluids for chemical processing. *AIChE J.* **2001**, *47*, 2384–2389. [CrossRef]
19. Cai, F.; Ibrahim, J.J.; Gao, L.; Wei, R.; Xiao, G. A study on the liquid–liquid equilibrium of 1-alkyl-3-methylimidazolium dialkylphosphate with methanol and dimethyl carbonate. *Fluid Phase Equilibria* **2014**, *382*, 254–259. [CrossRef]
20. Wen, G.; Geng, X.; Bai, W.; Wang, Y.; Gao, J. Ternary liquid-liquid equilibria for systems containing (dimethyl carbonate or methyl acetate + methanol + 1-methylimidazole hydrogen sulfate) at 298.15 K and 318.15 K. *J. Chem. Thermodyn.* **2018**, *121*, 49–54. [CrossRef]
21. Chen, S.; Dong, L.; Zhang, J.; Cheng, W.; Huo, F.; Su, Q.; Hua, W. Effects of imidazolium-based ionic liquids on the isobaric vapor–liquid equilibria of methanol + dimethyl carbonate azeotropic systems. *Chin. J. Chem. Eng.* **2020**, *28*, 766–776. [CrossRef]
22. Zhang, C.; Wu, J.; Wang, R.; Ma, E.; Wu, L.; Bai, J.; Wang, J. Study of the toluene absorption capacity and mechanism of ionic liquids using COSMO-RS prediction and experimental verification. *Green Energy Environ.* **2021**, *6*, 339–349. [CrossRef]

23. Jiang, C.; Cheng, H.; Qin, Z.; Wang, R.; Chen, L.; Yang, C.; Qi, Z.; Liu, X. COSMO-RS prediction and experimental verification of 1,5-pentanediamine extraction from aqueous solution by ionic liquids. *Green Energy Environ.* **2021**, *6*, 422–431. [CrossRef]
24. Zhao, X.; Wu, H.; Duan, M.; Hao, X.; Yang, Q.; Zhang, Q.; Huang, X. Liquid-liquid extraction of lithium from aqueous solution using novel ionic liquid extractants via COSMO-RS and experiments. *Fluid Phase Equilibria* **2018**, *459*, 129–137. [CrossRef]
25. Wang, Z.; Liu, S.; Jiang, Y.; Lei, Z.; Zhang, J.; Zhu, R.; Ren, J. Methyl chloride dehydration with ionic liquid based on COSMO-RS model. *Green Energy Environ.* **2021**, *6*, 413–421. [CrossRef]
26. Eckert, F.; Klamt, A. Fast solvent screening via quantum chemistry: COSMO-RS approach. *AIChE J.* **2002**, *48*, 369–385. [CrossRef]
27. Sellaoui, L.; Guedidi, H.; Masson, S.; Reinert, L.; Levêque, J.-M.; Knani, S.; Lamine, A.B.; Khalfaoui, M.; Duclaux, L. Steric and energetic interpretations of the equilibrium adsorption of two new pyridinium ionic liquids and ibuprofen on a microporous activated carbon cloth: Statistical and COSMO-RS models. *Fluid Phase Equilibria* **2016**, *414*, 156–163. [CrossRef]
28. Meng, H.; Ge, C.-T.; Ren, N.-N.; Ma, W.-Y.; Lu, Y.-Z.; Li, C.-X. Complex Extraction of Phenol and Cresol from Model Coal Tar with Polyols, Ethanol Amines, and Ionic Liquids Thereof. *Ind. Eng. Chem. Res.* **2014**, *53*, 355–362. [CrossRef]
29. Zhou, T.; Chen, L.; Ye, Y.; Chen, L.; Qi, Z.; Freund, H.; Sundmacher, K. An Overview of Mutual Solubility of Ionic Liquids and Water Predicted by COSMO-RS. *Ind. Eng. Chem. Res.* **2012**, *51*, 6256–6264. [CrossRef]
30. Bezold, F.; Weinberger, M.E.; Minceva, M. Assessing solute partitioning in deep eutectic solvent-based biphasic systems using the predictive thermodynamic model COSMO-RS. *Fluid Phase Equilibria* **2017**, *437*, 23–33. [CrossRef]
31. Bicak, N. A new ionic liquid: 2-Hydroxy ethylammonium formate. *J. Mol. Liq.* **2004**, *116*, 15. [CrossRef]
32. Iglesias, M.; Torres, A.; Gonzalez-Olmos, R.; Salvatierra, D. Effect of temperature on mixing thermodynamics of a new ionic liquid: {2-Hydroxy ethylammonium formate (2-HEAF)+ short hydroxylic solvents}. *J. Chem. Thermodyn.* **2008**, *40*, 119–133. [CrossRef]

Disclaimer/Publisher's Note: The statements, opinions and data contained in all publications are solely those of the individual author(s) and contributor(s) and not of MDPI and/or the editor(s). MDPI and/or the editor(s) disclaim responsibility for any injury to people or property resulting from any ideas, methods, instructions or products referred to in the content.

Article

Macrocyclic Ionic Liquids with Amino Acid Residues: Synthesis and Influence of Thiacalix[4]arene Conformation on Thermal Stability

Olga Terenteva [1], Azamat Bikmukhametov [1], Alexander Gerasimov [1], Pavel Padnya [1,*] and Ivan Stoikov [1,2,*]

[1] A.M. Butlerov Chemistry Institute, Kazan Federal University, 18 Kremlyovskaya Street, Kazan 420008, Russia
[2] Federal Center for Toxicological, Radiation and Biological Safety, 2 Nauchny Gorodok Street, Kazan 420075, Russia
* Correspondence: padnya.ksu@gmail.com (P.P.); ivan.stoikov@mail.ru (I.S.); Tel.: +7-843-233-7241 (I.S.)

Abstract: Novel thiacalix[4]arene based ammonium ionic liquids (ILs) containing amino acid residues (glycine and L-phenylalanine) in *cone*, *partial cone*, and *1,3-alternate* conformations were synthesized by alkylation of macrocyclic tertiary amines with N-bromoacetyl-amino acids ethyl ester followed by replacing bromide anions with bis(trifluoromethylsulfonyl)imide ions. The melting temperature of the obtained ILs was found in the range of 50–75 °C. The effect of macrocyclic core conformation on the synthesized ILs' melting points was shown, i.e., the ILs in *partial cone* conformation have the lowest melting points. Thermal stability of the obtained macrocyclic ILs was determined via thermogravimetry and differential scanning calorimetry. The onset of decomposition of the synthesized compounds was established at 305–327 °C. The compounds with L-phenylalanine residues are less thermally stable by 3–19 °C than the same glycine-containing derivatives.

Keywords: thiacalix[4]arenes; ionic liquids; amino acid; synthesis; thermal stability; melting point

Citation: Terenteva, O.; Bikmukhametov, A.; Gerasimov, A.; Padnya, P.; Stoikov, I. Macrocyclic Ionic Liquids with Amino Acid Residues: Synthesis and Influence of Thiacalix[4]arene Conformation on Thermal Stability. *Molecules* 2022, 27, 8006. https://doi.org/10.3390/molecules27228006

Academic Editors: Slavica Ražić, Aleksandra Cvetanović Kljakić and Enrico Bodo

Received: 1 November 2022
Accepted: 16 November 2022
Published: 18 November 2022

Publisher's Note: MDPI stays neutral with regard to jurisdictional claims in published maps and institutional affiliations.

Copyright: © 2022 by the authors. Licensee MDPI, Basel, Switzerland. This article is an open access article distributed under the terms and conditions of the Creative Commons Attribution (CC BY) license (https://creativecommons.org/licenses/by/4.0/).

1. Introduction

Ionic liquids (ILs) have been attracting the attention of researchers in the last two decades. Their unique properties, e.g., low vapor pressure, low toxicity, recyclability, high solvating ability, polarity, thermal and electrochemical stability and electrical conductivity can be explained by the structure [1–16]. ILs consist of bulky organic cations with low symmetry and inorganic or organic anions [17]. The physicochemical properties of ILs can also be affected by varying the cations and anions [18].

Despite the fact that the first examples of ILs are derivatives of ammonium salts, this class of compounds has not been fully investigated. Ammonium ILs were used for metal ions extraction, functional materials creation, as a reaction medium, battery electrolytes, components of pharmaceutical agents [19–23]. However, low biocompatibility and complexation selectivity limit the practical application of these compounds. A possible solution of this problem is the modification of ILs with various functional groups, e.g., amide, hydroxyl, carboxyl, amino acid fragments etc. ILs with amino acid fragments also increase the stability of biomolecules such as enzymes and DNA [24–29]. Prior works showed the rise of thermal stability and effect of such ILs on proteins activity and stability, their package changing and aggregation inhibition [30–33].

One of the actively developing classes of ILs are macrocyclic and polyionic liquids containing several cationic fragments [34–36]. The design of ILs based on supramolecular platforms, e.g., crown ethers [37], pillararenes [38–40], and (thia)calixarenes [41–47], was described. At this moment, the creation of such structures is a non-trivial synthetic problem. Only a few examples of their successful synthesis are known [48,49]. Our scientific group has shown that the introduction of quaternary ammonium, ester and amino acid fragments leads to obtain macrocyclic ILs in *cone* and *1,3-alternate* conformations [50]. There are no literature examples of the synthesis and properties study of macrocyclic ILs based on *partial cone* stereoisomer. The structure of *partial cone* is less symmetrical than *cone* and *1,3-alternate*, that can hypothetically decrease the melting point of such compounds. The introduction of different amino acid residues affects the thermal stability and melting point, which are important characteristics of ILs. In this work, *p-tert*-butylthiacalix[4]arenes tetrasubstituted at the lower rim with quaternary ammonium groups and amino acid fragments (glycine and L-phenylalanine) in *cone*, *partial cone*, and *1,3-alternate* conformations were synthesized for the first time. The influence of the conformation of macrocyclic core and the nature of the amino acid substituent on their thermal stability was investigated.

2. Results and Discussion

2.1. Synthesis of p-tert-butylthiacalix[4]arenes Containing Quaternary Ammonium Groups and Fragments of Glycine and L-phenylalanine

Previously, our scientific group developed an approach to the synthesis of macrocyclic ILs [50]. It consisted in the alkylation of *p-tert*-butylthiacalix[4]arene-based tertiary amines, followed by the replacement of bromide anions. Initially, we synthesized highly reactive alkylating agents containing amino acid residues. Glycine **1** and L-phenylalanine **2** were selected to evaluate the effect of planar π-aromatic ring systems on physical properties, e.g., melting point and thermal stability. The target compounds were obtained in two steps (Scheme 1). The first step was the synthesis of the amino acid esters **3** and **4**. The second step was the interaction of the obtained compounds **3** and **4** with bromoacetic acid bromide. *N*-Bromoacetyl-amino acid ethyl esters **5** and **6** were obtained in 92 and 90% yield.

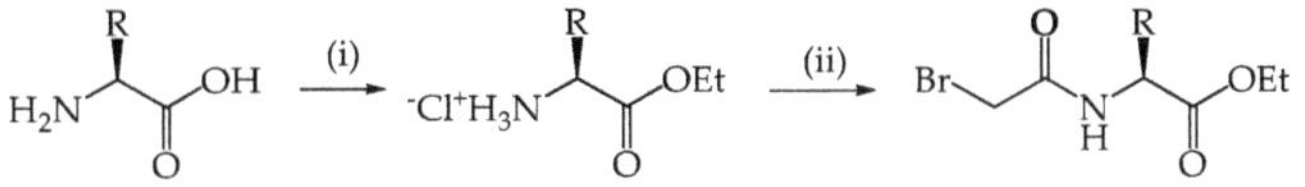

1 R=H
2 R=CH$_2$-Ph

3 R=H
4 R=CH$_2$-Ph

5 R=H (92%)
6 R=CH$_2$-Ph (90%)

Scheme 1. Reagents and reaction conditions: (i) SOCl$_2$, ethanol; (ii) BrCH$_2$C(O)Br, NaHCO$_3$, benzene/water.

The next stage of this work was the study of reaction of the obtained alkylating agents **5** and **6** with tetrasubstituted thiacalixarenes **7–9** containing terminal tertiary amino groups in *cone*, *partial cone* and *1,3-alternate* conformations (Scheme 2). Targeted bromides of macrocyclic quaternary ammonium salts **10–15** were obtained in high yields (93–95%). Previously, a significant decrease in the melting point by replacing halide ions with bis(trifluoromethylsulfonyl)imide ions (N(SO$_2$CF$_3$)$_2^-$, NTf$_2^-$) has also been shown [51]. This can be explained by the fact that the increase of anion size decreases symmetry of the molecule. Thus, the compounds **10–15** were reacted with lithium bis(trifluoromethylsulfonyl)imide in water at room temperature. The macrocyclic ILs **16–21** were obtained in yields close to quantitative (Scheme 2).

The structure and the composition of all synthesized compounds were confirmed by ^{1}H and ^{13}C NMR, IR spectroscopy, mass spectrometry and elemental analysis (Figures S1–S33). The conformation of modified *p-tert*-butylthiacalix[4]arene derivatives can be determined by one-dimensional ^{1}H NMR spectroscopy based on specific proton signals of *tert*-butyl group, aromatic ring, oxymethylene and amide group. Table 1 lists the values of the characteristic chemical shifts of the compounds **10–21**. The protons of oxymethylene and amide groups of the compound **15** (*1,3-alternate*) are located in the shielded zone of neighboring aromatic rings of the macrocycle. These signals in the ^{1}H NMR spectrum are recorded upfield (4.00 and 8.02 ppm, respectively) of those of the macrocycles **11** in *cone* conformation (4.81 and 8.50 ppm, respectively). The chemical shifts of the aromatic protons depend less on the conformation of the macrocyclic platform, shifting by only 0.20 ppm upfield from *cone* **11** (7.39 ppm) to *1,3-alternate* **15** (7.59 ppm) stereoisomers. This provides evidence of the shielding effect of neighboring aryl fragments in *cone* stereoisomer on the aryl protons of macrocycle ring. The *tert*-butyl groups proton signals of *cone* **11** were found upfield (1.07 ppm) in contrast to the same signals of *1,3-alternate* **15** (1.19 ppm). This effect was related to the spatial location of the *tert*-butyl groups of *1,3-alternate* stereoisomer shielded by neighboring fragments of the macrocycle. The proton signals in *partial cone* **13** differ from *cone* and *1,3-alternate* due to the asymmetric structure of the macrocycle. The *tert*-butyl groups proton signals of *partial cone* **13** were recorded upfield (1.00, 1.10 and 1.27 ppm) as singlets with 2:1:1 intensity ratio. The oxymethylene proton signals were located in 4.40–4.77 ppm as two singlets and an AB–system. The aromatic ring proton signals are recorded at 7.67, 7.75, 7.01–7.65 ppm as two singlets and an AB–system. The amide group proton signals appeared in 8.31–8.48 ppm as broadened triplets. This effect of AB–systems in the ^{1}H NMR spectra due to the asymmetry of *partial cone* stereoisomer structure shielded by neighboring aromatic fragments of the macrocycle.

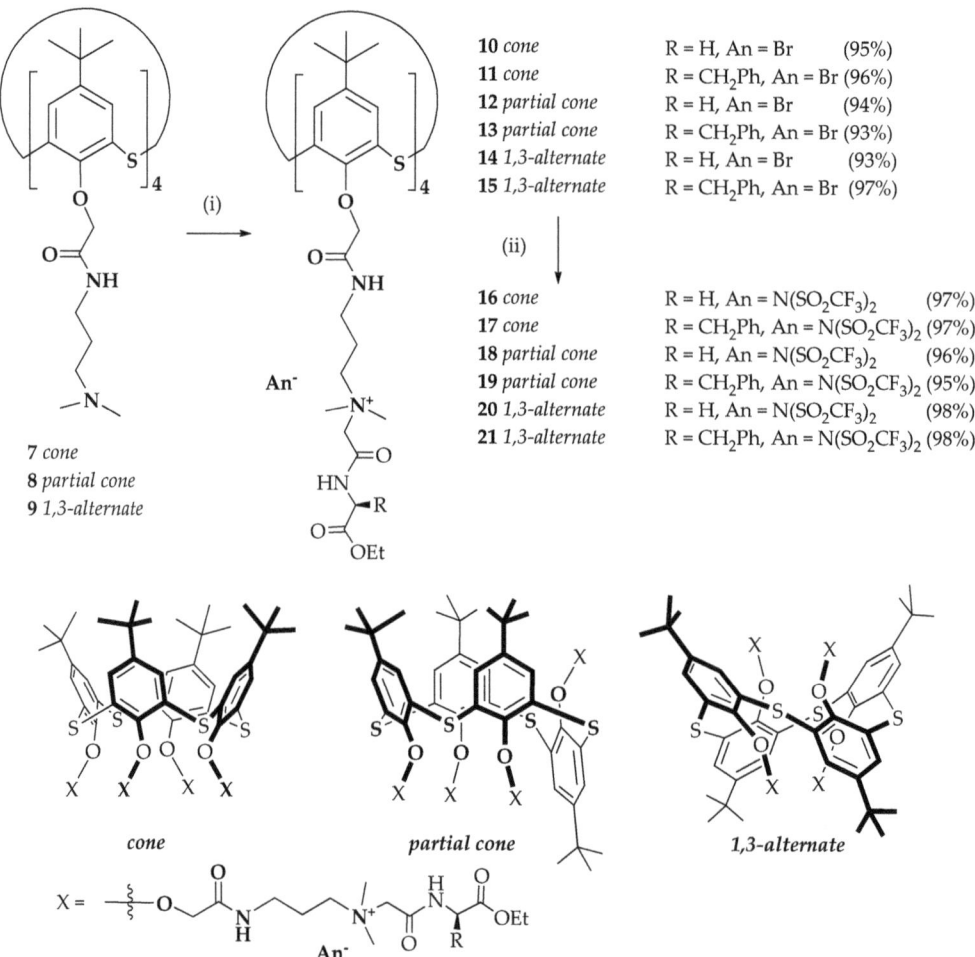

Scheme 2. Reagents and reaction conditions: (i) *N*-bromoacetyl-amino acid ethyl esters **5** or **6**, acetonitrile; and (ii) LiN(SO$_2$CF$_3$)$_2$, water.

Table 1. Chemical shifts (ppm) and spin-spin coupling constants (Hz) in the ^{1}H NMR spectra of the compounds **10–21** (DMSO-d_6, 298 K, 400 MHz).

Compounds	Amino Acid Fragments/Anion	^{t}Bu	OCH$_2$	ArH	NHCH$_2$CH$_2$CH$_2$N$^+$
10 (*cone*) *	Gly/Br$^-$	1.12	5.03	7.36	8.68
11 (*cone*)	L-Phe/Br$^-$	1.07	4.81	7.39	8.50
12 (*partial cone*)	Gly/Br$^-$	1.00, 1.17, 1.27	4.40 ($^2J_{HH}$ = 13.6 Hz), 4.49, 4.51, 4.80 ($^2J_{HH}$ = 13.6 Hz)	7.67, 7.01 ($^4J_{HH}$ = 2.4 Hz), 7.65 ($^4J_{HH}$ = 2.4 Hz), 7.75	8.36, 8.45, 8.50
13 (*partial cone*)	L-Phe/Br$^-$	1.00, 1.10, 1.27	4.42 ($^2J_{HH}$ = 13.5 Hz), 4.48, 4.51, 4.79 ($^2J_{HH}$ = 13.5 Hz)	7.67, 7.01 ($^4J_{HH}$ = 2.4 Hz), 7.65 ($^4J_{HH}$ = 2.4 Hz), 7.75	8.31, 8.41, 8.48
14 (*1,3-alternate*) *	Gly/Br$^-$	1.20	3.99	7.60	8.04
15 (*1,3-alternate*)	L-Phe/Br$^-$	1.19	4.00	7.59	8.02
16 (*cone*) *	Gly/NTf$_2^-$	1.11	4.89	7.35	8.48
17 (*cone*)	L-Phe/NTf$_2^-$	1.06	4.79	7.38	8.48
18 (*partial cone*)	Gly/NTf$_2^-$	1.00, 1.27, 1.30	4.39 ($^2J_{HH}$ = 13.6 Hz), 4.49, 4.51, 4.78 ($^2J_{HH}$ = 13.6 Hz)	7.67, 7.01 ($^4J_{HH}$ = 2.4 Hz), 7.65 ($^4J_{HH}$ = 2.4 Hz), 7.75	8.31, 8.41, 8.50
19 (*partial cone*)	L-Phe/NTf$_2^-$	1.00, 1.27, 1.28	4.40 ($^2J_{HH}$ = 13.5 Hz), 4.49, 4.52, 4.79 ($^2J_{HH}$ = 13.5 Hz)	7.67, 7.02 ($^4J_{HH}$ = 2.4 Hz), 7.65 ($^4J_{HH}$ = 2.4 Hz), 7.75	8.28, 8.40, 8.48
20 (*1,3-alternate*) *	Gly/NTf$_2^-$	1.20	3.99	7.59	8.03
21 (*1,3-alternate*)	L-Phe/NTf$_2^-$	1.19	3.99	7.59	8.00

* previously published data [50].

It should be noted that the proton signals in the ^{1}H NMR spectra of the compounds **10–15** containing halide anions, and the proton signals of salts **16–21** containing NTf$_2^-$ anions have identical multiplicity and exert very similar chemical shifts. This can be explained by the ability of compounds to form solvent-separated ion pairs. The quartet observed in the ^{13}C NMR spectra of compounds **16–21** at 120 ppm corresponds to the N(SO$_2$CF$_3$)$_2^-$ anion. The obtained salts **10–21** were characterized by ESI mass spectrometry. The mass spectra of the compounds **10–15** showed peaks corresponding to one-, two-, three-, and four-charged molecular ions without one, two, three, and four bromide anions. The obtained data also confirm the formation of solvate-separated ion pairs by the compounds **10–21**.

2.2. The Study of Thermal Stability of the Obtained Thiacalix[4]arene Based ILs

Melting point is one of the most important characteristics of ILs. Melting points of the synthesized thiacalix[4]arenes **10–21** are presented in Table 2. The replacement of halide ions by NTf$_2^-$ ions leads to significant decrease of the melting points of the thiacalix[4]arenes by 39–55 °C. All synthesized macrocycles **16–21** containing NTf$_2^-$ anions melt below 100 °C. It is well known that molecular packing density in the crystal lattice is a major factor affecting the melting point of the compound. More symmetrical molecules have denser packing in crystal and higher melting points. A comparison with the obtained results from previously published compounds with glycine residues [50] found that symmetrical *1,3-alternate* and *cone* stereoisomers showed higher melting points than asymmetrical *partial cone* structures. Thus, our hypothesis of lowering the melting point of *partial cone* stereoisomers (the compounds **12**, **13**, **18**, **19**) due to their molecular asymmetry was confirmed experimentally. However, the decrease of the melting point of the targeted compounds due to aromatic fragments in amino acid residues was not confirmed. The melting points of the macrocycles containing glycine fragments are lower by 1–11 °C compared to stereoisomers with L-phenylalanine fragments. Apparently, these results can be explained by the interaction of L-phenylalanine fragments with each other and the formation of additional hydrophobic and π-π interactions, which leads to a denser molecular packing in the crystal lattice and an increase in the melting point of the compounds as a result.

Table 2. Melting points (°C) of the macrocycles **10–21** containing amino acid fragments.

Amino Acid Fragments	Br⁻			N(SO$_2$CF$_3$)$_2$⁻		
	Cone	Partial Cone	1,3-alternate	Cone	Partial Cone	1,3-alternate
Gly	114 * (**10**)	105 (**12**)	112 * (**14**)	63 * (**16**)	50 (**18**)	73 * (**20**)
L-Phe	118 (**11**)	110 (**13**)	123 (**15**)	64 (**17**)	55 (**19**)	75 (**21**)

* previously published data [50].

High thermal stability is one of the characteristic properties of ILs [52,53]. The correlation between the obtained macrocyclic ILs structure and their thermal stability (influence of macrocycle conformation and amino acid residues) was investigated via thermogravimetric analysis (TG). Figure 1 shows the TG curves for the compounds **16–21**. All obtained macrocyclic ILs were thermally stable (decomposition temperature T_{onset} = 305–327 °C). Glycine containing compounds decomposed at a higher temperature (by 3–19 °C) compared to the compounds containing L-phenylalanine fragments. Many low molecular weight compounds for biofuel technology are obtained by pyrolysis (thermal lysis) of oligo- and polypeptides at a temperature of 300–350 °C [54]. However, thermal decomposition of proteins occurs at temperatures between 175 and 250 °C. In our case, the decomposition temperature of the amino acids containing compounds **16–21** was significantly higher. The obtained results are also consistent with the literature data on the thermal stability of macrocyclic ILs [38,50]. Temperature data $T_{10\%}$ and $T_{50\%}$ corresponding to 10% and 50% weight loss on decomposition are presented in Table 3. These characteristics are important in materials thermal stability research. The difference between T_{onset}, $T_{10\%}$ and $T_{50\%}$ is the measure of decomposition rate [52,55]. T_{onset} and $T_{10\%}$ of the studied compounds differ by 0–3 °C. The difference between T_{onset} and $T_{50\%}$ is more considerable, namely 41–109 °C. These results correspond to decomposition rate of non-macrocyclic ILs containing quaternary ammonium fragments [56,57]. The differential scanning calorimetry (DSC) heating curves of the synthesized compounds (Figure S34) in the temperature range of 300–375 °C clearly show endo effects corresponding to the first stage of decomposition. The values of endo effects for the compounds **16, 18, 20** are similar (38–41 J/g) (Figure S34a). These values are larger for the compounds **17, 19, 21** (53–55 J/g) (Figure S34b). The further decomposition of the obtained compounds at temperatures above 375 °C is accompanied by exo effects.

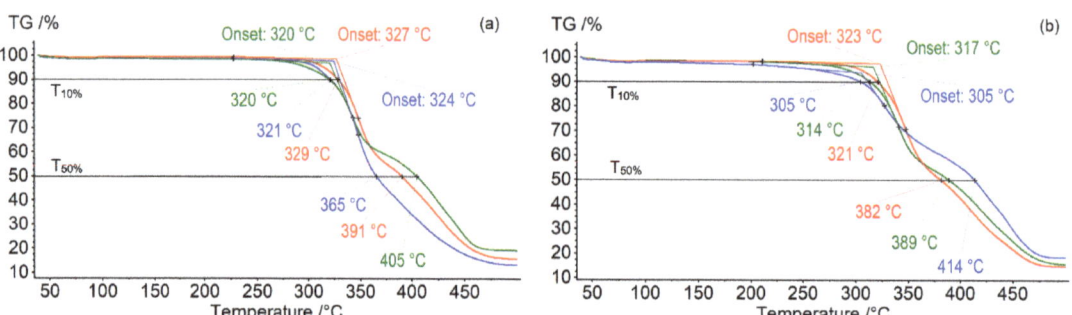

Figure 1. TG curves of the compounds with Gly: (**a**) **16** (green), **18** (blue), **20** (red), and L-Phe: (**b**) **17** (green), **19** (blue), **21** (red) fragments (dynamic argon atmosphere of 75 mL/min in the temperature range from 40 to 500 °C).

Table 3. TG data for the macrocyclic ILs **16–21** (°C).

Compounds	Amino Acid Fragments	T_{onset}	$T_{10\%}$	$T_{50\%}$
16 (cone)	Gly	320	320	405
18 (partial cone)		324	321	365
20 (1,3-alternate)		327	329	391
17 (cone)	L-Phe	317	314	389
19 (partial cone)		305	305	414
21 (1,3-alternate)		323	321	382

The experimental results and determined correlations between the synthesized macrocyclic ILs structure and their thermal characteristics showed that the introduction of aromatic amino acid (L-phenylalanine) fragments into ammonium salts based on p-tert-butylthiacalixarene reduces the thermal stability by 3–19 °C compared to glycine containing compounds. The melting points of the L-phenylalanine derivatives were higher than glycine ones by 1–11 °C. Oligo- and polypeptides thermal stability literature data [58–60] show that protein thermal stability increase is associated with rise of the number of charged amino acid fragments in their structures capable of electrostatic and cation–π interactions [61]. The presence of amino acids with uncharged polar fragments in protein structures reduces their thermal stability due to biomolecule packing efficiency decrease [62]. Thus, the obtained macrocyclic ILs are considered as biomimetic models of oligo- and polypeptides with the same structural patterns. The obtained results can also be applied to design of sensor systems capable for target substrate recognition.

3. Materials and Methods

3.1. General

All chemicals were purchased from Acros (Fair Lawn, NJ, USA), and most of them were used as received without additional purification. Organic solvents were purified by standard procedures. ^{1}H NMR and ^{13}C NMR spectra were obtained on the Bruker Avance-400 spectrometer (Bruker Corp., Billerica, MA, USA) (^{13}C{^{1}H} 100 MHz and ^{1}H 400 MHz). Chemical shifts were determined against the signals of residual protons of deuterated solvent (DMSO-d_6). The compounds concentration was equal to 3–5% by the weight in all records. The FTIR ATR spectra were recorded on the Spectrum 400 FT-IR spectrometer (Perkin–Elmer, Seer Green, Llantrisant, UK) with the Diamond KRS-5 attenuated total internal reflectance attachment (resolution 0.5 cm^{-1}, accumulation of 64 scans, recording time 16 s in the wavelength range 400–4000 cm^{-1}). Elemental analysis was performed on the Perkin–Elmer 2400 Series II instruments (Perkin–Elmer, Waltham, MA, USA). Melting points were determined using the Boetius block apparatus (VEB Kombinat Nagema, Radebeul, Germany). Mass spectra (ESI) were recorded on an AmaZonX mass spectrometer (Bruker Daltonik GmbH, Bremen, Germany). The drying gas was nitrogen at 300 °C. The capillary voltage was 4.5 kV. The samples were dissolved in acetonitrile (concentration ~ 10^{-6} g mL^{-1}). ESI HRMS experiments were performed at Agilent 6550 iFunnel Q-TOF LC/MS (Agilent Technologies, Santa Clara, CA, USA), equipped with Agilent 1290 Infinity II LC. Simultaneous thermogravimetry (TG) and differential scanning calorimetry (DSC) of solid samples were performed using the thermoanalyzer STA 449F1 Jupiter (Netzsch, Germany) at the temperature range of 40–500 °C. The measurements were carried out in aluminum crucibles in a dynamic argon atmosphere (75 mL/min) at a temperature scanning rate of 10 °C/min. The weights of sample were 4.9–10.2 mg.

N-Bromoacetyl-glycine ethyl ester **5** and thiacalix[4]arenes **7–10, 14, 16, 20** were synthesized according to the literature procedures [50,63–65].

3.2. Procedure for the Synthesis of the Compound 4

L–Phenylalanine **2** (1 g, 6.05 mmol) was dissolved in 10 mL of ethanol in a round-bottom flask equipped with a magnetic stirrer. Thionyl chloride (0.88 mL, 12.10 mmol) was added dropwise with stirring. The reaction mixture was left for 30 min at room temperature, then the reaction was carried out under heating for 4 h. The solvent was removed on a rotary evaporator. Diethyl ether was added to the residue, after which the formed precipitate was filtered off. The obtained product was dried in vacuum over phosphorus pentoxide.

L–Phenylalanine Ethyl Ester Hydrochloride (**4**) [66]

^{1}H NMR (DMSO-d_6, δ, ppm, J/Hz): 1.07 (t, $^3J_{HH}$ = 7.1 Hz, 3H, C$\underline{H_3}$CH$_2$O), 3.01–3.20 (m, 2H, C$\underline{H_2}$Ph), 4.08 (m, 2H, CH$_3$C$\underline{H_2}$O), 4.23 (m, 1H, NHC$\underline{H}$CO), 7.22–7.35 (m, 5H, P$\underline{h}$CH$_2$), 8.64 (br.s, 3H, NH$_3^+$).

3.3. Procedure for the Synthesis of the Compound 6

The solution of Na$_2$CO$_3$ (3.46 g, 32.66 mmol) in 50 mL of water was added to the suspension of L-phenylalanine ethyl ester hydrochloride **4** (3.41 g, 14.85 mmol) in benzene (50 mL). The reaction mixture was then cooled to 0 °C and bromoacetyl bromide (2.6 mL, 30 mmol) was added dropwise. The reaction mixture was allowed to warm to room temperature and stirred for 12 h with controlling pH = 6.5 by adding acetic acid. The organic phase was separated on a separating funnel and dried over anhydrous magnesium sulfate. Then the benzene was removed on a rotary evaporator. The obtained product **6** was dried in vacuum over phosphorus oxide.

N-Bromoacetyl-L-Phenylalanine Ethyl Ester (**6**) [66]

^{1}H NMR (DMSO-d_6, δ, ppm, J/Hz): 1.11 (t, $^3J_{HH}$ = 7.1 Hz, 3H, C$\underline{H_3}$CH$_2$O), 2.90–3.05 (m, 2H, C$\underline{H_2}$Ph), 3.86 (s, 2H, BrCH$_2$CO), 4.06 (q, $^3J_{HH}$ = 7.1 Hz, 2H, CH$_3$C$\underline{H_2}$O), 4.45 (m, 1H, NHC$\underline{H}$CO), 7.20–7.30 (m, 5H, P$\underline{h}$CH$_2$), 8.75 (d, $^3J_{HH}$ = 7.5 Hz, 1H, CON$\underline{H}$CH).

3.4. General Procedure for the Synthesis of the Compounds 10–15

The compounds **7–9** (0.30 g, 0.023 mmol) were dissolved in 5 mL of acetonitrile in a round-bottom flask equipped with a magnetic stirrer and reflux condenser. An equimolar amount per functional group (0.092 mmol) of the alkylating agent (N-bromoacetyl-glycine ethyl ester **5** or N-bromoacetyl-L-phenylalanine ethyl ester **6**) was added. The reaction mixture was refluxed for 18 h. Then the solvent was removed on a rotary evaporator. The obtained product was dried in vacuum over phosphorus oxide.

3.4.1. 5,11,17,23-Tetra-*tert*-butyl-25,26,27,28-tetrakis{N-[3′-(dimethyl{[(S)-ethoxycarbonylbenzylmethyl]aminocarbonylmethyl}ammonio)propyl]aminocarbonylmethoxy}-2,8,14,20-thiacalix[4]arene Tetrabromide in *cone* Conformation (**11**)

Yield: 0.57 g (96%). M.p. 118 °C. ^{1}H NMR (DMSO-d_6, δ, ppm, J/Hz): 1.06 (s, 36H, (CH$_3$)$_3$C), 1.11 (t, $^3J_{HH}$ = 7.1 Hz, 12H, C$\underline{H_3}$CH$_2$O), 1.88 (m, 8H, NHCH$_2$C$\underline{H_2}$CH$_2$N$^+$), 2.93 (m, 8H, C$\underline{H_2}$Ph), 3.08 (s, 24H, (CH$_3$)$_2$N$^+$), 3.20 (m, 8H, NHCH$_2$C$\underline{H_2}$CH$_2$N$^+$), 3.45 (m, 8H, NHC$\underline{H_2}$CH$_2$CH$_2$N$^+$), 4.02–4.13 (m, 16H, CH$_3$C$\underline{H_2}$O, N$^+$C$\underline{H_2}$CO), 4.55 (m, 4H, NHC$\underline{H}$CO), 4.80 (s, 8H, OCH$_2$CO), 7.21–7.30 (m, 20H, Ph), 7.38 (s, 8H, ArH), 8.51 (br.s, 4H, N$\underline{H}$CH$_2$CH$_2$CH$_2$N$^+$), 9.12. (d, $^3J_{HH}$ = 7.5 Hz, 4H, CON$\underline{H}$CH). ^{13}C NMR (DMSO-d_6, δ, ppm): 13.9, 22.6, 30.7, 33.9, 35.4, 36.5, 51.1, 53.7, 60.9, 61.7, 62.8, 74.2, 126.7, 128.1, 128.3, 129.1, 134.4, 136.6, 146.7, 157.9, 163.1, 168.3, 170.6. Elemental analysis. C$_{120}$H$_{168}$Br$_4$N$_{12}$O$_{20}$S$_4$ C, 56.60; H, 6.65; Br, 12.55; N, 6.60; S, 5.04. Found: C, 56.72; H, 6.85; Br, 12.23; N, 6.43; S, 4.79. MS (ESI), m/z: calculated for 556.5 [M–4 Br$^-$]$^{4+}$, 1192.5 [M–2 Br$^-$]$^{2+}$; found: 556.6 [M–4 Br$^-$]$^{4+}$, 1193.0 [M–2 Br$^-$]$^{2+}$. FTIR ATR (ν, cm^{-1}): 1095 (COC), 1677 (C=O), 3191 (N–H).

3.4.2. 5,11,17,23-Tetra-*tert*-butyl-25,26,27,28-tetrakis{*N*-[3′-(dimethyl{[ethoxycarbonylmethyl]aminocarbonylmethyl}ammonio)propyl]aminocarbonylmethoxy}-2,8,14,20-tetrathiacalix[4]arene Tetrabromide in *partial cone* Conformation (12)

Yield: 0.477 g (94%). M.p. 105 °C. ^{1}H NMR (DMSO-d_6, δ, ppm, J/Hz): 1.00 (s, 18H, (CH$_3$)$_3$C), 1.14–1.21 (m, 12H, C$\underline{H}_3$CH$_2$O), 1.27 (s, 9H, (CH$_3$)$_3$C), 1.29 (s, 9H, (CH$_3$)$_3$C), 1.94 (m, 8H, NHCH$_2$C$\underline{H}_2$CH$_2$N$^+$), 3.13 (s, 6H, (CH$_3$)$_2$N$^+$), 3.19 (s, 18H, (CH$_3$)$_2$N$^+$), 3.39 (m, 8H, NHCH$_2$CH$_2$C$\underline{H}_2$N$^+$), 3.53 (m, 8H, NHC$\underline{H}_2$CH$_2$CH$_2$N$^+$), 3.93 (d, $^3J_{HH}$ = 5.8 Hz, 8H, NHC$\underline{H}_2$CO), 4.07–4.13 (m, 16H, CH$_3$C$\underline{H}_2$O, N$^+$C$\underline{H}_2$CO), 4.40 (d, $^2J_{HH}$ = 13.6 Hz, 2H, OCH$_2$C(O)), 4.49 (s, 2H, OCH$_2$C(O)), 4.51 (s, 2H, OCH$_2$C(O)), 4.80 (d, $^2J_{HH}$ = 13.6 Hz, 2H, OCH$_2$C(O)), 7.01 (d, $^4J_{HH}$ = 2.4 Hz, 2H, ArH), 7.65 (d, $^4J_{HH}$ = 2.4 Hz, 2H, ArH), 7.67 (s, 2H, ArH), 7.75 (s, 2H, ArH), 8.32 (br.s, 2H, N$\underline{H}$CH$_2$CH$_2$CH$_2$N$^+$), 8.42 (br.s, 1H, N$\underline{H}$CH$_2$CH$_2$CH$_2$N$^+$), 8.50 (br.s, 1H, N$\underline{H}$CH$_2$CH$_2$CH$_2$N$^+$), 9.05. (m, 4H, CON$\underline{H}$CH). ^{13}C NMR (DMSO-d_6, δ, ppm): 14.1, 22.6, 30.7, 31.0, 33.8, 35.5, 40.8, 51.3, 60.8, 61.8, 62.7, 72.6, 126.4, 127.1, 127.6, 128.1, 133.7, 134.0, 135.1, 135.4, 145.3, 145.7, 146.5, 157.2, 159.4, 163.7, 166.9, 168.0, 168.8, 169.1. Elemental analysis. C$_{92}$H$_{144}$Br$_4$N$_{12}$O$_{20}$S$_4$ C, 50.55; H, 6.64; Br, 14.62; N, 7.69; S, 5.87; Found: C, 51.52; H, 6.25; Br, 14.26; N, 7.47; S, 5.89. HRMS (ESI), *m/z*: calculated for: 466.2370 [M−4 Br$^-$]$^{4+}$, 647.9557 [M−3 Br$^-$]$^{3+}$, 1012.3919 [M−2 Br$^-$]$^{2+}$; found: 466.2364 [M−4 Br$^-$]$^{4+}$, 647.9541 [M−3 Br$^-$]$^{3+}$, 1012.3937 [M−2 Br$^-$]$^{2+}$. FTIR ATR (ν, cm^{-1}): 1094 (COC), 1675 (C=O), 3207 (N−H).

3.4.3. 5,11,17,23-Tetra-*tert*-butyl-25,26,27,28-tetrakis{*N*-[3′-(dimethyl{[(*S*)-ethoxycarbonylbenzylmethyl]aminocarbonylmethyl}ammonio)propyl]aminocarbonylmethoxy}-2,8,14,20-thiacalix[4]arene Tetrabromide in *partial cone* Conformation (13)

Yield: 0.551 g (93%). M.p. 110 °C. ^{1}H NMR (DMSO-d_6, δ, ppm, J/Hz): 1.00 (s, 18H, (CH$_3$)$_3$C), 1.10 (t, $^3J_{HH}$ = 7.1 Hz, 12H, C$\underline{H}_3$CH$_2$O), 1.27 (s, 9H, (CH$_3$)$_3$C), 1.28 (s, 9H, (CH$_3$)$_3$C), 1.85 (m, 8H, NHCH$_2$C$\underline{H}_2$CH$_2$N$^+$), 2.90–2.93 (m, 8H, C$\underline{H}_2$Ph), 3.03 (s, 6H, (CH$_3$)$_2$N$^+$), 3.09 (s, 18H, (CH$_3$)$_2$N$^+$), 3.13–3.24 (m, 8H, NHCH$_2$CH$_2$C$\underline{H}_2$N$^+$), 3.46 (m, 8H, NHC$\underline{H}_2$CH$_2$CH$_2$N$^+$), 4.01–4.13 (m, 16H, CH$_3$C$\underline{H}_2$O, N$^+$C$\underline{H}_2$CO), 4.42 (d, $^2J_{HH}$ = 13.5 Hz, 2H, OCH$_2$C(O)), 4.48 (s, 2H, OCH$_2$C(O)), 4.51 (s, 2H, OCH$_2$C(O)), 4.55–4.60 (m, 4H, NHC$\underline{H}$CO), 4.79 (d, $^2J_{HH}$ = 13.5 Hz, 2H, OCH$_2$C(O)), 7.01 (d, $^4J_{HH}$ = 2.4 Hz, 2H, Ar-H), 7.22–7.30 (m, 20H, Ph), 7.65 (d, 2H, $^4J_{HH}$ = 2.4 Hz, ArH), 7.67 (s, 2H, ArH), 7.75 (s, 2H, ArH), 8.31 (br.s, 2H, N$\underline{H}$CH$_2$CH$_2$CH$_2$N$^+$), 8.41 (br.s, 1H, N$\underline{H}$CH$_2$CH$_2$CH$_2$N$^+$), 8.48 (br.s, 1H, N$\underline{H}$CH$_2$CH$_2$CH$_2$N$^+$), 9.12 (d, $^3J_{HH}$ = 7.5 Hz, 4H, CON$\underline{H}$CH). ^{13}C NMR (DMSO-d_6, δ, ppm): 14.0, 22.6, 28.8, 30.7, 31.0, 33.8, 34.0, 35.5, 36.5, 51.2, 53.7, 54.0, 60.7, 61.0, 61.6, 61.8, 62.8, 63.1, 72.6, 126.4, 126.7, 127.1, 127.6, 128.3, 129.2, 133.7, 134.1, 135.1, 135.4, 136.6, 145.3, 145.7, 146.5, 157.2, 159.4, 163.2, 166.9, 168.0, 168.8, 170.7. Elemental analysis. C$_{120}$H$_{168}$Br$_4$N$_{12}$O$_{20}$S$_4$ C, 56.60; H, 6.65; Br, 12.55; N, 6.60; S, 5.04; found: C, 56.52; H, 6.55; Br, 12.26; N, 6.47; S, 4.89. HRMS (ESI), *m/z*: calculated for: 556.5348 [M−4 Br$^-$]$^{4+}$, 768.3527 [M−3 Br$^-$]$^{3+}$, 1192.4858 [M−2 Br$^-$]$^{2+}$; found: 556.5339 [M−4 Br$^-$]$^{4+}$, 768.3522 [M−3 Br$^-$]$^{3+}$, 1192.4872 [M−2 Br$^-$]$^{2+}$. FTIR ATR (ν, cm^{-1}): 1094 (COC), 1672 (C=O), 3187 (N−H).

3.4.4. 5,11,17,23-Tetra-*tert*-butyl-25,26,27,28-tetrakis{*N*-[3′-(dimethyl{[(*S*)-ethoxycarbonylbenzylmethyl]aminocarbonylmethyl}ammonio)propyl]aminocarbonylmethoxy}-2,8,14,20-thiacalix[4]arene tetrabromide in *1,3-alternate* Conformation (15)

Yield: 0.574 g (97%). M.p. 123 °C. ^{1}H NMR (DMSO-d_6, δ, ppm, J/Hz): 1.12 (t, $^3J_{HH}$ = 7.1 Hz, 12H, C$\underline{H}_3$CH$_2$O), 1.19 (s, 36H, (CH$_3$)$_3$C), 1.90 (m, 8H, NHCH$_2$C$\underline{H}_2$CH$_2$N$^+$), 3.01 (m, 8H, C$\underline{H}_2$Ph), 3.09 (s, 24H, (CH$_3$)$_2$N$^+$), 3.15 (m, 8H, NHCH$_2$CH$_2$C$\underline{H}_2$N$^+$), 3.45 (m, 8H, NHC$\underline{H}_2$CH$_2$CH$_2$N$^+$), 4.00 (s, 8H, OCH$_2$CO), 4.01–4.14 (m, 16H, CH$_3$C$\underline{H}_2$O, N$^+$C$\underline{H}_2$CO), 4.59 (m, 4H, NHC$\underline{H}$CO), 7.20–7.30 (m, 20H, Ph), 7.59 (s, 8H, ArH), 8.02 (br.s, 4H, N$\underline{H}$CH$_2$CH$_2$CH$_2$N$^+$), 9.12 (d, $^3J_{HH}$ = 7.6 Hz, 4H, CON$\underline{H}$CH). ^{13}C NMR (DMSO-d_6, δ, ppm): 13.9, 22.6, 30.7, 33.9, 35.4, 36.5, 51.1, 53.7, 60.9, 61.7, 62.8, 74.2, 126.7, 128.1, 128.3, 129.1, 134.5, 136.6, 146.7, 157.9, 163.1, 168.3, 170.6. Elemental analysis. C$_{120}$H$_{168}$Br$_4$N$_{12}$O$_{20}$S$_4$ C, 56.60; H, 6.65; Br, 12.55; N, 6.60; S, 5.04; found: C, 56.78; H, 6.26; Br, 12.21; N, 6.31; S, 4.07. MS (ESI), *m/z*: calculated: 556.5 [M−4 Br$^-$]$^{4+}$, 768.3 [M−3 Br$^-$]$^{3+}$, 1192.5 [M−2 Br$^-$]$^{2+}$, 2463.9 [M−Br$^-$]$^+$;

found: 556.5 [M–4 Br⁻]⁴⁺, 768.9 [M–3 Br⁻]³⁺, 1193.5 [M–2 Br⁻]²⁺, 2464.0 [M–Br⁻]⁺. FTIR ATR (ν, cm⁻¹): 1086 (COC), 1675 (C=O), 3186 (N–H).

3.5. General Procedure for the Synthesis of the Compounds 16–21

The compounds **10–15** (0.10 g) were dissolved in 2 mL of water in a round-bottom flask equipped with a magnetic stirrer and reflux condenser. An equimolar amount per functional group of the lithium bis(trifluoromethylsulfonyl)imide was added. The reaction mixture was stirred for 24 h. The resulting precipitate was filtered off. The obtained product was dried in vacuum over phosphorus oxide.

3.5.1. 5,11,17,23-Tetra-*tert*-butyl-25,26,27,28-tetrakis{N-[3′-(dimethyl{[(S)-ethoxycarbonylbenzylmethyl]aminocarbonylmethyl}ammonio)propyl]aminocarbonylmethoxy}-2,8,14,20-thiacalix[4]arene tetra[bis(trifluoromethylsulfonyl)imide] in *cone* Conformation (**17**)

Yield: 0.127 g (97%). M.p. 64 °C. ^{1}H NMR (DMSO-d_6, δ, ppm, J/Hz): 1.06 (s, 36H, (CH$_3$)$_3$C), 1.12 (t, $^3J_{HH}$ = 7.1 Hz, 12H, C$\underline{H}_3$CH$_2$O), 1.86 (m, 8H, NHCH$_2$C$\underline{H}_2$CH$_2$N⁺), 2.89–295 (m, 8H, C$\underline{H}_2$Ph), 3.06 (s, 24H, (CH$_3$)$_2$N⁺), 3.20 (m, 8H, NHCH$_2$CH$_2$C$\underline{H}_2$N⁺), 3.42 (m, 8H, NHC$\underline{H}_2$CH$_2$CH$_2$N⁺), 3.98–4.09 (m, 16H, CH$_3$C$\underline{H}_2$O, N⁺C$\underline{H}_2$CO), 4.57 (m, 4H, NHC$\underline{H}$CO), 4.79 (s, 8H, OCH$_2$CO), 7.19–7.29 (m, 20H, Ph), 7.38 (s, 8H, ArH), 8.48 (br.s, 4H, N$\underline{H}$CH$_2$CH$_2$CH$_2$N⁺), 9.06 (d, $^3J_{HH}$ = 7.5 Hz, 4H, CON$\underline{H}$CH). ^{13}C NMR (DMSO-d_6, δ, ppm): 13.9, 22.6, 30.7, 33.9, 35.4, 36.6, 51.2, 53.5, 61.0, 61.7, 62.6, 74.2, 119.5 ($^1J_{CF}$ = 322 Hz), 126.8, 128.1, 128.4, 129.1, 134.5, 136.5, 146.8, 158.0, 163.1, 168.3, 170.6. Elemental analysis. C$_{128}$H$_{168}$F$_{24}$N$_{16}$O$_{36}$S$_{12}$ C, 45.93; H, 5.06; F, 13.62; N, 6.69; S, 11.49; found: C, 45.72; H, 5.85; F, 13.23; N, 6.43; S, 11.79. HRMS (ESI), m/z: calculated for: 556.5348 [M–4 NTf$_2^-$]⁴⁺, 835.3524 [M–3 NTf$_2^-$]³⁺; found: 556.5220 [M–4 NTf$_2^-$]⁴⁺, 835.3328 [M–3 NTf$_2^-$]³⁺. FTIR ATR (ν, cm⁻¹): 1094 (COC), 1668 (C=O), 3065 (N–H).

3.5.2. 5,11,17,23-Tetra-*tert*-butyl-25,26,27,28-tetrakis{N-[3′-(dimethyl{[ethoxycarbonylmethyl]aminocarbonylmethyl}ammonio)propyl]aminocarbonylmethoxy}-2,8,14,20-tetrathiacalix[4]arene tetra[bis(trifluoromethylsulfonyl)imide] in *partial cone* Conformation (**18**)

Yield: 0.131 g (96%). M.p. 50 °C. ^{1}H NMR (DMSO-d_6, δ, ppm, J/Hz): 1.00 (s, 18H, (CH$_3$)$_3$C), 1.14–1.20 (m, 12H, C$\underline{H}_3$CH$_2$O), 1.27 (s, 9H, (CH$_3$)$_3$C), 1.30 (s, 9H, (CH$_3$)$_3$C), 1.93 (m, 8H, NHCH$_2$C$\underline{H}_2$CH$_2$N⁺), 3.13 (s, 6H, (CH$_3$)$_2$N⁺), 3.18 (s, 18H, (CH$_3$)$_2$N⁺), 3.23–3.29 (m, 8H, NHCH$_2$CH$_2$C$\underline{H}_2$N⁺), 3.52 (m, 8H, NHC$\underline{H}_2$CH$_2$CH$_2$N⁺), 3.93 (d, $^3J_{HH}$ = 5.8 Hz, 8H, OC$\underline{H}_2$CH$_3$), 4.07 (m, 8H, N⁺C$\underline{H}_2$CO), 4.13 (m, 8H, NHC$\underline{H}_2$CO), 4.39 (d, $^2J_{HH}$ = 13.6 Hz, 2H, OCH$_2$C(O)), 4.49 (s, 2H, OCH$_2$C(O)), 4.51 (s, 2H, OCH$_2$C(O)), 4.78 (d, $^2J_{HH}$ = 13.6 Hz, 2H, OCH$_2$C(O)), 7.01 (d, $^4J_{HH}$ = 2.4 Hz, 2H, ArH), 7.65 (d, $^4J_{HH}$ = 2.4 Hz, 2H, ArH), 7.68 (s, 2H, ArH), 7.75 (s, 2H, ArH), 8.31 (br.s, 2H, N$\underline{H}$CH$_2$CH$_2$CH$_2$N⁺), 8.41 (br.s, 1H, N$\underline{H}$CH$_2$CH$_2$CH$_2$N⁺), 8.50 (br.s, 1H, N$\underline{H}$CH$_2$CH$_2$CH$_2$N⁺), 9.05. (m, 4H, CON$\underline{H}$CH). ^{13}C NMR (DMSO-d_6, δ, ppm): 14.1, 22.6, 30.7, 31.0, 33.8, 35.5, 40.8, 51.3, 60.8, 61.8, 62.7, 72.6, 119.5 ($^1J_{CF}$ = 322 Hz), 126.4, 127.1, 127.6, 128.1, 133.7, 134.0, 135.1, 135.4, 145.3, 145.7, 146.5, 157.2, 159.4, 163.7, 166.9, 168.0, 169.1. Elemental analysis. C$_{100}$H$_{144}$F$_{24}$N$_{16}$O$_{36}$S$_{12}$ C, 40.21; H, 4.86; F 15.26, N, 7.50; S, 12.88 found: C, 39.26; H, 4.46; F 15.08, N, 6.95; S, 12.49. HRMS (ESI), m/z: calculated for: 466.2370 [M–4 NTf$_2^-$]⁴⁺, 715.2898 [M–3 NTf$_2^-$]³⁺, 1212.8936 [M–2 NTf$_2^-$]²⁺; found: 466.2357 [M–4 NTf$_2^-$]⁴⁺, 715.2875 [M–3 NTf$_2^-$]³⁺, 1212.8914 [M–2 NTf$_2^-$]²⁺. FTIR ATR (ν, cm⁻¹):1094 (C–O–C), 1674 (C=O), 3207 (N–H).

3.5.3. 5,11,17,23-Tetra-*tert*-butyl-25,26,27,28-tetrakis{N-[3′-(dimethyl{[(S)-ethoxycarbonylbenzylmethyl]aminocarbonylmethyl}ammonio)propyl]aminocarbonylmethoxy}-2,8,14,20-thiacalix[4]arene tetra[bis(trifluoromethylsulfonyl)imide] in *partial cone* Conformation (**19**)

Yield: 0.125 g (95%). M.p. 55 °C. ^{1}H NMR (DMSO-d_6, δ, ppm, J/Hz): 1.00 (s, 18H, (CH$_3$)$_3$C), 1.10 (t, $^3J_{HH}$ = 7.1 Hz, 12H, C$\underline{H}_3$CH$_2$O), 1.27 (s, 9H, (CH$_3$)$_3$C), 1.28 (s, 9H, (CH$_3$)$_3$C), 1.86 (m, 8H, NHCH$_2$C$\underline{H}_2$CH$_2$N⁺), 2.90–2.95 (m, 8H, C$\underline{H}_2$Ph), 3.02 (s, 6H, (CH$_3$)$_2$N⁺), 3.08 (s, 18H, (CH$_3$)$_2$N⁺), 3.13–3.24 (m, 8H, NHCH$_2$CH$_2$C$\underline{H}_2$N⁺), 3.42–3.45 (m,

8H, NHC$\underline{H_2}$CH$_2$CH$_2$N$^+$), 4.04–4.07 (m, 16H, CH$_3$C$\underline{H_2}$O, N$^+$C$\underline{H_2}$CO), 4.40 (d, $^2J_{HH}$ = 13.5 Hz, 2H, OCH$_2$C(O)), 4.49 (s, 2H, OCH$_2$C(O)), 4.52 (s, 2H, OCH$_2$C(O)), 4.56–4.61 (m, 4H, NHC$\underline{H}$CO), 4.79 (d, $^2J_{HH}$ = 13.5 Hz, 2H, OCH$_2$C(O)), 7.02 (d, $^4J_{HH}$ = 2.4 Hz, 2H, ArH), 7.21–7.28 (m, 20H, Ph), 7.65 (d, $^4J_{HH}$ = 2.4 Hz, 2H, ArH), 7.67 (s, 2H, ArH), 7.75 (s, 2H, ArH), 8.28 (br.s, 2H, N$\underline{H}$CH$_2$CH$_2$CH$_2$N$^+$), 8.40 (br.s, 1H, N$\underline{H}$CH$_2$CH$_2$CH$_2$N$^+$), 8.48 (br.s, 1H, N$\underline{H}$CH$_2$CH$_2$CH$_2$N$^+$), 9.09. (m, 4H, CON$\underline{H}$CH). ^{13}C NMR (DMSO-d_6, δ, ppm): 13.9, 22.6, 30.7, 31.0, 33.8, 35.5, 36.6, 51.3, 53.6, 61.0, 61.7, 62.6, 72.6, 119.5 ($^1J_{CF}$ = 322 Hz), 126.3, 126.8, 127.7, 128.4, 129.2, 133.7, 134.1, 135.2, 135.4, 136.6, 145.4, 145.7, 146.6, 157.3, 163.1, 168.0, 168.8, 170.7. Elemental analysis. C$_{128}$H$_{168}$F$_{24}$N$_{16}$O$_{36}$S$_{12}$ C, 45.93; H, 5.06; F, 13.62; N, 6.69; S, 11.49; found: C, 45.79; H, 5.00; F, 13.35; N, 6.43; S, 11.47. HRMS (ESI), m/z: calculated for: 556.5348 [M−4 NTf$_2^-$]$^{4+}$, 835.3524 [M−3 NTf$_2^-$]$^{3+}$; found: 556.5206 [M−4 NTf$_2^-$]$^{4+}$, 835.3283 [M−3 NTf$_2^-$]$^{3+}$. FTIR ATR (ν, cm$^{−1}$): 1096 (COC), 1668 (C=O), 3064 (N–H).

3.5.4. 5,11,17,23-Tetra-*tert*-butyl-25,26,27,28-tetrakis{*N*-[3′-(dimethyl{[(S)-ethoxycarbonylbenzylmethyl]aminocarbonylmethyl}ammonio)propyl]aminocarbonylmethoxy}-2,8,14,20-thiacalix[4]arene tetra[bis(trifluoromethylsulfonyl)imide] in *1,3-alternate* Conformation (21)

Yield: 0.129 g (98%). M.p. 75 °C. ^{1}H NMR (DMSO-d_6, δ, ppm, J/Hz): 1.12 (t, $^3J_{HH}$ = 7.1 Hz, 12H, C$\underline{H_3}$CH$_2$O), 1.19 (s, 36H, (CH$_3$)$_3$C), 1.89 (m, 8H, NHCH$_2$C$\underline{H_2}$CH$_2$N$^+$), 2.91–2.96 (m, 8H, C$\underline{H_2}$Ph), 3.08 (s, 24H, (CH$_3$)$_2$N$^+$), 3.15 (m, 8H, NHCH$_2$CH$_2$C$\underline{H_2}$N$^+$), 3.43 (m, 8H, NHC$\underline{H_2}$CH$_2$CH$_2$N$^+$), 3.99 (s, 8H, OCH$_2$CO), 4.03–4.07 (m, 16H, CH$_3$C$\underline{H_2}$O, N$^+$C$\underline{H_2}$CO), 4.59 (m, 4H, NHC$\underline{H}$CO), 7.23–7.28 (m, 20H, Ph), 7.59 (s, 8H, ArH), 8.00 (br.s, 4H, N$\underline{H}$CH$_2$CH$_2$CH$_2$N$^+$), 9.07 (d, $^3J_{HH}$ = 7.6 Hz, 4H, CON$\underline{H}$CH). ^{13}C NMR (DMSO-d_6, δ, ppm): 13.9, 22.6, 30.8, 33.9, 35.8, 36.6, 51.2, 53.6, 61.0, 61.7, 62.5, 71.0, 119.5 ($^1J_{CF}$ = 322 Hz), 126.7, 127.6, 128.4, 129.2, 133.1, 136.6, 146.1, 157.2, 163.1, 167.4, 170.7. Elemental analysis. C$_{128}$H$_{168}$F$_{24}$N$_{16}$O$_{36}$S$_{12}$ C, 45.93; H, 5.06; F, 13.62; N, 6.69; S, 11.49; found: C, 45.22; H, 5.05; F, 13.53; N, 6.09; S, 11.09. HRMS (ESI), m/z: calculated for: 556.5348 [M−4 NTf$_2^-$]$^{4+}$, 835.3524 [M−3 NTf$_2^-$]$^{3+}$; found: 556.5198 [M−4 NTf$_2^-$]$^{4+}$, 835.3278 [M−3 NTf$_2^-$]$^{3+}$. FTIR ATR (ν, cm$^{−1}$): 1093 (COC), 1669 (C=O), 3064 (N–H).

4. Conclusions

Novel macrocyclic quaternary ammonium ILs containing amino acid fragments (glycine and L-phenylalanine) based on *p-tert*-butylthiacalix[4]arene in *cone*, *partial cone*, and *1,3-alternate* conformations were synthesized. The melting temperature of the obtained ILs was found in the range of 50–75 °C. Replacement of the bromide anion with bis(trifluoromethylsulfonyl)imide led to a decrease in the melting point by 39–55 °C. The ILs in *partial cone* conformation had the lowest melting points among all stereoisomers. Thermal stability of the obtained macrocyclic ILs was determined via thermogravimetry and differential scanning calorimetry. The onset of decomposition of the synthesized compounds was established at 305–327 °C. The obtained results can be applied to the design of sensor systems capable for target substrate recognition. These compounds can also be used as synthetic biomimetic models of oligo- and polypeptides.

Supplementary Materials: The following are available online at https://www.mdpi.com/article/10.3390/molecules27228006/s1, Figures S1–S10: ^{1}H NMR spectra of the compounds **4**, **6**, **11–13**, **15**, **17–19**, **21**; Figures S11–S18: ^{13}C NMR spectra of the compounds **4**, **6**, **11–13**, **15**, **17–19**, **21**; Figures S19–S26: FT-IR spectra of the compounds **4**, **6**, **11–13**, **15**, **17–19**, **21**; Figures S27–S33: HRMS spectra of the compounds **4**, **6**, **11–13**, **15**, **17–19**, **21**; Figure S34: DSC curves of the compounds **16–21**.

Author Contributions: Conceptualization, writing—review and editing, supervision, I.S.; Project administration, writing—review and editing, funding acquisition and visualization, P.P.; writing—original draft preparation, investigation, O.T., A.B.; resources, formal analysis, A.G. All authors have read and agreed to the published version of the manuscript.

Funding: This work was financially supported by Russian Science Foundation, Russian Federation (grant № 19-73-10134, https://rscf.ru/en/project/19-73-10134/) (accessed on 1 November 2022).

Institutional Review Board Statement: Not applicable.

Informed Consent Statement: Not applicable.

Data Availability Statement: The data presented in this study are available in Supplementary Materials.

Acknowledgments: High resolution mass-spectrometry (ESI) has been carried out by the Kazan Federal University Strategic Academic Leadership Program ('PRIORITY-2030').

Conflicts of Interest: The authors declare no conflict of interest. The funders had no role in the design of the study; in the collection, analyses, or interpretation of data; in the writing of the manuscript, or in the decision to publish the results.

Sample Availability: Samples of all obtained compounds are available from the authors.

References

1. Patil, K.R.; Surwade, A.D.; Rajput, P.J.; Shaikh, V.R. Investigations of Solute–Solvent Interactions in Aqueous Solutions of Amino Acids Ionic Liquids Having the Common Nitrate as Anion at Different Temperatures. *J. Mol. Liq.* **2021**, *329*, 115546. [CrossRef]
2. Fabre, E.; Murshed, S.M.S. A Review of the Thermophysical Properties and Potential of Ionic Liquids for Thermal Applications. *J. Mater. Chem. A* **2021**, *9*, 15861–15879. [CrossRef]
3. Nikfarjam, N.; Ghomi, M.; Agarwal, T.; Hassanpour, M.; Sharifi, E.; Khorsandi, D.; Ali Khan, M.; Rossi, F.; Rossetti, A.; Nazarzadeh Zare, E.; et al. Antimicrobial Ionic Liquid-Based Materials for Biomedical Applications. *Adv. Funct. Mater.* **2021**, *31*, 2104148. [CrossRef]
4. Xu, C.; Yang, G.; Wu, D.; Yao, M.; Xing, C.; Zhang, J.; Zhang, H.; Li, F.; Feng, Y.; Qi, S.; et al. Roadmap on Ionic Liquid Electrolytes for Energy Storage Devices. *Chem. Asian J.* **2021**, *16*, 549–562. [CrossRef] [PubMed]
5. Minea, A.A.; Sohel Murshed, S.M. Ionic Liquids-Based Nanocolloids—A Review of Progress and Prospects in Convective Heat Transfer Applications. *Nanomaterials* **2021**, *11*, 1039. [CrossRef]
6. Buettner, C.S.; Cognigni, A.; Schröder, C.; Bica-Schröder, K. Surface-Active Ionic Liquids: A Review. *J. Mol. Liq.* **2022**, *347*, 118160. [CrossRef]
7. Correia, D.M.; Fernandes, L.C.; Fernandes, M.M.; Hermenegildo, B.; Meira, R.M.; Ribeiro, C.; Ribeiro, S.; Reguera, J.; Lanceros-Méndez, S. Ionic Liquid-Based Materials for Biomedical Applications. *Nanomaterials* **2021**, *11*, 2401. [CrossRef]
8. Maculewicz, J.; Świacka, K.; Stepnowski, P.; Dołżonek, J.; Białk-Bielińska, A. Ionic Liquids as Potentially Hazardous Pollutants: Evidences of Their Presence in the Environment and Recent Analytical Developments. *J. Hazard. Mater.* **2022**, *437*, 129353. [CrossRef]
9. Zhuang, W.; Hachem, K.; Bokov, D.; Javed Ansari, M.; Taghvaie Nakhjiri, A. Ionic Liquids in Pharmaceutical Industry: A Systematic Review on Applications and Future Perspectives. *J. Mol. Liq.* **2022**, *349*, 118145. [CrossRef]
10. Niu, H.; Wang, L.; Guan, P.; Zhang, N.; Yan, C.; Ding, M.; Guo, X.; Huang, T.; Hu, X. Recent Advances in Application of Ionic Liquids in Electrolyte of Lithium Ion Batteries. *J. Energy Storage* **2021**, *40*, 102659. [CrossRef]
11. Rauber, D.; Hofmann, A.; Philippi, F.; Kay, C.W.M.; Zinkevich, T.; Hanemann, T.; Hempelmann, R. Structure-Property Relation of Trimethyl Ammonium Ionic Liquids for Battery Applications. *Appl. Sci.* **2021**, *11*, 5679. [CrossRef]
12. Cho, C.-W.; Pham, T.P.T.; Zhao, Y.; Stolte, S.; Yun, Y.-S. Review of the Toxic Effects of Ionic Liquids. *Sci. Total Environ.* **2021**, *786*, 147309. [CrossRef] [PubMed]
13. El Seoud, O.A.; Keppeler, N.; Malek, N.I.; Galgano, P.D. Ionic Liquid-Based Surfactants: Recent Advances in Their Syntheses, Solution Properties, and Applications. *Polymers* **2021**, *13*, 1100. [CrossRef] [PubMed]
14. Gonçalves, A.R.P.; Paredes, X.; Cristino, A.F.; Santos, F.J.V.; Queirós, C.S.G.P. Ionic Liquids—A Review of Their Toxicity to Living Organisms. *Int. J. Mol. Sci.* **2021**, *22*, 5612. [CrossRef] [PubMed]
15. Himani; Pratap Singh Raman, A.; Babu Singh, M.; Jain, P.; Chaudhary, P.; Bahadur, I.; Lal, K.; Kumar, V.; Singh, P. An Update on Synthesis, Properties, Applications and Toxicity of the ILs. *J. Mol. Liq.* **2022**, *364*, 119989. [CrossRef]
16. Chen, Y.; Han, X.; Liu, Z.; Li, Y.; Sun, H.; Wang, H.; Wang, J. Thermal Decomposition and Volatility of Ionic Liquids: Factors, Evaluation and Strategies. *J. Mol. Liq.* **2022**, *366*, 120336. [CrossRef]
17. Holbrey, J.D.; Seddon, K.R. The Phase Behaviour of 1-Alkyl-3-Methylimidazolium Tetrafluoroborates; Ionic Liquids and Ionic Liquid Crystals. *J. Chem. Soc., Dalton Trans.* **1999**, *13*, 2133–2140. [CrossRef]
18. Shmukler, L.E.; Fedorova, I.V.; Fadeeva, Y.A.; Safonova, L.P. The Physicochemical Properties and Structure of Alkylammonium Protic Ionic Liquids of $R_nH_{4-n}NX$ (n = 1–3) Family. A Mini–Review. *J. Mol. Liq.* **2021**, *321*, 114350. [CrossRef]
19. Vereshchagin, A.N.; Frolov, N.A.; Egorova, K.S.; Seitkalieva, M.M.; Ananikov, V.P. Quaternary Ammonium Compounds (QACs) and Ionic Liquids (ILs) as Biocides: From Simple Antiseptics to Tunable Antimicrobials. *Int. J. Mol. Sci.* **2021**, *22*, 6793. [CrossRef]
20. Zhuravlev, O.E.; Voronchikhina, L.I.; Gorbunova, D.V. Comparative Characteristics of Thermal Stability of Quaternary Ammonium and Pyridinium Tetrachloroferrates. *Russ. J. Gen. Chem.* **2022**, *92*, 348–354. [CrossRef]
21. Egorova, K.S.; Gordeev, E.G.; Ananikov, V.P. Biological Activity of Ionic Liquids and Their Application in Pharmaceutics and Medicine. *Chem. Rev.* **2017**, *117*, 7132–7189. [CrossRef] [PubMed]

22. Khachatrian, A.A.; Mukhametzyanov, T.A.; Yakhvarov, D.G.; Sinyashin, O.G.; Garifullin, B.F.; Rakipov, I.T.; Mironova, D.A.; Burilov, V.A.; Solomonov, B.N. Intermolecular Interactions between Imidazolium- and Cholinium-Based Ionic Liquids and Lysozyme: Regularities and Peculiarities. *J. Mol. Liq.* **2022**, *348*, 118426. [CrossRef]
23. Melo, C.I.; Bogel-Łukasik, R.; Nunes da Ponte, M.; Bogel-Łukasik, E. Ammonium Ionic Liquids as Green Solvents for Drugs. *Fluid Phase Equilib.* **2013**, *338*, 209–216. [CrossRef]
24. Bisht, M.; Jha, I.; Venkatesu, P. Comprehensive Evaluation of Biomolecular Interactions between Protein and Amino Acid Based-Ionic Liquids: A Comparable Study between [Bmim][Br] and [Bmim][Gly] Ionic Liquids. *ChemistrySelect* **2016**, *1*, 3510–3519. [CrossRef]
25. Schindl, A.; Hagen, M.L.; Muzammal, S.; Gunasekera, H.A.D.; Croft, A.K. Proteins in Ionic Liquids: Reactions, Applications, and Futures. *Front. Chem.* **2019**, *7*, 347. [CrossRef]
26. Ossowicz, P.; Klebeko, J.; Roman, B.; Janus, E.; Rozwadowski, Z. The Relationship between the Structure and Properties of Amino Acid Ionic Liquids. *Molecules* **2019**, *24*, 3252. [CrossRef]
27. Miao, S.; Atkin, R.; Warr, G. Design and Applications of Biocompatible Choline Amino Acid Ionic Liquids. *Green Chem.* **2022**, *24*, 7281–7304. [CrossRef]
28. Shukla, S.K.; Mikkola, J.-P. Use of Ionic Liquids in Protein and DNA Chemistry. *Front. Chem.* **2020**, *8*, 598662. [CrossRef]
29. Egorova, K.S.; Posvyatenko, A.V.; Larin, S.S.; Ananikov, V.P. Ionic Liquids: Prospects for Nucleic Acid Handling and Delivery. *Nucleic Acids Res.* **2021**, *49*, 1201–1234. [CrossRef]
30. Patel, R.; Kumari, M.; Khan, A.B. Recent Advances in the Applications of Ionic Liquids in Protein Stability and Activity: A Review. *Appl. Biochem. Biotechnol.* **2014**, *172*, 3701–3720. [CrossRef]
31. Egorova, K.S.; Seitkalieva, M.M.; Posvyatenko, A.V.; Ananikov, V.P. An Unexpected Increase of Toxicity of Amino Acid-Containing Ionic Liquids. *Toxicol. Res.* **2015**, *4*, 152–159. [CrossRef]
32. Yadav, R.; Kahlon, N.K.; Kumar, S.; Devunuri, N.; Venkatesu, P. Biophysical Study on the Phase Transition Behaviour of Biocompatible Thermoresponsive Polymer Influenced by Tryptophan-Based Amino Acid Ionic Liquids. *Polymer* **2021**, *228*, 123871. [CrossRef]
33. Sahoo, D.K.; Jena, S.; Tulsiyan, K.D.; Dutta, J.; Chakrabarty, S.; Biswal, H.S. Amino-Acid-Based Ionic Liquids for the Improvement in Stability and Activity of Cytochrome c: A Combined Experimental and Molecular Dynamics Study. *J. Phys. Chem. B* **2019**, *123*, 10100–10109. [CrossRef]
34. Durga, G.; Kalra, P.; Kumar Verma, V.; Wangdi, K.; Mishra, A. Ionic Liquids: From a Solvent for Polymeric Reactions to the Monomers for Poly(Ionic Liquids). *J. Mol. Liq.* **2021**, *335*, 116540. [CrossRef]
35. Yang, B.; Yang, G.; Zhang, Y.-M.; Zhang, S.X.-A. Recent Advances in Poly(Ionic Liquid)s for Electrochromic Devices. *J. Mater. Chem. C* **2021**, *9*, 4730–4741. [CrossRef]
36. Barrulas, R.V.; Zanatta, M.; Casimiro, T.; Corvo, M.C. Advanced Porous Materials from Poly(Ionic Liquid)s: Challenges, Applications and Opportunities. *Chem. Eng. J.* **2021**, *411*, 128528. [CrossRef]
37. Thapaliya, B.P.; Puskar, N.G.; Slaymaker, S.; Feider, N.O.; Do-Thanh, C.-L.; Schott, J.A.; Jiang, D.; Teague, C.M.; Mahurin, S.M.; Dai, S. Synthesis and Characterization of Macrocyclic Ionic Liquids for CO_2 Separation. *Ind. Eng. Chem. Res.* **2021**, *60*, 8218–8226. [CrossRef]
38. Ogoshi, T.; Ueshima, N.; Yamagishi, T.; Toyota, Y.; Matsumi, N. Ionic Liquid Pillar[5]Arene: Its Ionic Conductivity and Solvent-Free Complexation with a Guest. *Chem. Commun.* **2012**, *48*, 3536. [CrossRef]
39. Zhou, J.; Rao, L.; Yu, G.; Cook, T.R.; Chen, X.; Huang, F. Supramolecular Cancer Nanotheranostics. *Chem. Soc. Rev.* **2021**, *50*, 2839–2891. [CrossRef]
40. Wang, M.; Fang, X.; Yang, S.; Li, Q.; Khashab, N.M.; Zhou, J.; Huang, F. Separation of Ethyltoluene Isomers by Nonporous Adaptive Crystals of Perethylated and Perbromoethylated Pillararenes. *Mater. Today Chem.* **2022**, *24*, 100919. [CrossRef]
41. Selivanova, N.; Gubaidullin, A.; Padnya, P.; Stoikov, I.; Galyametdinov, Y. Phase Behaviour, Structural Properties and Intermolecular Interactions of Systems Based on Substituted Thiacalix[4]Arene and Nonionic Surfactants. *Liq. Cryst.* **2018**, *46*, 415–421. [CrossRef]
42. Selivanova, N.M.; Zimina, M.V.; Padnya, P.L.; Stoikov, I.I.; Gubaidullin, A.T.; Galyametdinov, Y.G. Development of Efficient Luminescent Soft Media by Incorporation of a Hetero-Ligand Macrocyclic Terbium Complex into a Lyomesophase. *Russ. Chem. Bull.* **2020**, *69*, 1763–1770. [CrossRef]
43. Shurpik, D.N.; Padnya, P.L.; Stoikov, I.I.; Cragg, P.J. Antimicrobial Activity of Calixarenes and Related Macrocycles. *Molecules* **2020**, *25*, 5145. [CrossRef] [PubMed]
44. Yang, F.; Guo, H.; Jiao, Z.; Li, C.; Ye, J. Calixarene Ionic Liquids: Excellent Phase Transfer Catalysts for Nucleophilic Substitution Reaction in Water. *J. Iran. Chem. Soc.* **2012**, *9*, 327–332. [CrossRef]
45. Padnya, P.L.; Porfireva, A.V.; Evtugyn, G.A.; Stoikov, I.I. Solid Contact Potentiometric Sensors Based on a New Class of Ionic Liquids on Thiacalixarene Platform. *Front. Chem.* **2018**, *6*, 594. [CrossRef]
46. Padnya, P.L.; Terenteva, O.S.; Akhmedov, A.A.; Iksanova, A.G.; Shtyrlin, N.V.; Nikitina, E.V.; Krylova, E.S.; Shtyrlin, Y.G.; Stoikov, I.I. Thiacalixarene Based Quaternary Ammonium Salts as Promising Antibacterial Agents. *Bioorg. Med. Chem.* **2021**, *29*, 115905. [CrossRef]
47. Podyachev, S.N.; Zairov, R.R.; Mustafina, A.R. 1,3-Diketone Calix[4]Arene Derivatives—A New Type of Versatile Ligands for Metal Complexes and Nanoparticles. *Molecules* **2021**, *26*, 1214. [CrossRef]

48. Iampolska, A.D.; Kharchenko, S.G.; Voitenko, Z.V.; Shishkina, S.V.; Ryabitskii, A.B.; Kalchenko, V.I. Synthesis of Thiacalix[4]Arene Task-Specific Ionic Liquids. *Phosphorus Sulfur Silicon Relat. Elem.* **2016**, *191*, 174–179. [CrossRef]
49. Alishahi, N.; Mohammadpoor-Baltork, I.; Tangestaninejad, S.; Mirkhani, V.; Moghadam, M.; Kia, R. Calixarene Based Ionic Liquid as an Efficient and Reusable Catalyst for One-Pot Multicomponent Synthesis of Polysubstituted Pyridines and Bis-pyridines. *ChemistrySelect* **2019**, *4*, 5903–5910. [CrossRef]
50. Padnya, P.L.; Andreyko, E.A.; Gorbatova, P.A.; Parfenov, V.V.; Rizvanov, I.K.; Stoikov, I.I. Towards Macrocyclic Ionic Liquids: Novel Ammonium Salts Based on Tetrasubstituted *p-Tert*-Butylthiacalix[4]Arenes. *RSC Adv.* **2017**, *7*, 1671–1686. [CrossRef]
51. Zhang, S.; Sun, N.; He, X.; Lu, X.; Zhang, X. Physical Properties of Ionic Liquids: Database and Evaluation. *J. Phys. Chem. Ref. Data* **2006**, *35*, 1475–1517. [CrossRef]
52. Xu, C.; Cheng, Z. Thermal Stability of Ionic Liquids: Current Status and Prospects for Future Development. *Processes* **2021**, *9*, 337. [CrossRef]
53. Villanueva, M.; Coronas, A.; García, J.; Salgado, J. Thermal Stability of Ionic Liquids for Their Application as New Absorbents. *Ind. Eng. Chem. Res.* **2013**, *52*, 15718–15727. [CrossRef]
54. Moldoveanu, S.C. Pyrolysis of Amino Acids and Small Peptides. In *Pyrolysis of Organic Molecules*, 2nd ed.; Moldoveanu, S.C., Ed.; Elsevier: London, UK, 2019; pp. 555–633. [CrossRef]
55. Efimova, A.; Pfützner, L.; Schmidt, P. Thermal Stability and Decomposition Mechanism of 1-Ethyl-3-Methylimidazolium Halides. *Thermochim. Acta* **2015**, *604*, 129–136. [CrossRef]
56. Lorenzo, M.; Vilas, M.; Verdía, P.; Villanueva, M.; Salgado, J.; Tojo, E. Long-Term Thermal Stabilities of Ammonium Ionic Liquids Designed as Potential Absorbents of Ammonia. *RSC Adv.* **2015**, *5*, 41278–41284. [CrossRef]
57. Kurnia, K.A.; Wilfred, C.D.; Murugesan, T. Thermophysical Properties of Hydroxyl Ammonium Ionic Liquids. *J. Chem. Thermodyn.* **2009**, *41*, 517–521. [CrossRef]
58. Pucci, F.; Rooman, M. Improved Insights into Protein Thermal Stability: From the Molecular to the Structurome Scale. *Phil. Trans. R. Soc. A.* **2016**, *374*, 20160141. [CrossRef]
59. Wingreen, N.S.; Li, H.; Tang, C. Designability and Thermal Stability of Protein Structures. *Polymer* **2004**, *45*, 699–705. [CrossRef]
60. Maheshwari, A.S.; Archunan, G. Distribution of Amino Acids in Functional Sites of Proteins with High Melting Temperature. *Bioinformation* **2012**, *8*, 1176–1181. [CrossRef]
61. Jelesarov, I.; Karshikoff, A. Defining the Role of Salt Bridges in Protein Stability. *Methods Mol. Biol.* **2008**, *490*, 227–260. [CrossRef]
62. Kumar, S.; Tsai, C.-J.; Nussinov, R. Factors Enhancing Protein Thermostability. *Protein Eng. Des. Sel.* **2000**, *13*, 179–191. [PubMed]
63. Padnya, P.L.; Andreyko, E.A.; Mostovaya, O.A.; Rizvanov, I.K.; Stoikov, I.I. The Synthesis of New Amphiphilic *P-Tert*-Butylthiacalix[4]Arenes Containing Peptide Fragments and Their Interaction with DNA. *Org. Biomol. Chem.* **2015**, *13*, 5894–5904. [CrossRef] [PubMed]
64. Andreyko, E.A.; Padnya, P.L.; Stoikov, I.I. Supramolecular Self-Assembly of Water-Soluble Nanoparticles Based on Amphiphilic *p-Tert*-Butylthiacalix[4]Arenes with Silver Nitrate and Fluorescein. *Colloids Surf. A* **2014**, *454*, 74–83. [CrossRef]
65. Andreyko, E.A.; Padnya, P.L.; Daminova, R.R.; Stoikov, I.I. Supramolecular "Containers": Self-Assembly and Functionalization of Thiacalix[4]Arenes for Recognition of Amino- and Dicarboxylic Acids. *RSC Adv.* **2014**, *4*, 3556–3565. [CrossRef]
66. Nazarova, A.; Shurpik, D.; Padnya, P.; Mukhametzyanov, T.; Cragg, P.; Stoikov, I. Self-Assembly of Supramolecular Architectures by the Effect of Amino Acid Residues of Quaternary Ammonium Pillar[5]Arenes. *Int. J. Mol. Sci.* **2020**, *21*, 7206. [CrossRef]

MDPI
St. Alban-Anlage 66
4052 Basel
Switzerland
www.mdpi.com

Molecules Editorial Office
E-mail: molecules@mdpi.com
www.mdpi.com/journal/molecules

Disclaimer/Publisher's Note: The statements, opinions and data contained in all publications are solely those of the individual author(s) and contributor(s) and not of MDPI and/or the editor(s). MDPI and/or the editor(s) disclaim responsibility for any injury to people or property resulting from any ideas, methods, instructions or products referred to in the content.

www.ingramcontent.com/pod-product-compliance
Lightning Source LLC
LaVergne TN
LVHW070415100526
838202LV00014B/1463